P9-ELF-716

THE EARTH'S NATURAL FORCES

THE EARTH'S NATURAL FORCES

GENERAL EDITOR

Professor K.J. Gregory

New York
OXFORD UNIVERSITY PRESS
1990

CONSULTANT EDITOR
Professor Peter Haggett, University of Bristol

Professor Peter Beaumont, University of Wales, Lampeter
The Middle East

Dr Michael J. Clark, University of Southampton
Canada and the Arctic

Professor Ian Douglas, University of Manchester
Southeast Asia

Dr V. Gardiner, University of Leicester
The Low Countries, The Indian Subcontinent

Professor Ken J. Gregory, University of Southampton
The Forces that Shape the Earth, The Nordic Countries

Dr J. Grzybowski, Polish Academy of Sciences, Warsaw
Eastern Europe

Dr J.E. Hobbs, University of New England, Armidale, NSW
Australasia, Oceania and Antarctica

Dr David K. C. Jones, London School of Economics and Political Science
Italy and Greece

Professor John Lewin, University College of Wales, Aberystwyth
The British Isles

Professor Leonid Serebryanny, USSR Academy of Sciences, Moscow
The Soviet Union

Professor M.F. Thomas, University of Stirling
Central Africa

Professor Colin R. Thorne, University of Nottingham
The United States

Professor John Thornes, University of Bristol
South America, Spain and Portugal

Professor J. Tricart, University of Strasbourg
France and its neighbors

Dr A. Trilsbach, University of Durham
Northern Africa

Dr Lance Tufnell, Huddersfield Polytechnic
Central Europe

Professor D.E. Walling, University of Exeter
China and its neighbors

Dr R.P.D. Walsh, University College, Swansea
Central America

Dr J.R. Whitlow, University of Zimbabwe, Harare
Southern Africa

Dr Michael Witherick, University of Southampton
Japan and Korea

AN EQUINOX BOOK

Planned and produced by
Andromeda Oxford Limited
11–15 The Vineyard, Abingdon,
Oxfordshire, England OX14 3PX

Published in the United States of America by
Oxford University Press, Inc.,
200 Madison Avenue,
New York, N.Y. 10016

Oxford is a registered trademark of
Oxford University Press

Library of Congress
Cataloging-in-Publication Data

Earth's natural forces / general editor, Kenneth J. Gregory.
p. cm.
"An Equinox book."
Includes bibliographical references and index.
ISBN 0-19-520860-9
1. Physical geography. I. Gregory, K. J. (Kenneth John)
GB54.5.E27 1990
910'.02—dc20 90-7885
CIP

Volume editor Bill MacKeith
Designers Jerry Goldie, Bob Gordon, Jaqueline Palmer
Cartographic manager Olive Pearson
Cartographic editors Alison Dickinson, Zoë Goodwin
Picture manager Alison Renney
Picture researcher David Pratt

Project editor Candida Hunt
Art editor Steve McCurdy

ISBN 0-19-520860-9

Printing (last digit): 9 8 7 6 5 4 3 2

Printed in Spain by Heraclio Fournier SA, Vitoria

INTRODUCTORY PHOTOGRAPHS
Half title: *Limestone pavement, Yorkshire, UK (Bruce Coleman/G. Doré)*
Half title verso: *Anvers Island, Antarctica (Mountain Camera/C. Monteath)*
Title page: *Monument Valley, Utah, USA (Art Directors)*
This page: *The Maldives, Indian Ocean (Ace Photo Agency/P. Steel)*

Contents

PREFACE

MANY OF US HAVE SEEN THE EARTH BENEATH US FROM THE WINDOW of an aircraft. We are also familiar with the surface of the land on which we build our towns and cities, roads and communication systems, and with the mountain ranges, rivers and lakes we exploit for commerical and leisure purposes. Yet in comparison with other branches of the physical and natural sciences, scientific study of the Earth's surface is comparatively new – barely more than a century old. For a long time research was dominated by studies of how the Earth had developed; only recently has attention turned to how the Earth works – to its present processes.

Since we have been able to view the Earth from space we have come to appreciate in much more detail the way our planet functions. Awareness has grown of its resilience and its vulnerability, and of the effects of human activity on the land, the oceans and the atmosphere. At the same time, technological advances in the last 25 years, providing us with more and more information at greater speed than ever before, have made it possible to study more closely how the Earth's surface is changing, and how it is likely to change in the future. Such techniques include the observation of Earth from satellites and aircraft to track weather systems, including the paths of hurricanes. Changes in climatic history over very long periods of time, and their effect on the processes that have shaped the Earth, can be discovered by analyzing cores taken from the ice sheets of Antarctica and from the deep ocean floor. Sophisticated methods have also been devised to assess the extent to which human activity may accelerate natural processes.

Energy from the Sun and from deep inside the Earth play a major part in shaping the surface of the planet. Their power is demonstrated through winds and ocean currents, through volcanoes, earthquakes and other forms of tectonic activity. Landscapes have developed gradually over the geological past, particularly as the result of the dramatic changes of climate and of sea level associated with the sequence of ice ages during the last 2 million years. The present environment continues to be shaped by natural processes – the cycles of water, of sediment, of chemical elements and of soil development.

The aim of this book is to explore the way the Earth works, both globally and locally within each region, with an awareness of the new techniques and discoveries of recent years. In many respects the Earth's environments are sensitive and fragile. We all have a responsibility to ensure the future well-being of the planet, and we can begin to do this by acquiring a greater understanding of the Earth.

Professor K.J. Gregory
UNIVERSITY OF SOUTHAMPTON

The Dolomites in northeast Italy which once lay beneath the ocean floor.

The Namib Desert, southern Africa, photographed from space *(overleaf)*.

THE FORCES THAT SHAPE THE EARTH

THE EARTH SYSTEM

DEVELOPMENT OF THE EARTH

CYCLES AT WORK ON EARTH

PEOPLE IN THE LANDSCAPE

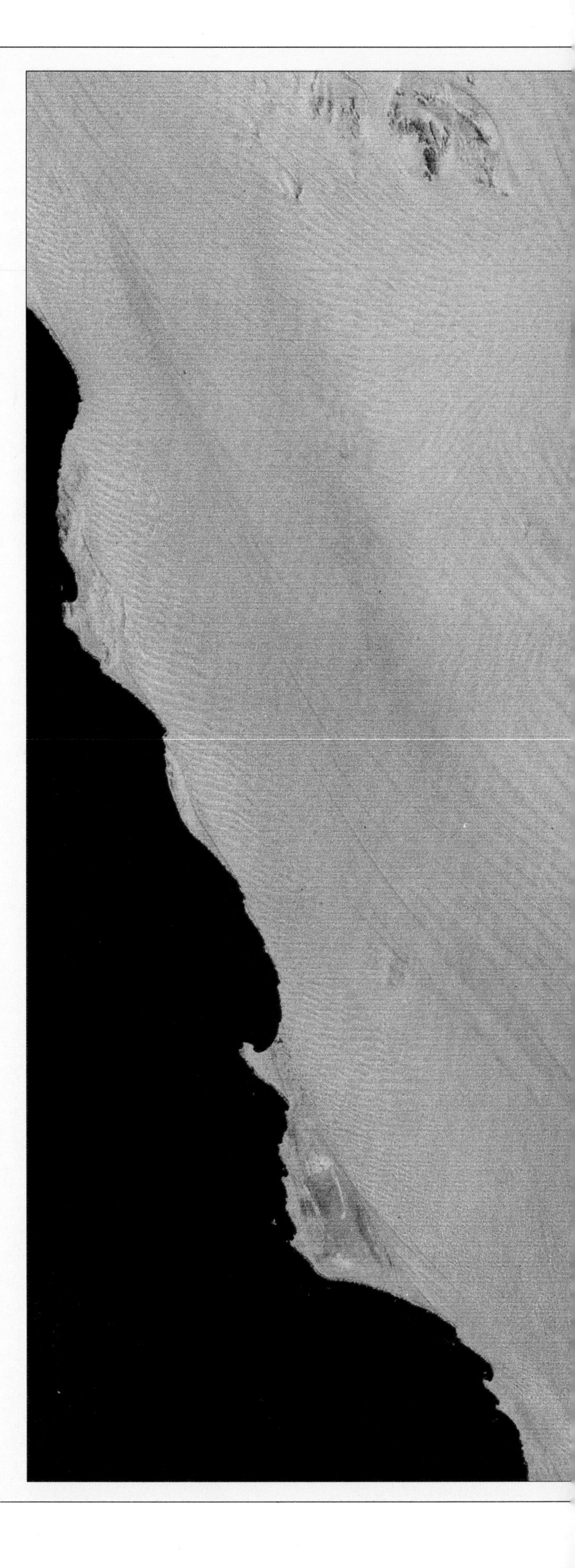

The Earth and the Sun

The fate of the Earth and of all the life that inhabits it depends primarily on the heat and light of the Sun. Energy from the Sun and from within the Earth has been responsible for creating continents, mountains, valleys, plains and oceans. Over millions of years that energy is still changing the world.

Energy from the Sun

The Sun generates its energy from nuclear fusion. At its center, matter is a hundred times more dense than water, and at temperatures of 15–20 million degrees centigrade particles collide and combine, releasing large amounts of energy. This energy radiates out into space. A tiny proportion reaches the Earth, 150 million km (93 million mi) away, where it is strong enough to heat the atmosphere, the land and the oceans.

The Earth intercepts only one part in 2 billion (2,000 million) of the energy radiated by the Sun. If all even of that small portion heated the Earth, the planet's surface would become too hot for life. However, not all the radiation is absorbed: a third of it is reflected straight back into space by an envelope of gases surrounding the Earth (the atmosphere), by clouds and by the Earth's surface. The ability of surfaces to reflect radiation, their albedo, depends on their color and texture. Clouds and snow have a high albedo, reflecting more radiation than forests, which have a low albedo. The albedo of the oceans, which cover nearly three-quarters of the surface of the Earth, varies depending on the angle at which the Sun's rays strike.

About half the incoming solar radiation is absorbed by the land surface and the oceans. Energy from the Sun interacts with the atmosphere, and it is the driving force of processes such as the water cycle, which plays an important part in sculpting the surface of the Earth.

Energy is radiated from the Sun at a wide range of wavelengths. Shortwave radiation, with a wavelength the same as or less than that of visible light, causes little heating of the atmosphere, but some is absorbed by the Earth's surface and some is re-radiated back into the atmosphere from the surface as lower-energy long waves. The atmosphere, particularly its carbon dioxide (just 0.05 percent of the atmosphere), is opaque to longwave radiation. The waves cannot pass through it, but are absorbed by the atmosphere or reflected back toward the Earth's surface, so the lower atmosphere and the surface of the Earth are warmer than they otherwise would be. This process is called the greenhouse effect.

The amount of energy reaching the surface of the Earth at any point also depends on the angle at which the Sun's rays strike the surface; this affects the amount of atmosphere they have to pass through. Rays strike the Earth almost vertically in the tropics, which receive three times as much radiation as the polar regions. Movement of the atmosphere and of water in the oceans constantly transfers energy from one area to another. Were this not so, the tropical areas would be much hotter and the polar regions much colder than they are.

Energy inside the Earth

Energy deep within the Earth also plays a major part in shaping the continents and mountains on the surface. The Earth becomes hotter with depth. Near the surface the temperature increases by about 3°C (5°F) with every 100m (330 ft), although the rate of increase may be different in cold continental or hot volcanic areas. The source of most of this heat near the surface of the Earth is the decay of long-lived radioactive isotopes of elements such as uranium. Some of the energy reaching the surface in the form of heat is almost certainly primordial, generated during or soon after the formation of the Earth.

THE ORIGIN OF THE EARTH

The Earth is one of nine planets orbiting the Sun in our solar system. The Sun is one of millions of stars orbiting the center of the galaxy known as the Milky Way. The Milky Way is just one of millions of galaxies in the universe, which originated 15 billion years ago when the "Big Bang" created pure energy, probably from a previous universe collapsing in on itself – the "Big Crunch".

The Milky Way was a cloud of gas revolving in space 10 billion years ago. Parts of the cloud began to condense; as they were pulled together by gravity the central mass became hot enough for thermonuclear reactions (fusion) to start. Stars, including our Sun, were created.

The solar system – the planets, asteroids, comets and debris that orbit the Sun – is about 5 billion years old.

The origin of the planets is not known for certain. They may be the remains of a second sun that exploded, or may have formed from gases and solids revolving around our Sun.

The Earth itself dates back some 4.6 billion years. At some stage in its early life it was completely molten, but as the surface cooled a crust was formed. Volcanic activity released vast quantities of carbon dioxide and steam. The steam condensed into water to form oceans, which were prevented from freezing by the insulating effect of the carbon dioxide. About 3,800 million years ago, at about the same time as the oldest known rocks were formed, conditions were right for life to start to evolve. The ancestors of humans evolved some 3 million years ago; our own species, *Homo sapiens*, dates back only some 350,000 years.

Weather in the making This view of the Hawaiian island chain in the Pacific Ocean, seen from the space shuttle, shows how substantial trails of cloud can form in the lee of islands as they intercept the prevailing northeasterly winds.

Over billions of years the tremendous energy within the Earth has caused the continental landmasses to move across the surface of the planet. Today's continents have not only experienced vastly different climates in the past, but the division and collision of continents has created major features of the Earth's surface: the rifts and oceans, and the great mountain ranges. The energy is expressed in the active belts of mountain-building and ocean-floor spreading across the world, in earthquakes and in volcanic activity. The classic example of continental drift is India, now welded to Asia, with the mountain range of the Himalayas still being formed.

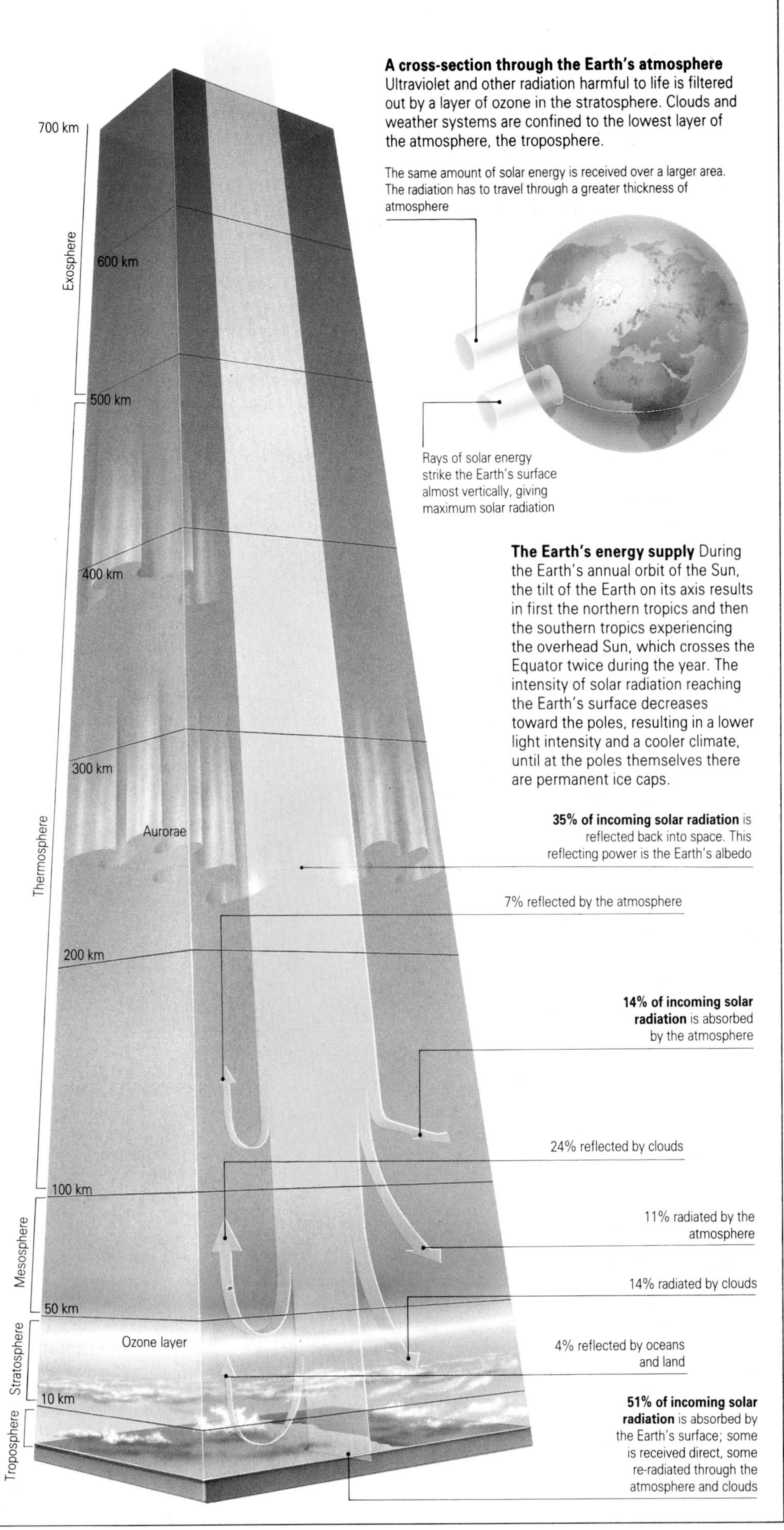

A cross-section through the Earth's atmosphere Ultraviolet and other radiation harmful to life is filtered out by a layer of ozone in the stratosphere. Clouds and weather systems are confined to the lowest layer of the atmosphere, the troposphere.

The Earth's energy supply During the Earth's annual orbit of the Sun, the tilt of the Earth on its axis results in first the northern tropics and then the southern tropics experiencing the overhead Sun, which crosses the Equator twice during the year. The intensity of solar radiation reaching the Earth's surface decreases toward the poles, resulting in a lower light intensity and a cooler climate, until at the poles themselves there are permanent ice caps.

Climates and Oceans

Hurricane Elena, 1985 A hurricane forms when the ocean surface is heated to over 27°C (81°F). This creates an area of very low pressure, which is made to spin by the rotation of the Earth. Water vapor is sucked into this vortex, creating towering clouds.

THE EARTH IS PACKAGED IN TWO FLUID envelopes, the atmosphere and the oceans. Powered by the Sun's energy and by the rotation of the Earth, they create the planet's broad climatic patterns.

The atmosphere is a protective layer of gases surrounding the Earth. It is composed mainly of nitrogen (on average 75 percent by mass) and oxygen (23 percent). Most of its layers are very thin; 90 percent of the mass of the atmosphere is in the lowest layer, called the troposphere. Climate and weather patterns are largely determined by movements in this layer, which varies in depth from about 18 km (11 mi) over the Equator to 7 km (4 mi) over the poles.

The oceans cover nearly three-quarters of the surface of the Earth. The constant motion of the ocean currents, and of air masses in the atmosphere above the oceans and continents, redistributes heat received from the Sun, providing the energy that drives the processes shaping the landscape.

World climates

The broad pattern of movement in the atmosphere is reflected in the Earth's climatic patterns. The Earth receives the greatest amount of solar radiation in the tropics, a zone that lies between the Tropic of Cancer and Tropic of Capricorn and is centered on the Equator.

The pattern of movement in the atmosphere begins in the equatorial zone with a circulation of air known as the Hadley cell. Warm air expands, becomes less dense and rises, creating an area of low pressure at the Earth's surface. As it rises the air cools, and some of the moisture condenses and falls as rain. Once aloft, the cool, heavier air spreads out toward the tropics and then sinks again, creating belts of high pressure in subtropical latitudes bordering the tropics. There is very little rain in this belt because the air warms as it sinks and is thus able to absorb moisture. The subtropical high-pressure belts of both the northern and southern hemispheres correspond to the major areas of desert – the Sahara in North Africa, the Kalahari in southern Africa and the deserts of the Australian interior.

From the subtropical zones most of the air is drawn back toward the equatorial belt, as the northeast trade winds in the northern hemisphere and the southeast trade winds in the southern hemisphere.

The rotation of the Earth prevents the air currents from moving due north and south. Winds deflect to the right in the northern hemisphere and to the left in the southern hemisphere. This is called the Coriolis effect.

The zone where the trade winds meet is known as the intertropical convergence zone (ITCZ). This "meteorological equator", a region of towering cumulus clouds and rain, moves north and south of the true Equator, following the latitude where the Sun is overhead at midday.

Some of the air in the subtropical high-pressure zones moves toward the poles as the westerly winds, and meets cool air moving away from the polar areas. The two air masses meet along the polar front, whose north–south position (latitude) again migrates with the seasons. At high altitudes of 10–12 km (6–7.5 mi) in the upper reaches of the troposphere, the westerlies form the polar front jet streams, winds blowing at speeds up to 350–450 k/h (220–280 mph). Westerly jet streams also form above the subtropical high-pressure belts.

The amount of moisture in the atmosphere at any one time is the equivalent of only about 25 mm (1 in) of rainfall spread over the surface of the Earth. The average rainfall in a year over the Earth as a whole is 1,000 mm (40 in). Moisture constantly moves between the surface and the atmosphere in the water (or hydrological) cycle. Moisture can be carried long distances in the atmosphere before the air cools sufficiently for the moisture to condense and return to the surface in the form of precipitation – mist, rain, sleet, snow or hail.

Patterns of precipitation relate to the global patterns of air movement. The southeast coast of Nigeria in West Africa, for example, lies in the equatorial belt and receives over 3,000 mm (120 in) of rain a

THE OCEANS

The oceans cover some 368 million sq km (142 million sq mi) – 70.8 percent of the Earth's surface. The average depth is 3,800 m (12,500 ft) and the total volume of water 1.4 billion cu km (330 million cu mi). The oceans provide some 0.02 percent of the Earth's total weight.

The most important way in which the oceans affect climate and weather is through evaporation. When water evaporates heat is transferred from the water into the moisture in the atmosphere. When the water vapor condenses heat is released and remains in the upper air. Through evaporation and condensation energy is transferred both vertically and from one place to another. In this way the oceans play a vital part in driving the water cycle, which has a major influence on the appearance of the land surface, through the action of rain and rivers.

The oceans store heat. The top 3 m (10 ft) of water stores as much energy as the entire atmosphere. They can store and release great quantities of heat with only small changes in temperature, and this can have a major effect on air temperatures. In the middle of a large continent in the mid-latitudes the difference between summer and winter temperatures can be as much as 80°C (144°F). On the other hand the seasonal variation of sea temperatures in the same latitude is rarely more than 10°C (18°F).

Heat is transported by movement in the waters of the ocean. Cold currents flow from the polar areas toward the Equator and warm currents away from the Equator, in huge circular motions driven by the wind and the Earth's rotation. The warm North Atlantic Drift moves more water across the North Atlantic than all the rivers in the world. The prevailing westerly winds carry this heat over northwestern Europe, producing a warmer climate there than would be expected for its latitude. St John's in Newfoundland, Canada has an average temperature in January of about –5°C (23°F); on the other side of the Atlantic Ocean and a few degrees farther north, the January temperature at Kew Gardens near London, England averages about 4°C (39°F).

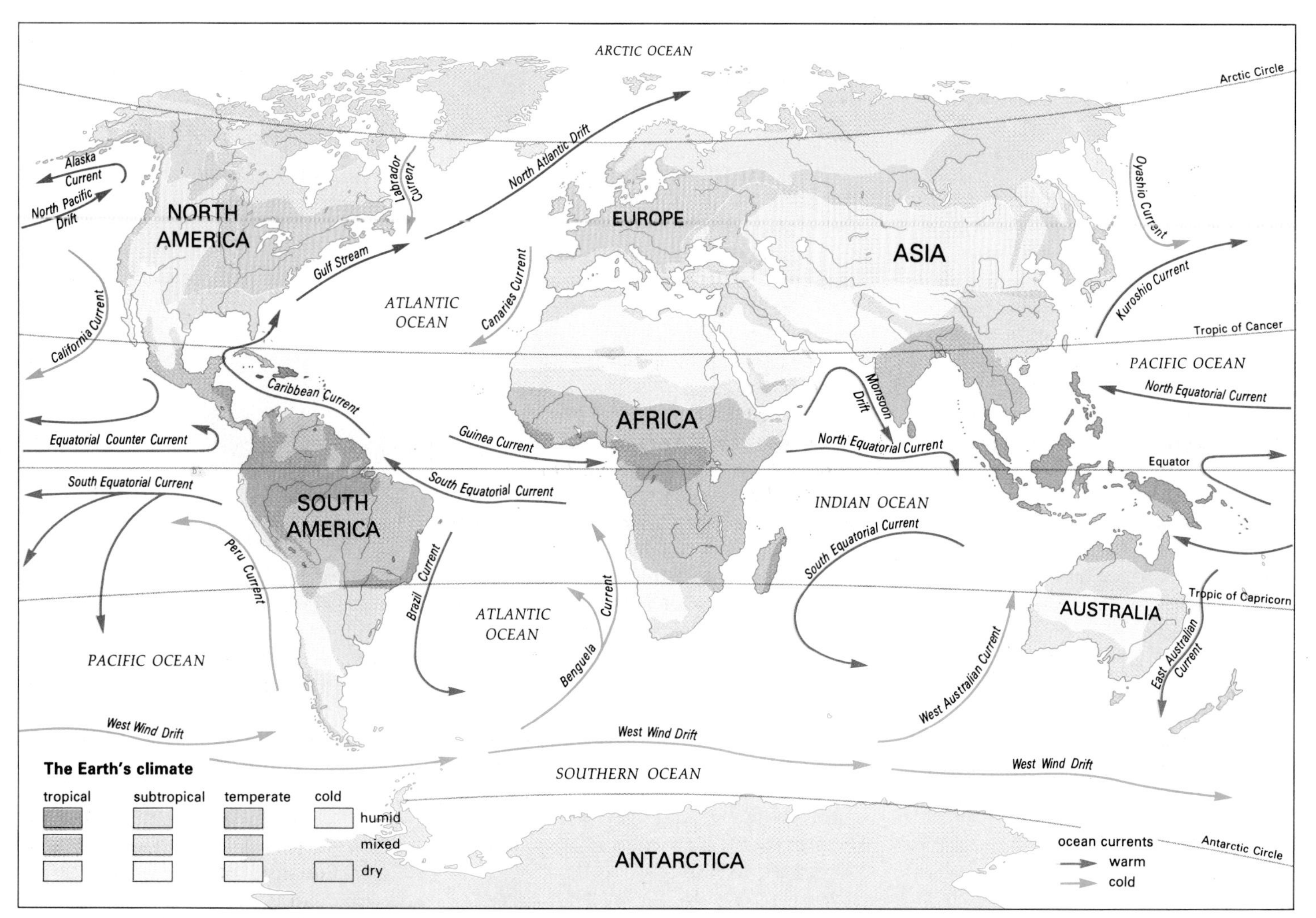

Climate zones and ocean currents Climate is affected by latitude, prevailing winds, topography, distance from the sea and ocean currents. The currents are caused by a combination of wind, differences in water density due to temperature and salinity, and the Earth's rotation.

year; the northern part of the country lies in the subtropical high-pressure belt and receives only 250 mm (10 in).

Climate and land-shaping processes

While climate provides the overall context within which land-shaping processes operate, climatic averages do not always provide the best indicators of the major influences on the landscape. Exceptional conditions may be more significant. In recent years droughts in the Sahel, the semidesert southern fringe of the Sahara, have killed much of the vegetation that once held the soil in place. When it does rain, erosion is much greater than before and deep gullies can be formed.

Climates are also not the same now as they have been in the past. In many areas the scenery has been determined by processes in the past under very different conditions. Ice sheets and glaciers have extended to lower latitudes during several ice ages, and many of today's landscapes – for example in parts of northern Europe and around the Great Lakes of North America – reflect processes of glacial erosion and deposition in the past.

Global wind system

The Earth's rotation deflects winds to the right in the northern hemisphere and to the left in the southern hemisphere

Polar jet stream

Tropic of Cancer high pressure belt

Intertropical convergence zone (ITCZ)

Hadley cell

Subtropical jet stream

Westerlies

Northeast trades

Southeast trades

The circulation of the atmosphere is driven by energy from the Sun. Warm air is less dense than cold air; it rises, creating areas of low pressure to which colder air is drawn. These effects are complicated by the Earth's rotation, the differential warming of land and sea, and the topography of the land. Rossby waves, undulations in the high-altitude westerly jet streams above the polar fronts, play a major role in climate.

Formation of Rossby waves

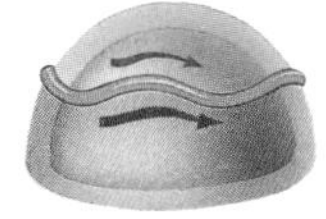

An undulation hundreds of kilometers long develops in the high-altitude jet stream above the polar front as a result of temperature gradients and the effects of the Earth's rotation

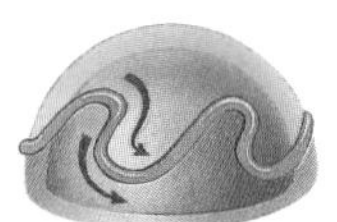

The undulation increases in amplitude, and is also modified by the relief of the land below. It affects the distribution of warm and cold air in the lower atmosphere

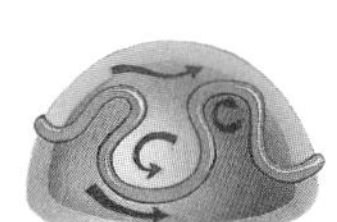

Masses of cold and warm air enclosed in the loops of pronounced waves become detached, to form low-pressure cells (cyclones) and high-pressure cells (anti-cyclones) that dominate climate in mid-latitude regions

Inside the Earth

ENERGY FROM THE SUN DRIVES THE PROCESSES that sculpture the Earth; energy from within the Earth makes the basic structures such as the continents and mountain ranges, and determines their positions on the Earth's surface.

The Earth is a near-sphere, with a diameter at the Equator of 12,756 km (7,921 mi) and at the poles of 12,714 km (7,895 mi). At some stage in its evolution it was a molten mass, but as it cooled from the outside a solid skin was formed with a series of concentric layers beneath.

At the center is the core, made of iron with a smaller amount of nickel. The temperature may be as high as 5,500°C (9,900°F), but because the pressure is 3 to 4 million times greater than atmospheric pressure at sea level the central part, or inner core, cannot melt and is solid. The outer layer of the core, under less pressure, is like a very thick liquid.

Around the core is the mantle, a layer of more or less solid rock rich in iron. Within the upper mantle there are two distinct layers. The asthenosphere consists of rocks close enough to their melting points to be malleable or partly molten; above it is another, more rigid layer of rock.

The topmost layer of the solid Earth is the thin skin of the Earth's crust, consisting of relatively light rocks, largely granite beneath the continents and basalt beneath the oceans. The oceanic crust is about 6 km (3.7 mi) thick on average. Its material is heavy compared with that of the continental crust, which averages some 33 km (20 mi) in thickness and therefore sinks much deeper into the mantle. The boundary between mantle and crust is known as the Mohorovicic discontinuity, or Moho, after the Yugoslav seismologist whose work helped in its discovery.

The Earth's heat

The heat inside the Earth has three main sources. Some of it remains from when the Earth was first formed, some is caused by the enormous pressure the rocks are under, and some comes from radioactive chemicals that release a lot of heat when they break down. Near the center the Earth's original heat is the most significant source, while at depths of some 10 km (6 mi) radioactivity contributes about three-quarters of the heat, and fuels volcanic outbreaks at the surface. All this heat from geothermal sources amounts to less than 1 part in 3000 of the total energy reaching the Earth from the Sun.

Continental drift

Early in the 20th century the German meteorologist Alfred Wegener suggested that the continents were once all part of one supercontinent, Pangea, which split into sections and drifted apart. He supported his idea by showing that the outlines of the continents could be fitted together like a jigsaw. Further evidence for the theory was that similar rocks and patterns of geological history were found in widely separated places. Traces of glaciation in what is now India revealed that past climates were very different from those of today.

The idea was received with much skepticism and generated a lot of debate until the 1950s, when paleomagnetism could be measured. The magnetic alignment in old rocks follows the Earth's slowly changing magnetic field at the time the rocks were formed. As the alignment in rocks of the same age on different continents often does not match, the continents must have moved in relation to each other.

Plate tectonics

How the continents gradually change their position is explained by the theory of plate tectonics. The outer shell of the Earth consists of moving slabs called tectonic or lithospheric plates. Each plate comprises a proportion of oceanic crust or continental crust, or both, attached to a portion of the underlying rigid layer of the upper mantle. The plates, which bear the surface of the Earth, move over the malleable asthenosphere underneath.

There are seven major tectonic plates and several smaller ones. They are all constantly and very slowly moving in relation to each other at rates of from a few millimeters to a few centimeters each year. Where plates move apart molten rock comes to the surface and forms new oceanic crust. Where they collide, one plate plunges beneath the other, creating ocean trenches and crumpled mountain ridges, often associated with earthquakes and volcanoes.

Plate tectonics has explained why the continents have moved and accounts for the distribution of earthquakes and volcanic activity. No one is quite sure where the forces driving the plates come from; they are probably driven by energy from within the Earth.

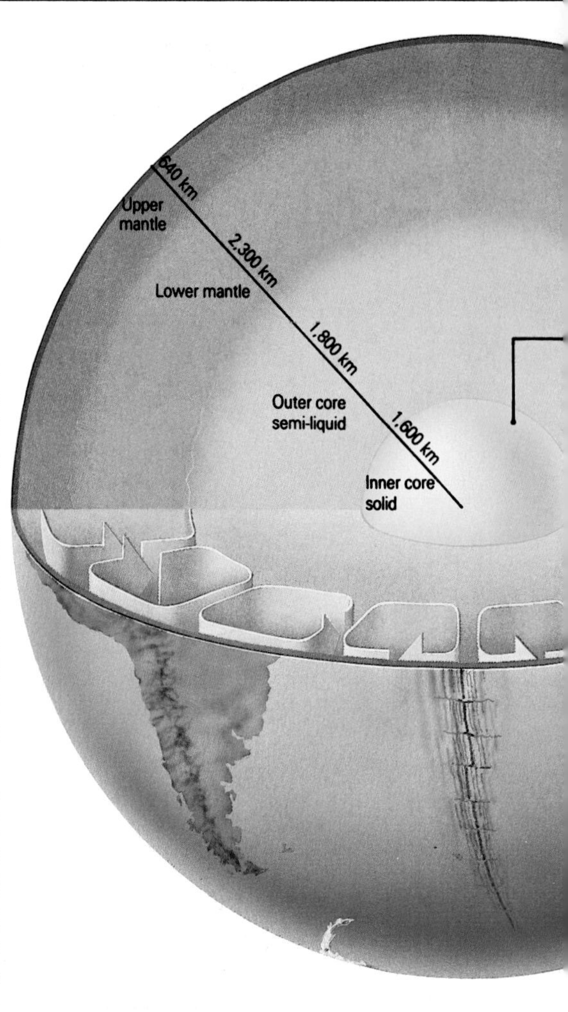

EARTHQUAKES

Lisbon (in 1755), San Francisco (1906), Quetta (1935) and Armenia (1988) have all suffered earthquakes that reduced cities to rubble and killed thousands of people. They lie on the boundaries of major plates, where 95 percent of all earthquakes occur.

As the plates move, pressure builds up in the Earth's crust, eventually causing it to fracture and find a new position. Not all earthquakes are so devastating. The plates move constantly, and in some places frequent small tremors go almost unnoticed.

Earthquakes set up waves that radiate out from the center of movement like ripples in a pool of water. The point at the surface above the focus of the earthquake is called the epicenter. In the areas of most recent mountain-building the land can be raised by as much as 20 m (66 ft) in a thousand years, but the activity may be concentrated into very brief periods, each one associated with an earthquake. In other areas, such as the Rift Valley in East Africa, the land has subsided, and has been flooded by a series of lakes.

It is often possible to tell from the landscape where earthquakes have taken place, as rock strata and river courses may fracture and be displaced. The line along which the rocks fracture is called a faultline.

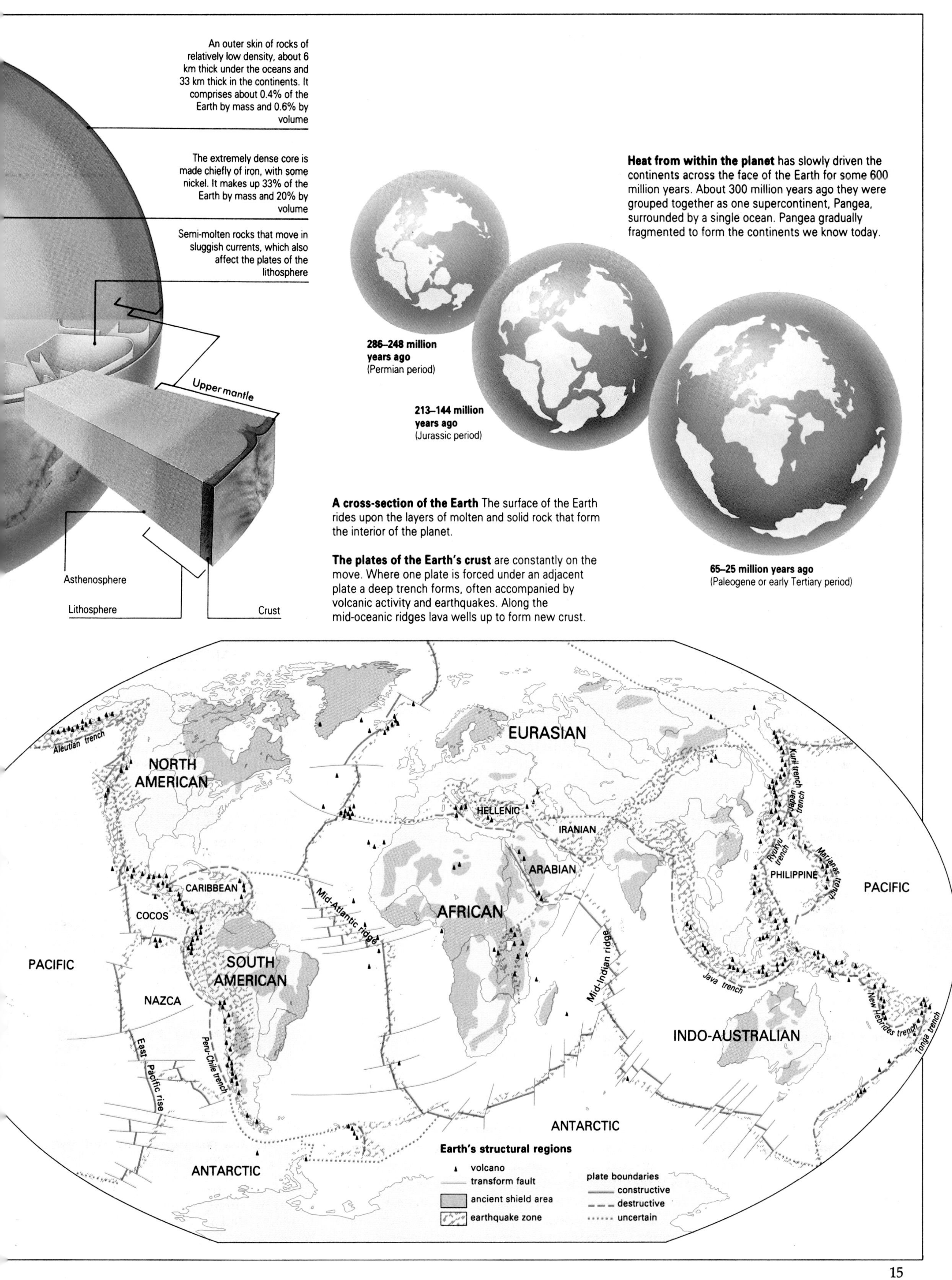

Heat from within the planet has slowly driven the continents across the face of the Earth for some 600 million years. About 300 million years ago they were grouped together as one supercontinent, Pangea, surrounded by a single ocean. Pangea gradually fragmented to form the continents we know today.

A cross-section of the Earth The surface of the Earth rides upon the layers of molten and solid rock that form the interior of the planet.

The plates of the Earth's crust are constantly on the move. Where one plate is forced under an adjacent plate a deep trench forms, often accompanied by volcanic activity and earthquakes. Along the mid-oceanic ridges lava wells up to form new crust.

The Earth's Crust

THE MOVEMENTS OF THE PORTIONS OF THE Earth's outer shell known as tectonic plates account both for the movements of the continents and to a great extent for the patterns and processes of volcanic activity and mountain-building.

Plate boundaries

Areas where tectonic plates move apart are called constructive plate boundaries. They are usually found beneath the oceans, where molten rock from inside the Earth rises to fill the gap and form new oceanic crust. The molten material wells up to form a ridge, and the sea floor spreads. The Atlantic Ocean is widening by about 2 cm (0.8 in) a year on both sides of the Mid-Atlantic ridge, which curves down the center of the Atlantic Ocean for 16,000 km (10,000 mi). It is studded with live volcanoes; occasionally its peaks rise above the sea as islands, such as the Azores and Iceland. The ridge has been formed as the American, Eurasian and African plates have moved apart over the past 190 million years.

Although the divergence of plates is most common beneath the oceans, there are two major zones of crustal spreading on the continents. The deeper parts of one rift system, in Siberia, northern Asia, are occupied by the world's deepest lake, Lake Baikal. In Africa a series of rifts extends south from the Red Sea and through the East African highlands, where the Rift Valley contains several lakes including Tanganyika and Malawi. These continental rift systems are not the margins of major plates today; they can be thought of as large faults or incipient plate boundaries.

When plates move toward each other and collide, one plate slides below the other at an angle of 30–60°. The sinking plate, carrying the crust with it, melts and is drawn down (subducted) into the mantle to become part of the asthenosphere. The zone along this margin is known as a subduction zone.

The structures resulting from the collision depend upon the types of crust that are brought together.Where oceanic crust meets oceanic crust, island arcs such as that of the eastern Philippine islands are formed. Volcanoes are formed by molten rock (magma) rising from the melting tectonic plate. Where oceanic crust meets less dense continental crust, the oceanic crust plunges downward and creates deep ocean trenches. The Marianas trench in the western Pacific is the deepest, at 11,033 m (36,198 ft) in Challenger Deep. The continental crust is buckled by the impact, and mountain ranges such as the Andes of South America are formed. Parts of the continental crust may melt and become magma; because it is less dense than the surrounding material it tends to rise, and may break through the overlying rocks to erupt and build volcanoes.

Mountains are also formed where continental crust meets continental crust, as a result of continental plates colliding over millions of years. The Himalayas, for example, were formed by the collision of the plates carrying India and Siberia. At these plate boundaries there are very few volcanoes.

Plates sliding past each other are neither destructive nor constructive, and are known as conservative plates. Deep vertical fractures, or transform faults, along which portions of the lithosphere slide past each other, are usually associated with oceanic ridges, but a well-known example on land is the San Andreas fault in California. Here the margins of the northward-moving Pacific plate meet the North American plate. Movement along the fault is irregular, making this a zone of intense earthquake activity.

Mountain-building

The two main forces building mountains are volcanic and tectonic activity. Volcanoes are built up of lava and ash, which originate under the ground as magma. In tectonic activity the ground is buckled and broken by the movements in the Earth. The two kinds of activity often combine to produce a mountain range.

The chief periods of mountain-building include the Caledonian, which about 400 million years ago affected both Scandinavia, the northern British Isles, and eastern North America; the Hercynian/Appalachian, at its most active some 300 million years ago; and the Alpine, which began about 30 million years ago and is still active in many parts of the world. It is one of the two narrow belts of active mountain-building today. It lies mainly along continental margins, and extends from the Atlas Mountains in North Africa, through Europe and Asia to the Philippine Islands. Here it joins a second belt circling the Pacific, known as the Ring of Fire. Off the continent of Asia the belt takes the form of island arcs running through the Philippines, Japan and the Aleutian Islands. In the Americas this mountain belt lies mainly on the continents, forming the Rockies and other ranges of western North America, and the Andes in South America.

The San Andreas fault in California, USA, forms a distinctive feature of the landscape. In this area the Pacific plate is moving north past the North American plate; friction between the margins of the two plates is powerful enough to fracture the rocks.

The crust of the Earth is less dense than the mantle beneath it, and behaves as if it were floating. Mountain ranges on the surface are matched by deep depressions in the mantle caused by the weight of the crust pressing down. Erosion wears away mountains; as the weight of the crust is reduced it starts to rise out of the mantle. The sedimentary rocks derived from eroded material are usually deposited near the coast. When they build up they too start to press down into the mantle. The state of balance between crust and mantle is called isostasy.

Precambrian shields

At the heart of every continent are vast tracts of ancient rock dating back to before the Cambrian period, which began some 600 million years ago. These "shield" areas were once subjected to the stresses of mountain-building, but the mountains have long since been worn away. The shields are so thick and solid that they cannot be part of any further mountain-building, and form the stable, rigid core of the continents.

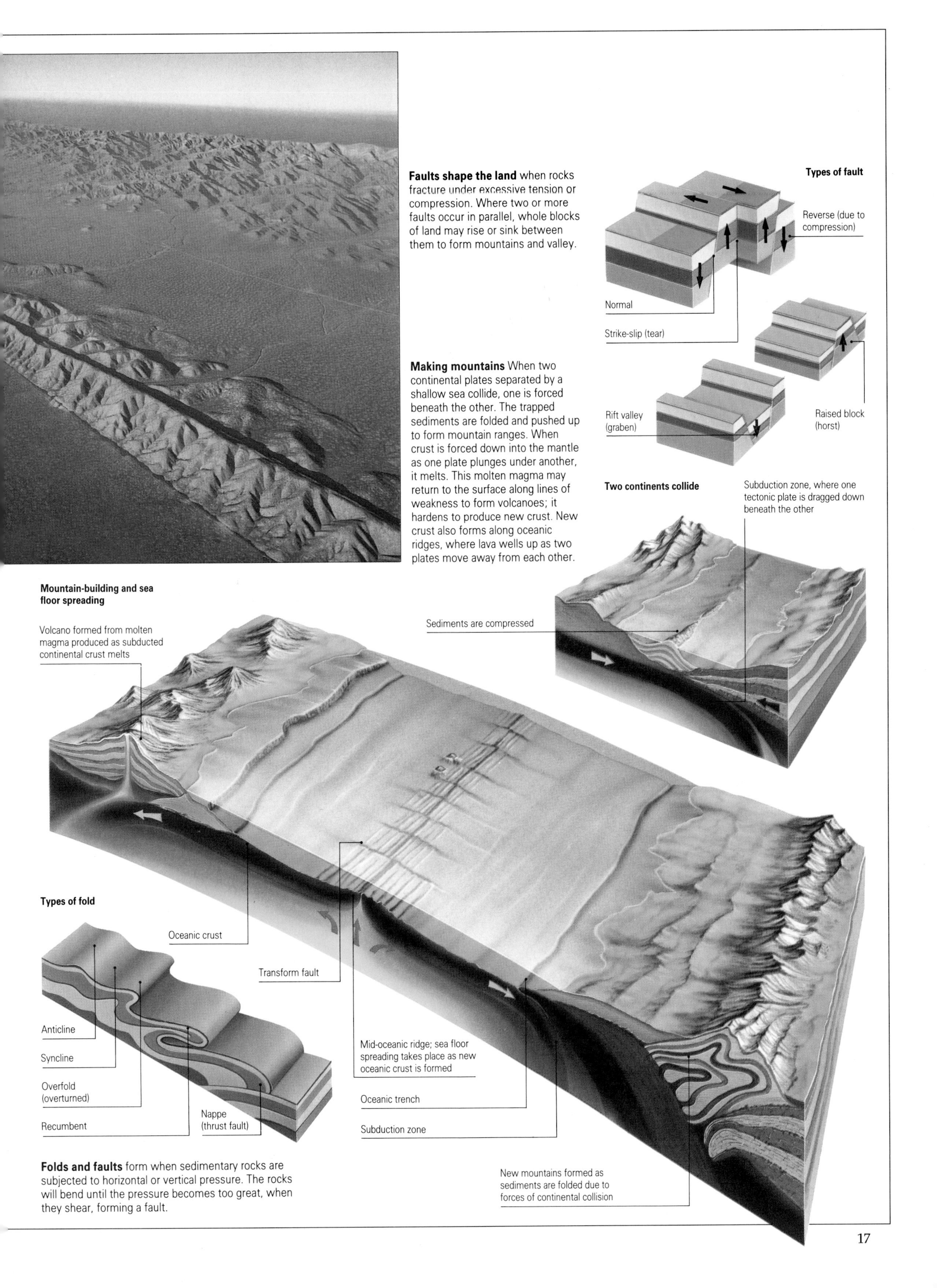

Faults shape the land when rocks fracture under excessive tension or compression. Where two or more faults occur in parallel, whole blocks of land may rise or sink between them to form mountains and valley.

Making mountains When two continental plates separated by a shallow sea collide, one is forced beneath the other. The trapped sediments are folded and pushed up to form mountain ranges. When crust is forced down into the mantle as one plate plunges under another, it melts. This molten magma may return to the surface along lines of weakness to form volcanoes; it hardens to produce new crust. New crust also forms along oceanic ridges, where lava wells up as two plates move away from each other.

Folds and faults form when sedimentary rocks are subjected to horizontal or vertical pressure. The rocks will bend until the pressure becomes too great, when they shear, forming a fault.

History through Rocks

THE HISTORY OF THE EARTH CAN BE followed from the characteristics of the rocks that form the crust. Some hard granites have remained unchanged for 3,800 million years, and geologists can learn about the Earth's origins from them. Other rocks are being formed today, sometimes from a spectacular volcanic eruption - but usually from very slow processes, such as the accumulation of fine-grained silt in a river estuary. These new rocks will help geologists of the future to find out about today's world.

New rocks can be formed only if old ones are destroyed. The minerals of the Earth's crust are recycled over millions of years to create a great variety of rock types, though all rocks are formed in one of three ways.

Rock formation

Molten rock (magma) deep underground provides the basis for igneous rocks. Some magma pushed toward the surface cools and solidifies before reaching it, forming intrusive rocks; granite is a major example. Molten magma that reaches the surface is known as lava, and the rocks formed from it are called extrusive or volcanic rocks. Over nine-tenths of volcanic rock is basalt, which is dark in color and fine-grained.

As soon as rocks are exposed at the surface the weather starts to break them down. Heat and cold, water, ice and chemicals split, dissolve or rot the rock until it crumbles. Broken fragments are carried by ice, rivers and the wind and deposited elsewhere; most end up on the sea bed fringing a continent. The deposits are compressed and bound together by natural cements to form layers of new rock. The heavier fragments, including pebbles, settle out nearest the coast and become breccias and conglomerates, natural concretes with sharp-edged and rounded fragments respectively, embedded in finer material such as sand and clay. Much of the finer material is carried farther out from the coast before it settles and becomes sandstone and shale.

Coal and some limestones – also sedimentary rocks – are formed from the compressed remains of plants and animals. Other sedimentary rocks, such as halite (rock salt) and gypsum, are formed from chemicals dissolved in water and precipitated when the water evaporates.

Metamorphic rocks are formed when great heat or pressure alters igneous or sedimentary rocks to give new minerals and textures. Rocks may be metamorphosed by the great weight of rocks lying over them, by being close to hot igneous rocks, or by the intense pressures caused during mountain-building. Igneous rocks such as granite may be changed into gneiss; limestone becomes marble.

Rocks and landscape

The recycling of the minerals of the Earth's crust does not take place at a constant rate. During the Earth's long history there have been in different areas stable periods of erosion and deposition separated by periods of mountain-building. Today's landscapes are the result of rock movements during the last 65 million years, the Cenozoic era. These created the structures on which the processes of weathering, erosion and deposition are operating today.

The characteristics of a rock affect its shape in the landscape, and help the history of that landscape to be understood. Resistant rocks normally form the highest areas and the steepest slopes. More easily eroded rocks form lowlands with less dramatic outlines. Where hard and soft rocks have been folded or faulted to form alternate bands at the surface, a succession of ridges and vales may be formed.

MINERAL FORMATION

The Earth has 92 naturally occurring elements. Eight of these account for 98 percent of the weight of the Earth's crust: they are (in order, by mass) oxygen, silicon, aluminum, iron, calcium, manganese, sodium and potassium. Only five elements are found alone in a natural state – sulfur, carbon, and three metals, gold, silver and copper. The other elements join together in various combinations to make compounds such as copper sulfide, which is formed when copper and sulfur atoms combine.

Elements and compounds found in the Earth are called minerals. There are about 2,000 minerals, but only 30 of them are common. Most rocks are composed of two or more minerals. Granite, for example, is composed mainly of quartz, orthoclase, muscovite and biotite. Quartz is one of the most widespread minerals in the Earth's crust. It is made of the elements silicon and oxygen, which are the basic ingredients of most rocks.

PRINCIPAL ROCK TYPES

Rock	Type
Igneous	
intrusive	
plutonic (deep)	gabbro, granite
hyperbyssal (in cracks in rocks)	dolerite
extrusive	lava, (basalt, obsidian, pumice), ash
Sedimentary	
cemented fragments (detrital)	clay (consolidated as shale) sand (consolidated as sandstone) silt (consolidated as siltstone) gravel (may become conglomerate)
from chemicals dissolving out of seas and lakes	some limestone, rock salt (halite)
from living things	some limestone (chalk, coral), peat, coal
Metamorphic	
layered (foliated)	slate (from shale), schist (from slate or basalt), gneiss (e.g. from granite)
not layered	quartzite (from sandstone), marble (from limestone)

The Grand Canyon of the Colorado river in the United States is 1.5 km (0.93 mi) deep. For 10 million years the river has been cutting through a rising tableland of rocks of different ages; the oldest date back over 4 billion years. Weathering of the rocks has produced the dramatic shapes of the canyon walls.

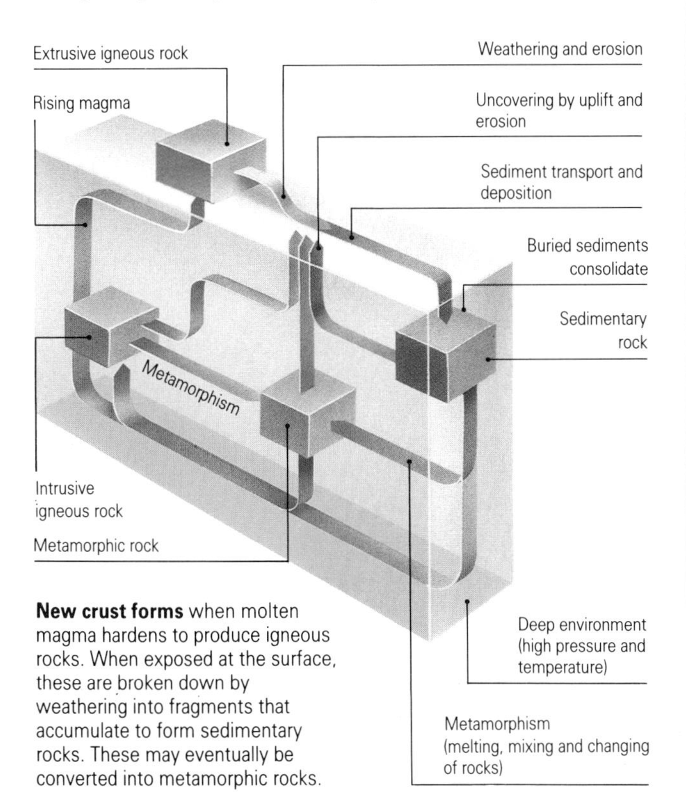

New crust forms when molten magma hardens to produce igneous rocks. When exposed at the surface, these are broken down by weathering into fragments that accumulate to form sedimentary rocks. These may eventually be converted into metamorphic rocks.

THE GEOLOGICAL TIME-SCALE

Eon	Era	Period	Event
			15 billion (15,000 m) years ago **"Big Bang" origin of the Universe** 10 bya **Milky Way a cloud of gas** 5 bya **Solar System originated**
Archean 4.6 bya	Pre– cambrian		4.6 bya **Origin of the Earth** 3.8 bya **Oldest known rocks** 3.5 bya Bacteria, algae
Proterozoic 2.5 bya			2.4 bya **Large continents develop** 2.3 bya Glaciation 1.9, 1.2 and 0.7 bya **Mountain-building** 0.8–0.7 bya Glaciations widespread in all continents (except South America) 0.7 bya Oxygen-dependent life in seas
Phanerozoic 590 mya	Paleozoic	Cambrian	Marine animals with skeletons
		505 mya Ordovician	500 mya **Antarctica near Equator** 470–440 mya Glaciation in Sahara
		438 mya Silurian	Coral reefs First land plants
		408 mya Devonian	400 mya **Caledonian mountain-building in NW Europe and NE America**
		360 mya Carboniferous	Coal forests 320–250 Gondwanaland glaciation
		286 mya Permian	300 mya **Mountain-building in central Europe (Hercynian) and Appalachians**
	Mesozoic	248 mya Triassic	250 mya **Single continent, Pangea, formed** Deserts widespread
		213 mya Jurassic	190 mya **Pangea starts to break up, North America moving away from Eurasia**
		144 mya Cretaceous	140–130 mya South Atlantic opening up 120 mya **Gondwanaland splitting up** Flowering plants evolve 100 mya **India splits from Antarctica**
		Epoch	
	Cenozoic	Tertiary 65 mya Paleocene	65 mya **Rocky Mountains formed**
		55 mya Eocene	55 mya **Africa touches Eurasia** 45 mya **Australasia separates from Antarctica**
		38 mya Oligocene	40–30 mya **India collides with Asia**
		25 mya Miocene	30 mya **Alpine–Himalayan mountain-building in progress**
		5 mya Pliocene	3–2 mya Human species
		Quaternary 2 mya	2 mya **North and South America joined**
		Pleistocene	2 mya recent ice age begins
		0.01 mya Holocene	18,000 ya maximum extent of ice 10,000 ya beginning of agriculture 6,000 ya postglacial rise in sea levels

Changing Climates

Orbit stretch 96,000-year cycle. The orbit of the Earth around the Sun alternates gradually between elliptical and circular. Thus the average distance of the Earth from the Sun, at present 149.6 million km (92.8 million mi), varies by up to 18.3 million km (11.4 million mi). This variation affects the amount of solar radiation reaching the Earth

A DROP OF ABOUT 4.5°C (8°F) IN THE AVERAGE temperature is all that is necessary to plunge the Earth into an ice age in which nearly three-tenths of the land would be covered by ice, in places 3 km (2 mi) thick. In its long history the Earth has experienced many ice ages, some lasting for 100 million years. During an ice age there are periods when it is very cold (glacials or glaciations) separated by warmer periods (interglacials).

The most recent ice age started about 2 million years ago. Since then the ice sheets and glaciers have advanced and retreated several times. At present the Earth is in one of the temporary interglacial periods, and only one-tenth of its land surface is covered by ice.

A glacial period begins with the glaciers in mountain areas reaching farther down the valleys before they begin to melt, and the great ice sheets spreading out farther from the polar areas. Glaciers and ice sheets eventually join up, and then extend farther toward the middle latitudes (45° N or S) as the ice age tightens its grip. As more and more water is stored in the ice instead of the oceans the sea level falls, and the outlines of the continents change. The major climatic belts and vegetation zones all shift toward the Equator. Only small areas of tropical forest remain unaffected.

Causes of climatic change

It is not known exactly why climates fluctuate enough to cause ice ages. Many factors influence the climate, and the time-scale over which they operate varies enormously. This makes it difficult to assess their relative impact. One major factor is probably very long-term variation in the amount of radiation emitted by the Sun. Changes in the relative positions of the poles, oceans and landmasses as a result of continental drift help to explain the ice ages of about 280 million years ago. Other factors may include variations in the intensity of the Earth's magnetic field, volcanic eruptions and the effects of major periods of mountain-building.

The most popular theory to explain colder and warmer phases in the most recent ice age links cycles in the motion and position of the Earth as it orbits the Sun to changes in the amount and the distribution of solar energy the planet receives. The cyclical variations were first calculated by the Yugoslav astronomer Milankovitch, who identified three cycles, each of different length. The Earth orbits the Sun once a year along an elliptical path that elongates and shrinks, a feature known as eccentricity, in a 96,000-year cycle. As a result, there are periods when the average distance between Sun and Earth is greater than at others, and the Earth therefore receives proportionately less solar radiation.

A second cycle, known as the precession of the equinoxes, takes place over a period of 26,000 years. The time of year (at present on about 2 January) when the Earth is closest to the Sun varies; 13,000 years ago it was on 21 June, and northern hemisphere summers would have been warmer. The tilt of the Earth's axis in relation to the plane of its orbit (the ecliptic) also changes, in a 40,000-year cycle, and seasonal contrasts are magnified. Together these cycles modify the amount and distribution of solar radiation received, substantially altering the Earth's temperature.

Recent changes of climate

The climatic changes of the Quaternary period, which coincides with the most recent ice age, have had the most effect on today's landscape and are the easiest to record accurately. The period is divided into the Pleistocene epoch, which started about 2 million years ago, and the present Holocene epoch, beginning about 10,000 years ago. It was once thought that there were four main glacial periods during the Pleistocene, but recent research shows that there were more. Samples taken from the bottom of the Pacific Ocean indicate that there were eight glacial periods over about 1 million years.

The Pleistocene's last glaciation was at its maximum about 18,000 years ago, when large areas of the polar and middle latitudes were covered with ice. Beyond the ice margins there were great areas of open tundra and frozen soils, similar to those of northern Canada and the Soviet Union today. The lower latitudes appear to have been drier than they are now, and the great tropical deserts were much more extensive.

The Holocene epoch started with a marked increase in temperatures around the world, and a retreat of the ice sheets. However, within this short period the climate has still fluctuated: in Europe particularly severe winters and a growth in glaciers were recorded between the 15th and early 18th centuries, which is sometimes known as the Little Ice Age.

CHANGES IN SEA LEVEL

The level of the sea in relation to the land is never static. Seashores are covered and uncovered twice a day as the tide ebbs and flows, drawn by the gravitational pulls of the Sun and the Moon. The difference between high and low tide in the Bay of Fundy on Canada's Atlantic coast can be as much as 12 m (40 ft), and tides of over 20 m (65 ft) have been recorded there. Waves driven by the wind, and others caused by earthquakes in the oceanic crust, reach even greater heights, and can cause tremendous damage to many coastal areas.

These short-term changes do not alter the average sea level in relation to the land, and are small in comparison with the effects of ice ages. During an ice age a greater proportion of moisture evaporated from the sea is held as ice. Sea levels fell by some 170 m (560 ft) during the last great advance of the ice. There is still enough water in the icecaps and glaciers to raise sea level by another 65–70 m (215–230 ft) should they melt. These changes in level resulting from changes in the volume of the oceans are known as eustatic changes.

Changes of sea level in relation to the land can also be caused by the rising and falling of the land itself. Glaciers and ice sheets place a heavy load on the Earth's crust, and the land is depressed. As the ice melts the land gradually rises again to reach a new equilibrium (isostasy). Old shorelines are often found on hillsides in areas that were once under ice.

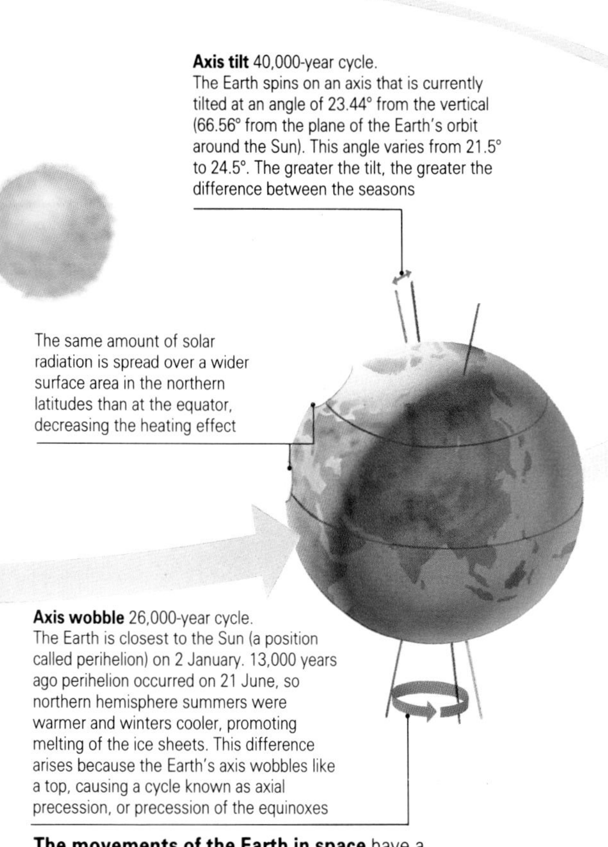

The movements of the Earth in space have a powerful effect on climate. The distance of the Earth from the Sun determines the total amount of solar radiation it receives, while the tilt and wobble of its axis affect the relative temperatures of summer and winter.

The Devil's Hole Golf Course in Death Valley, California, USA. The crust of salt was left behind when a lake evaporated as the climate warmed following the last ice age. The salts are still being added to when flash floods refill the lake.

The world's vegetation 18,000 years ago, at the height of the last glaciation. With so much water trapped in snow and ice, sea level was much lower and the climate drier. Deserts were more extensive and rainforests greatly reduced.

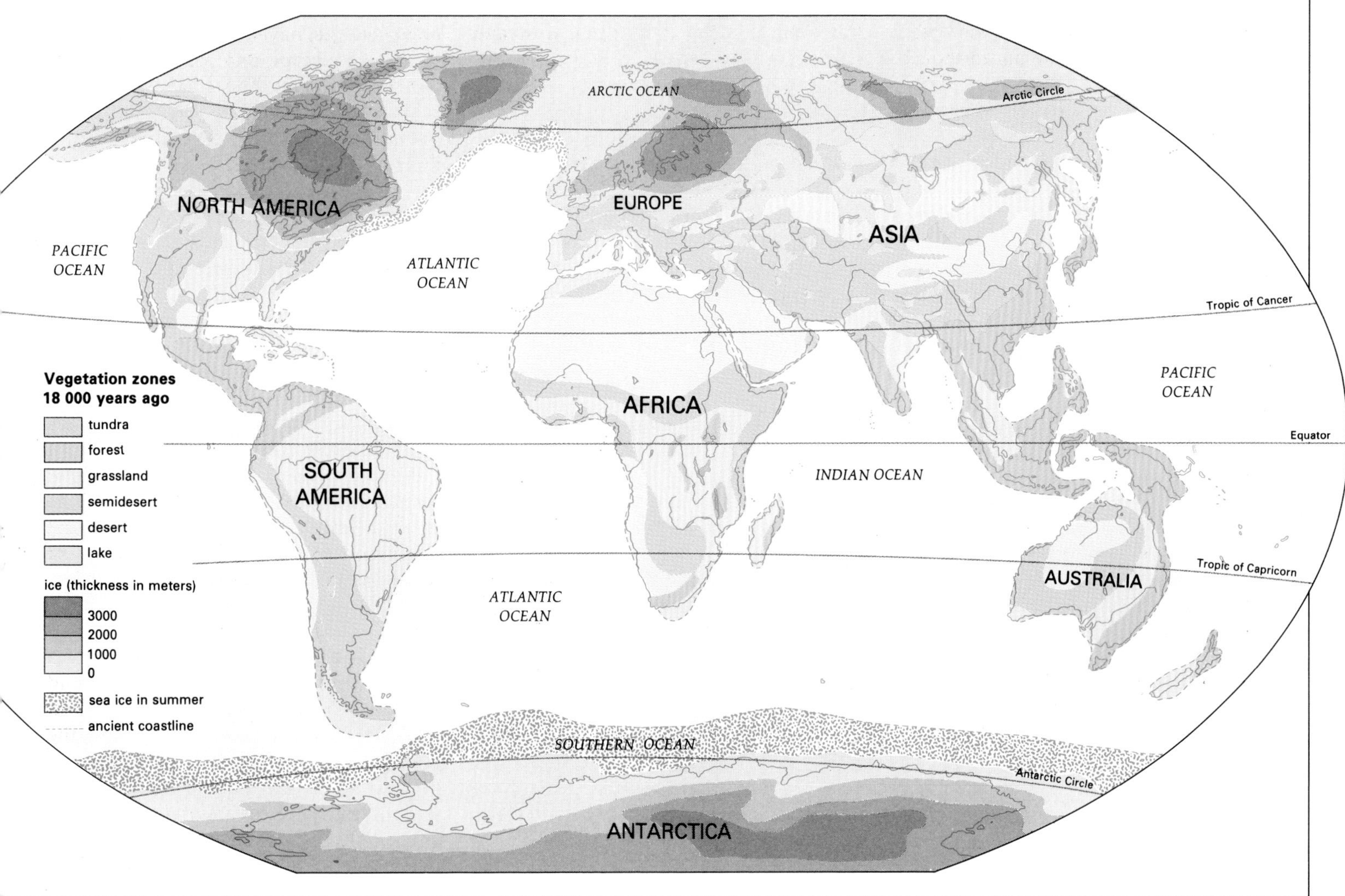

Legacies of the Ice

IN THE PAST ICE HAS COVERED ABOUT A THIRD of the land surface of the Earth. The landscape in those parts can be understood only by knowing how glaciers and ice sheets work, which may be learned from the ice that still covers 10 percent of the land. Glacier ice erodes the land; sediment contained within the ice or associated with it is deposited on the land. In the areas adjacent to glaciers and ice sheets alternate freezing and thawing of water, both in cracks in the rocks and beneath the surface, can create a series of distinctive features. Ice transports debris of all sizes very effectively. Eroded material can be carried hundreds of kilometers from its original source, and provides clues about the direction and extent of ice movements.

Erosion and weathering

Moving ice erodes the land effectively only if fragments of rock are embedded in it. A glacier plucks or scrapes its load from the rock and other loose material lying under the ice, and also from pieces of rock broken off the slopes above the ice by the freeze–thaw process. In mountain areas erosion by ice is responsible for enlarging the hollows known as cirques found at the head of river valleys. Where two cirques cut back into a mountain a steep-sided, knife-edged ridge called an arête is formed. If three or more cirques converge, the arêtes meet to form a pyramidal peak such as the Matterhorn in the Swiss–Italian Alps.

Glacial erosion can take place wherever ice comes into contact with the surface. The original valley is widened, deepened and straightened to produce a glacial trough. A glacially eroded valley or trough has a U-shaped cross section, whereas a valley eroded by river and slope processes is typically V-shaped in cross section. The lower ends of tributary valleys are truncated, and when the ice retreats are left hanging above the main valley, with waterfalls from one to the other. Fjords, old glacial troughs that have been flooded by the sea, are found in Norway and other areas of glacial erosion.

A glaciated valley often has a series of steps along it. These are caused by variations in the strength of the erosion or the strength of the underlying rock. *Roches moutonnées* are hills of harder rock found on the floor of a valley. The upstream edge has a gentle slope scraped smooth by the ice, while the trailing edge has a steeper, rougher surface from which rock has been plucked by the ice.

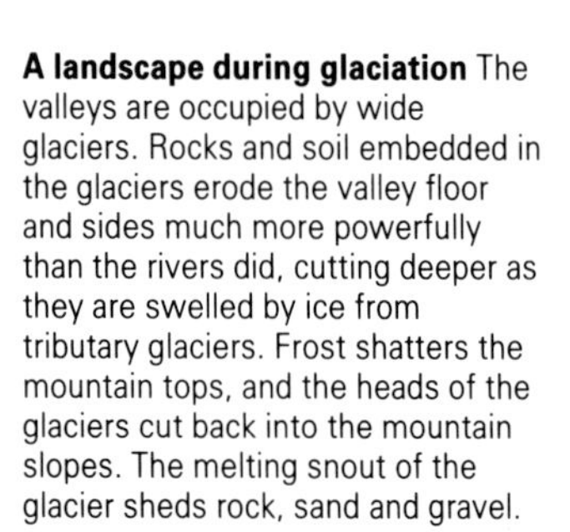

A landscape before glaciation Water is the main agent of erosion. Rivers scour their beds and banks, carving the landscape into hills and V-shaped valleys. Weathering and erosion shape hills, widen valleys and round hilltops. As the gradient decreases rivers flow more slowly, their erosive power decreases, and the valleys gradually become wider and shallower.

A landscape during glaciation The valleys are occupied by wide glaciers. Rocks and soil embedded in the glaciers erode the valley floor and sides much more powerfully than the rivers did, cutting deeper as they are swelled by ice from tributary glaciers. Frost shatters the mountain tops, and the heads of the glaciers cut back into the mountain slopes. The melting snout of the glacier sheds rock, sand and gravel.

Features of deposition

Material deposited by the ice is known as till. It is made of rock fragments of many shapes and sizes embedded in a matrix of sand or clay – hence its alternative name, boulder clay. The most common landscape effect of till is a cover of drift deposit or ground moraine, found over large areas of lowland in northern Europe, Asia and North America. Such lowlands are not always flat, because where the ice margin has halted for a time mounds of till have built up to form end (terminal) moraines. Lateral moraines are formed along the sides of valley glaciers; where two valley glaciers meet, the two inner moraines join to form a medial moraine.

Drumlins are another feature made of till that add variety to glacial deposition landscapes. They are elongated mounds, usually found in groups on a valley floor or on till plains. The largest mounds in this "basket of eggs topography" may be up to 1.6 km (1 mi) long and 90 m (300 ft) wide, but most are much smaller.

Other deposits have been laid down by water flowing in, under, or issuing from the margin of the ice. They are known as fluvioglacial deposits, and are easily recognized because the material has often been rounded by the action of the water, and has been sorted by it into bands of similar-sized fragments.

Eskers are long ridges of deposits running across lowlands, along valley floors or sometimes across valley sides. They are either the casts of former river channels in or under the ice, left behind on melting, or were left behind by rivers that flowed across the area in front of the ice margin. Deposits laid down in crevasses in the ice are known as kames. Beyond the margin of the glacier or ice sheet, the streams of meltwater in this proglacial zone form outwash plains of gravel, sands and clays. The heavier fragments are deposited close to the ice, the finer material farther away.

Beyond the ice

Areas near an ice sheet experience very cold climates. Most areas of northern Europe, Asia and North America have had periglacial conditions similar to those in tundra areas today. The principal influence on the landscape in these areas is the permafrost, the layer of ground beneath the surface that remains permanently frozen. The surface layer melts each summer, and is described as an active layer. It will be only 1–3 m (3.3–10ft) thick. Because there is still permafrost beneath, the water cannot drain into the ground, and the surface can become saturated and creep downhill over the frozen layer by a process known as solifluction. Alternate freezing and thawing leaves a legacy of shattered rock debris, and frost-shattered rocks accumulate as scree at the foot of slopes.

Alternate freezing and thawing is also responsible for convoluted patterns in layers of soil. During the most recent ice age little vegetation grew in the cold zones around the ice sheets, and the strong winds blowing out over the ice picked up quantities of fine sand and silt, called loase. The fine sediment was released by the slowing winds and formed deposits, often very thick, of a fine-grained, friable material known as loess (German "loose"). Loess deposits can produce fertile soils, and are extensive in the middle latitudes of Eurasia and North and South America, and in New Zealand's South Island.

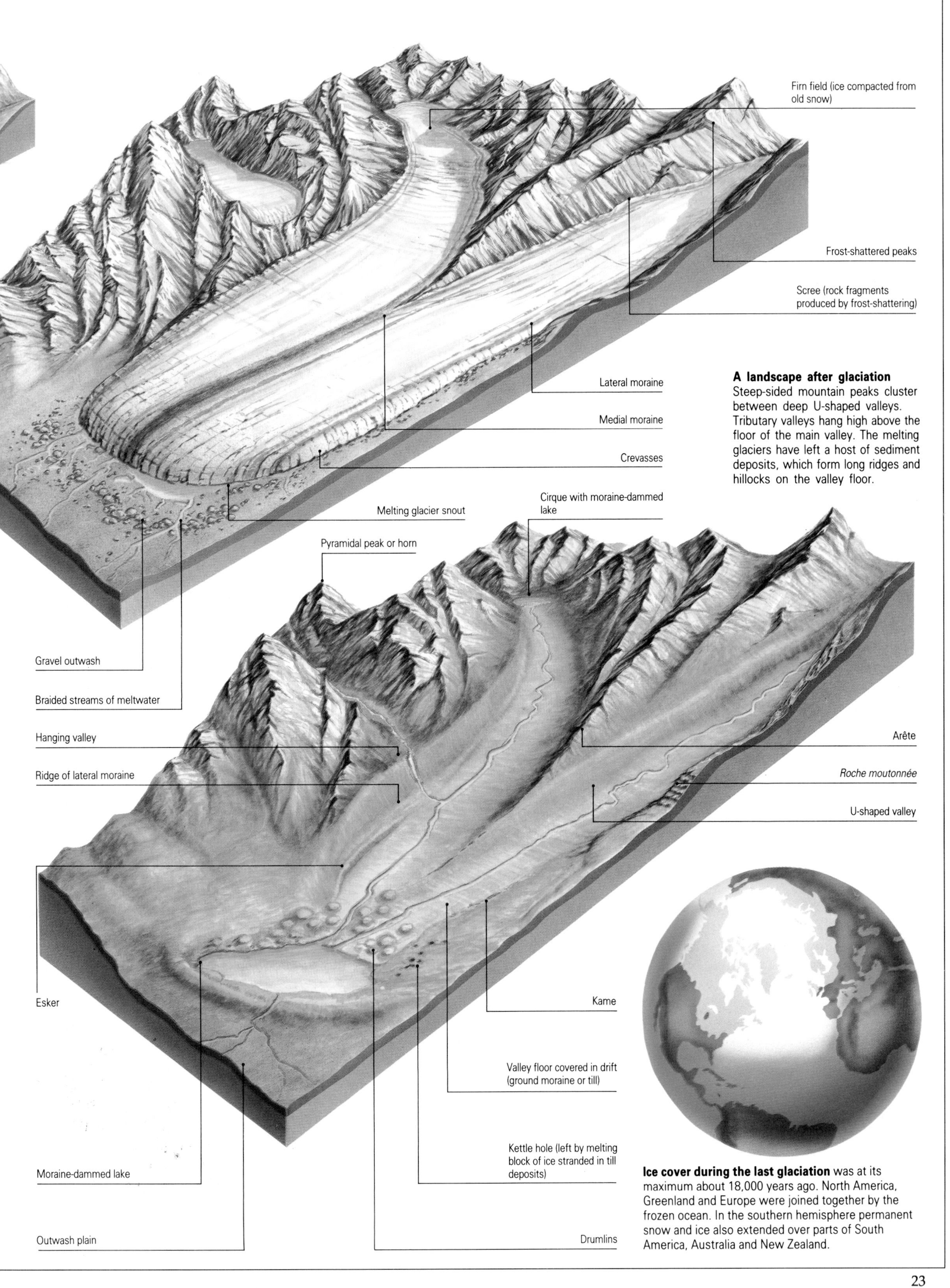

A landscape after glaciation
Steep-sided mountain peaks cluster between deep U-shaped valleys. Tributary valleys hang high above the floor of the main valley. The melting glaciers have left a host of sediment deposits, which form long ridges and hillocks on the valley floor.

Ice cover during the last glaciation was at its maximum about 18,000 years ago. North America, Greenland and Europe were joined together by the frozen ocean. In the southern hemisphere permanent snow and ice also extended over parts of South America, Australia and New Zealand.

Legacies in the Tropics

THE WHOLE OF THE EARTH WAS AFFECTED BY changes in climate during the glacial periods of the most recent ice age. Temperatures fell by as much as 17°C (30°F) near the polar ice sheets, and by about 4°C (7°F) nearer to the Equator.

Deserts and forests

During glacial periods the climate of areas close to the Equator was much drier, and the tropical and subtropical sandy deserts were much larger than they are now. Deserts are estimated to have covered as much as half the land area between latitudes 30°N and 30°S, compared with one-tenth today.

The extent of the deserts has been worked out from the existence of old (fossil) sand dunes beyond the margins of today's deserts. Sand dunes form in deserts where the rainfall is insufficient for vegetation to grow. The dunes grow, and migrate slowly downwind as sand is blown over the top and around the edges. Once the rainfall exceeds 100–300 mm (4–12 in), vegetation is able to fix the dune, which is prevented from migrating farther and becomes a fossil dune.

The presence of fossil dune fields in western Africa in areas where the annual rainfall is now 1,000 mm (40 in) indicates that the Sahara extended 600 km (370 mi) farther south during the last glaciation 18,000 years ago. In southern Africa, the Kalahari Desert must have been much larger because fossil dunes are found in areas where the annual rainfall is now 600 mm (24 in).

Fossil dune fields are also found in tropical South America, far from today's deserts. One such area is in the tropical grasslands of the Llanos in Venezuela, where rainfall can now be as high as 1,400 mm (55 in) a year; another is in the state of Bahia in north Brazil. This evidence, together with studies of plants and animals in these areas, suggests a picture of the Amazon basin and the surrounding tropical areas very different from what is seen today.

Tropical forests were much less extensive 18,000 years ago. It was once thought that all areas of tropical forest had continued to evolve for millions of years, uninterrupted by the glacial periods. This now seems to be true only for very small areas such as the Korup rainforest in Cameroon, western Africa. These islands of forest surrounded by vast tracts of savanna and dry land can be recognized today because they have a greater variety of plant and animal species than the newer forests around them. Many of their species are not found in any other area.

Changing lake levels

Other clues can help to piece together the patterns of climatic changes in the tropics. Lakes expand and contract with changes in temperature, which affect evaporation rates, and in rainfall, which affect the amount of water flowing into the lake. Higher lake levels can be traced from such features as old shorelines above the present ones, and lower levels deduced from flooded dune fields.

Lake Chad must at one time have been much smaller than it is now because sand dunes are covered by the lake. It has also been much larger: old shorelines are found in northeast Nigeria, more than 160 km (100 mi) from the shoreline today. In general, lake levels in most parts of tropical Africa seem to have been much lower during the last glaciation, and higher about 9,000 to 8,000 years ago.

FOSSIL EVIDENCE

Fossils are the remains of organisms found in rocks or sediments. They are produced when plants or animals are buried in sediments. Most of the organism decays, but the harder parts may be preserved and hardened by minerals from the surrounding rock. Sometimes the organism decays and leaves a space, which fills with minerals that make a cast of it. Earth movements and erosion can bring the fossils to the surface, where they provide clues to the Earth's history.

Fossils are used to establish the main divisions of geological time. For example, the Phanerozoic eon marks the appearance of animals whose hard shells and skeletons provided the first widespread fossils. They reveal changes in plant and animal life, and can also indicate the climatic and environmental conditions of the past.

The fossil skeletons of tiny sea creatures called foraminiferans yield information about fluctuating ocean temperatures. It is possible to measure the proportions of the isotopes oxygen-18 and oxygen-16 in oxygen they absorbed when they were alive. The light isotope oxygen-16 is more readily evaporated from oceans in a glacial climate, so relatively high oxygen-18 levels in their skeletons reveal a colder climate.

Fossilized teeth, scales and bones of crocodiles have been found in sediments in Lake Edward in Uganda in eastern Africa, though crocodiles are no longer found in the area. It is thought that they died out when the lake shrank during a drier period. Fossil finds of related and identical plant and animal species in South America, southern Africa, Antarctica and Australia provided impressive evidence of the existence of the supercontinent of Gondwanaland, between about 550 and 100 million years ago; fossils of sea creatures found in the high Andes offered startling evidence of mountain-building.

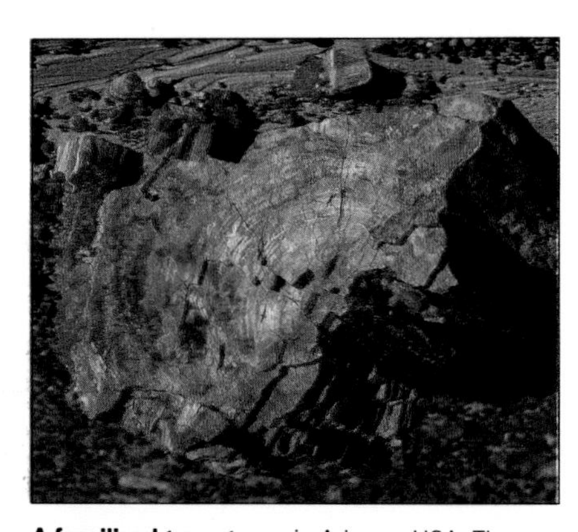

A fossilized tree stump in Arizona, USA. The tissues have been replaced by silica, which reproduces the finest details of the structure.

Huge dunes in the Namib Desert of southwest Africa shift with the wind as sand is blown off the windward slopes and deposited on the leeward slopes. Where vegetation can gain a hold the dunes are stabilized, and soil begins to form. Water is available for plants from the damp ground of old river courses, and from the moist coastal fogs that frequently envelop the Namib.

Retreating sand dunes 18,000 years ago sand dunes were much more widespread than they are today, and were found in areas farther to the north and to the south of the Equator. The climate of the low latitudes was drier then because so much moisture was trapped in ice sheets and glaciers.

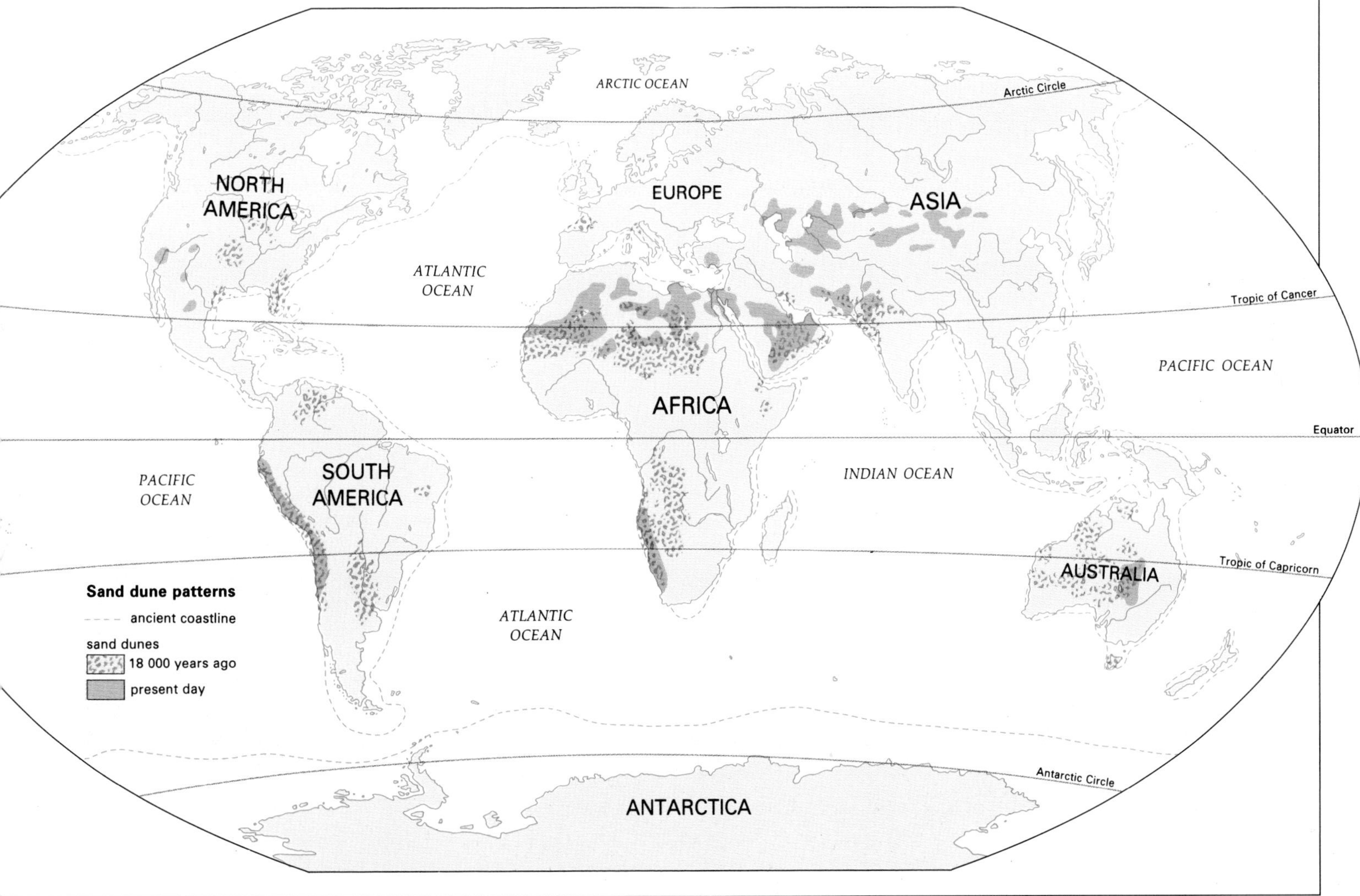

Waters of the Earth

THE PRESENCE OF HUGE AMOUNTS OF WATER distinguishes our planet from every other planet in the solar system. There are some 1.4 billion cu km (330 million cu mi) of water on Earth, and all life on Earth depends on it. As the circulation of blood in our veins and arteries is vital to human life, so water circulation is vital to many of the processes that fashion the surface of the Earth. Today's landscapes are the product of the continuous recycling of materials and energy through the environment. Water is required for most weathering and erosion processes; it also carries away eroded material and deposits it as sediment elsewhere.

The circulation of water through the Earth's environments is called the water (or hydrological) cycle. Water exists in three states: solid (crystalline ice), liquid (water) and gaseous (water vapor). About 96.5 percent of the Earth's water is in the oceans; only 3.5 percent is fresh water. Most of the fresh water (69 percent) is stored in ice sheets, glaciers and snow cover, and in rocks below the surface as groundwater (30 percent). Lakes represent 0.25 percent, while only about 0.04 percent is in the atmosphere.

As little as 0.006 percent of the total quantity of fresh water on Earth is in rivers. Despite this tiny percentage, rivers have a profound effect on the way landscapes are shaped, because the water has great power to erode rocks, to carry and to deposit the eroded material. Water thus helps gradually both to destroy landscapes and to build them.

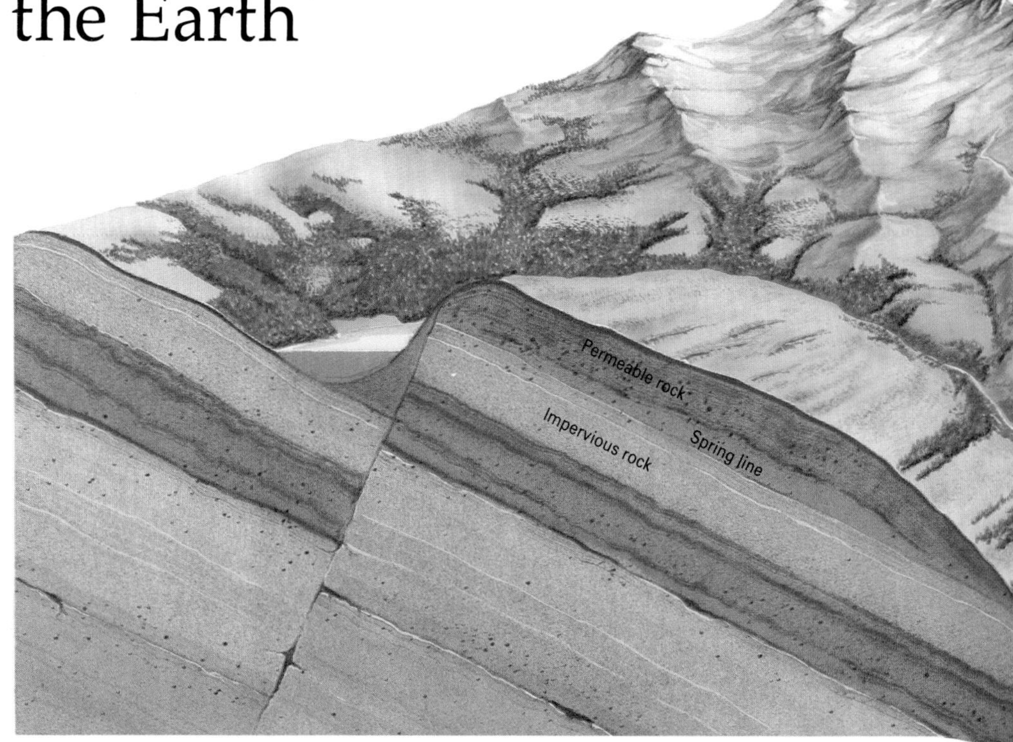

Underground reservoirs Rainwater falling on the surface of the Earth follows many different routes in its passage to the sea. Where the surface rock is impervious and does not allow water through, the water runs off to form puddles and ponds in natural hollows or to join rivers. Where the rock is permeable the water seeps through until it reaches a layer of impervious rock. It then flows under gravity along the surface of this rock until it emerges from the hillside as a spring. Some rocks allow water to percolate through them until they are saturated; then they channel water along their bedding planes.

Changing states of water

Water can change from one state to another: ice melts, water freezes and also evaporates, water vapor condenses. The driving force behind these changes is the energy from the Sun.

Most water vapor comes from the ocean. Evaporation from water and land surfaces is greatest where the amount of radiation received from the Sun is greatest, the air is dry, and the wind is strong. Where there is moisture on land more water evaporates from bare ground than from vegetated areas. In arid areas more water is lost through evapotranspiration (evaporation from the land and transpiration from plants) than is gained in precipitation.

When air cools beyond a certain level (its dew point) or becomes saturated with moisture, some of the water it carries condenses from a vaporous to a liquid state, as rain, mist, fog or dew or, if the dew point is below 0°C (32°F), as snow, sleet or hail. All these are forms of precipitation. Over the land precipitation exceeds evaporation, so on a world scale evaporation and precipitation are maintained roughly in balance.

The water cycle

Almost all the Earth's water is stored in the oceans, or as ice or groundwater. Yet water is constantly moving through the hydrological system, evaporating from oceans and lakes to provide the moisture for clouds and for precipitation.

Not all the precipitation falling toward the Earth reaches the surface. Condensed droplets of rain smaller than 0.2 mm (0.008 in) evaporate. Some precipitation lands on plants and then evaporates. The amount of rain reaching the ground varies according to the density of the vegetation and the intensity of rainfall. The water reaching the ground may flow over the surface, seep into the soil, or collect in puddles or small ponds. The route it takes depends both on the intensity of rainfall and on how quickly the soil can soak up water.

Water that infiltrates into the soil may percolate through it into the underlying rock. Water beneath the surface moves much more slowly than surface water. The movement of water in rocks depends on how easily water passes through them (the rock's permeability) and on how much water a rock can hold (its porosity). The groundwater that passes through and is retained by rocks replenishes the water table, which is the level below which permeable rocks are saturated. Rocks that both store water and allow it to flow are known as aquifers.

Where the water table reaches the surface, springs appear. Many of the world's great rivers have such beginnings; oases in deserts depend on springs to water their vegetation. Groundwater can also be reached by digging or drilling wells down to the water table.

The time that water spends in any one stage of the hydrological cycle varies considerably. In rivers it is often less than 4 days, and in the soil between 5 and 20 days. In the lower atmosphere the period is 4–8 days, increasing in the upper atmosphere (the stratosphere) to 100–500 days. Close to the ocean surface between 6 and 200 days may pass before evaporation takes place, whereas water in the deeper ocean may remain for up to 1,500 years. The water in ice sheets may be locked up for as much as 100,000 years. Groundwater may have been stored for 10,000 years: in some parts of the world wells supply water that fell as rain many thousands of years ago.

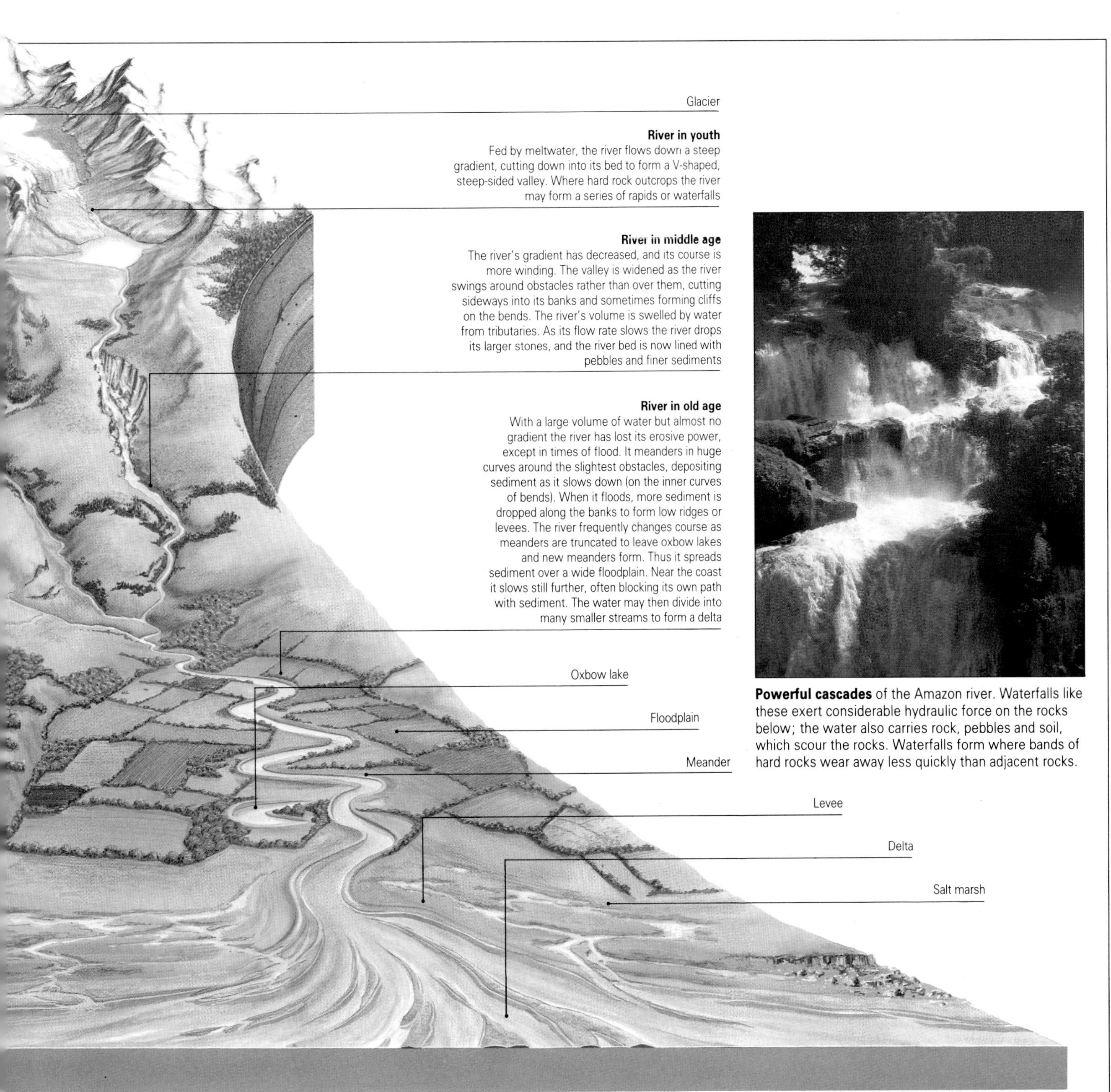

Powerful cascades of the Amazon river. Waterfalls like these exert considerable hydraulic force on the rocks below; the water also carries rock, pebbles and soil, which scour the rocks. Waterfalls form where bands of hard rocks wear away less quickly than adjacent rocks.

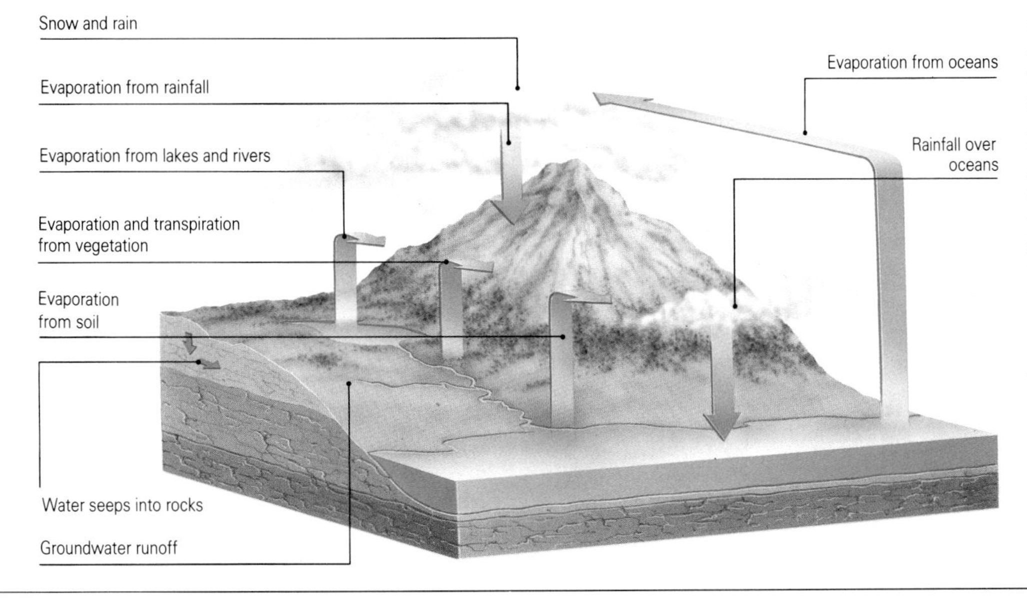

Water carves a landscape Below the frost-shattered peaks, glaciers gouge out deep valleys. The river carves the land into valleys as it flows toward the sea. In its upper reaches it flows swiftly down a steep gradient, splashing over rocks that interrupt its course. Here it has great erosive power, carrying along boulders that scour the rocks. As it slows down on leaving the hills the river sheds its load of sediment. It drops the largest particles first, and meanders around the slightest obstacle. The smaller pebbles, sand and silt are carried out to sea, where they accumulate on the sea bed and eventually form new sedimentary rocks.

The hydrological cycle As it absorbs energy from the Sun, water evaporates from the surface of lakes, rivers, seas and plants. On rising into the atmosphere it cools and condenses into droplets, losing energy as it does so. Eventually it falls back to the Earth's surface as rain, sleet, snow or hail. This water then flows under the influence of gravity in rivers and through rocks until it reaches the sea once more.

The Cycle of Erosion

THE LANDSCAPE IS COMPOSED OF LANDFORMS such as hills, valleys, escarpments, gorges and plains. The shapes of these landforms are constantly changing as the rocks are broken down by weathering, worn away by erosion, carried away and finally redeposited to become the raw material for new rocks. This cycle of rock material forms an important part of the larger rock cycle.

Weathering processes

Weather can disintegrate rock at or near the surface in two ways – physically or mechanically, and chemically. Physical weathering breaks rock into smaller fragments. Sharp changes in temperature can cause the outside of rocks to heat up and cool down faster than the inside. The surface layer eventually becomes weak and peels off (exfoliation). In cool climates water gets into cracks or other spaces during the day, and freezes at night. As it freezes it expands and the pressure weakens the rock; eventually pieces break away or the rock disintegrates. Dissolved salts can have a similar effect, because as the salt crystallizes expansion takes place, and the rock is weakened or disintegrates (salt weathering). Tree roots can also open up cracks in the rocks.

A number of chemical processes destroy rock. Limestone is particularly vulnerable to such chemical weathering. It can be dissolved by rainwater, which is a dilute acid containing dissolved carbon dioxide from the air. This process of carbonation carries chemicals away in solution, often leaving bare limestone pavements, or spectacular gorges or towers. Below ground the same process results in limestone caverns.

Oxidation takes place when oxygen in the atmosphere combines with compounds in the rock. The reddish crust on sandstone is formed in this way. Plants produce organic acids, and these too can attack rocks (organic weathering).

Hydrolysis involves reactions between water and mineral compounds to make new compounds: feldspar in granite is changed into clay, to produce kaolin or china clay. Some minerals take up water and expand. This normally affects the surface layer of the rock, which peels off.

Erosion and deposition

Fragments of weathered rock are moved by water, ice, wind and gravity. As they ve, the rock fragments can be further broken down by colliding with other fragments and with the surrounding rock. When the energy in the water or other vehicle transporting the eroded material is insufficient to keep it moving, it is deposited. Landforms can therefore be divided into those carved out by erosion and those built by deposition.

Erosional landforms such as mountain peaks, gorges and river valleys are created when material is moved away from an area. Depositional landforms are created when sediment is laid down. They include such features as drumlins deposited under the ice, floodplains and deltas associated with rivers.

Movement of eroded material

Eroded material moves through the landscape in one or more stages. The movement does not take place continuously because sediments will move only when there is a driving force such as a flash flood. There are six different stages of movement (sediment cascades).

On hill slopes mass movements take place in which weathered material, often lubricated by water in the ground, moves down slopes rapidly or gradually under the force of gravity. Typical features include earthflows and mudflows, rockfalls, landslides and soil creep.

Material is then moved by river (fluvial) processes through a drainage basin to the coast. The balance between erosion and deposition changes along the course of a river. Features that are caused by erosion, such as gorges and steep-sided river valleys, are found in the headwater parts of the basin, while farther downstream depositional features such as floodplains and mud flats are more common.

Coastal processes also create features of both erosion and deposition. Waves may erode cliffs, the material be carried away by waves and currents and deposited elsewhere as bars, spits and beaches.

In dry areas wind (eolian) processes dominate. Larger fragments can erode rocks by a sandblasting type of action as they are bounced along the ground by the wind. Coarse-grained material is usually not carried far, and is deposited as sand dunes. Very fine particles can be carried long distances before they are deposited forming a sediment known as loess.

Ice is responsible for two stages of movement. Glacier ice erodes rock, transports the material and then deposits it creating sharp mountain peaks and ridges, and large areas of drift deposits (till) and features such as moraines.

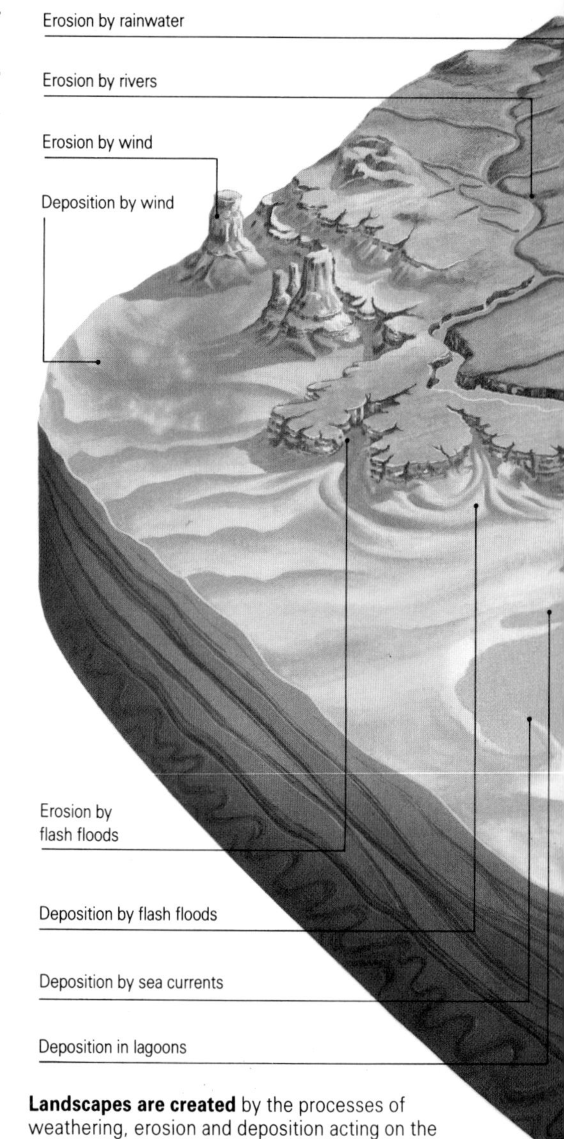

Landscapes are created by the processes of weathering, erosion and deposition acting on the surface of the Earth. Rock fragments eroded from the hills are deposited on lowland plains and the coast.

The freeze–thaw action of ice in the ground is also associated with movement of material. When water in rocks alternately freezes and thaws fragments o rock break away and fall onto the ice. Frost action may also cause soil to move downslope (solifluction), and may sort stones on the surface into patterns (patterned ground). Large mounds known as pingos are produced when underground water repeatedly freezes.

Measuring erosion

Rivers dump about 20 billion tonnes of eroded material into the sea every year. That is enough to lower the level of the land by about 3 cm (1.2 in) every 1,000 years. The rate at which erosion is taking

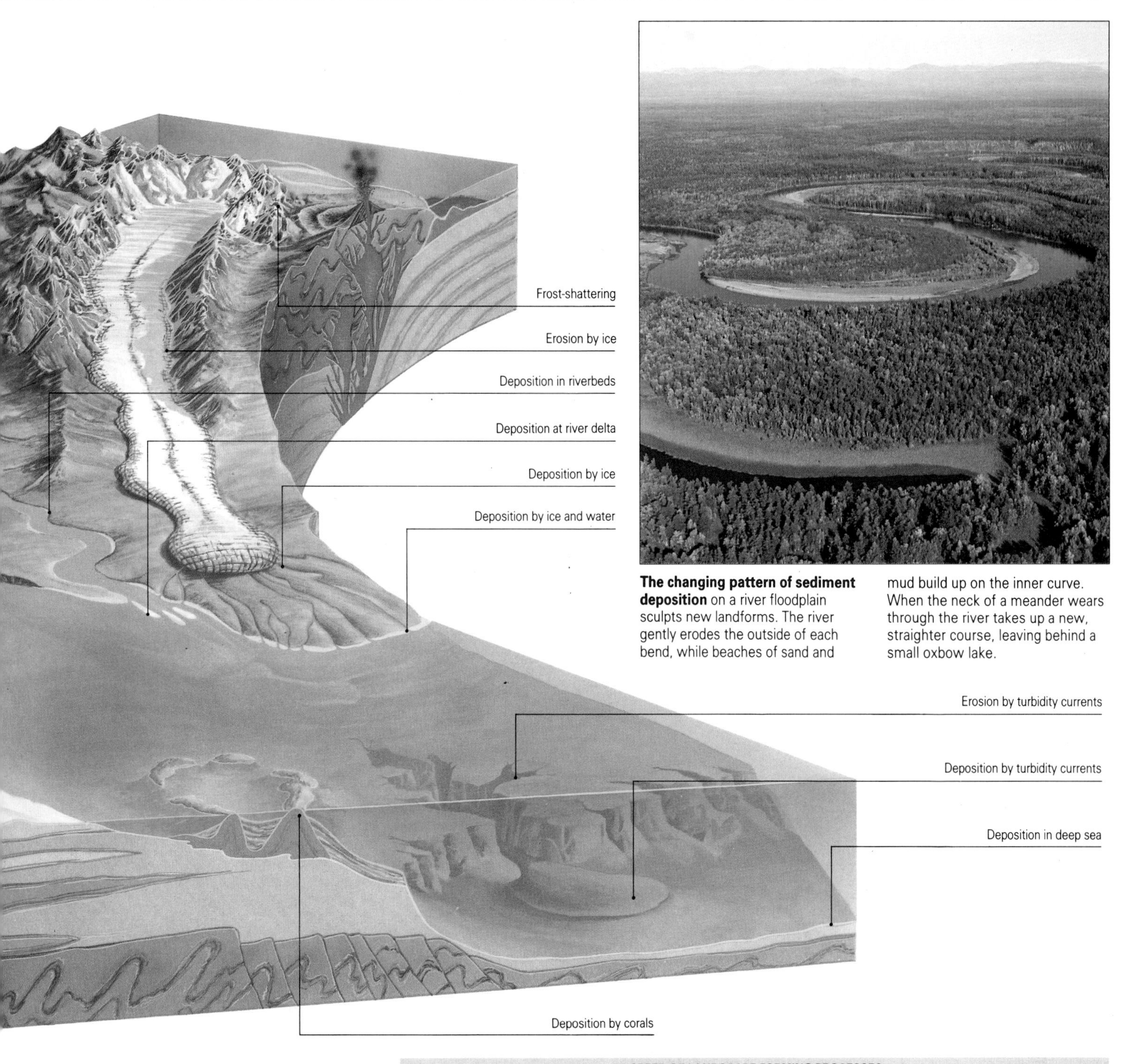

The changing pattern of sediment deposition on a river floodplain sculpts new landforms. The river gently erodes the outside of each bend, while beaches of sand and mud build up on the inner curve. When the neck of a meander wears through the river takes up a new, straighter course, leaving behind a small oxbow lake.

place can be calculated by measuring the amount of sediment being carried at any one point on the course of a river. This amount is divided by the area of the land drained by the river (its catchment area or drainage basin).

The highest rates of erosion take place in semiarid areas and in mountains, and where the land surface is disturbed by human activity. As humans have increasingly come to dominate the landscape over the past few thousand years, the impact of our activities is also growing. The lowest rates are found in low-lying areas where the rocks are very resistant, such as the Canadian Shield. To compare rates of erosion a standard measure, the Bubnoff unit, has been devised, which represents the lowering of the surface by one meter in a million years.

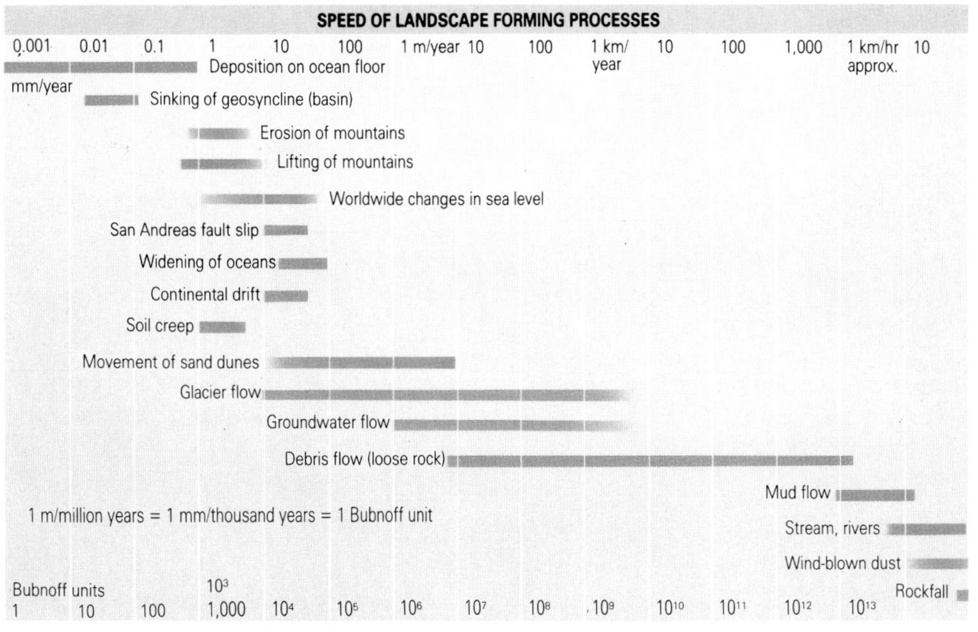

The Earth's Nutrient Cycles

THE TOTAL MASS OF ORGANIC LIFE ON EARTH is known as the biomass. It includes bacteria, fungi, plants and animals. Plants make up 97 percent of the biomass, of which 99 percent is found on or near the surface of the land. The Earth's biomass represents only about one-millionth of the mass of the oceans, but nevertheless has a quite dramatic influence both on the environment itself and on the processes that shape it.

Nutrient cycles

The elements and compounds (minerals) that make up the rocks, rivers, seas and atmosphere are the Earth's chemical resources. Many of these are taken in from the air or absorbed in solution by living organisms as nutrients. Key elements such as nitrogen, carbon, oxygen, sulfur and phosphorus are transformed from an inorganic to an organic state and back again through the nitrogen, carbon and other nutrient cycles.

These great global cycles are well understood as processes, but there are uncertainties about the quantities of nutrients involved in the cycles. Vegetation can act as a reservoir of some elements, and impede their removal from a drainage basin. In parts of the United States, for example, the deciduous forest takes up 20 times more of the element potassium than is carried away by streams. For other elements the reverse can be true. Four times more sodium is carried away in streams than is used by the forest. Experiments have shown that forest clearance speeds up the removal of certain elements, such as calcium, from a drainage basin.

Scientists are analyzing data on the amounts of nutrients carried in solution by rivers. Preliminary results show that the highest amounts are in Asia, where rainfall and runoff are high, and in Europe, whose sedimentary rocks are easily eroded. The lowest rates are in Australia and Africa, where the hard rocks are very resistant to weathering and erosion. A recent or continuing phase of mountain-building will tend to increase rates of erosion in a locality, and consequently the removal of nutrients by streams and rivers as well.

Plants, fungi and bacteria play an important part in nutrient recycling. The organisms themselves are sensitive to, and can be devastated by, changes that affect these cycles. The effect of acid rain pollution on the forests and on the plant and animal life of lakes and streams in northern Europe and North America in recent years is just one example.

Organisms and landforms

The energy to drive the nutrient cycles comes from the Sun; energy used in these cycles is not available to drive other processes that change the landscape. When plants absorb energy they slow down the rate of change in the landscape, especially changes caused by the hydrological cycle, such as erosion.

Vegetation protects the land surface in tropical areas, preventing it from being baked hard by the Sun and swept away by sudden rains. When rain falls the water is intercepted by the vegetation. Some of it evaporates back into the atmosphere, some flows down the trunks and stems of plants, and some drips onto the ground, where it soaks in rather than running over the surface and causing erosion. Where the forests have been cleared, as in the Himalayas where the river Ganges rises, for example, erosion of the land has increased, and the river carries more sediment than before. Much of the sediment load is deposited on the riverbed, raising the level of the river. When the heavy monsoon rains come, more water and sediment run off the land and the river overflows its banks, causing floods such as those that are becoming more frequent in Bangladesh. Reforestation has become an important way of controlling erosion, especially in tropical and arid zones where vegetation cover is so vital in retaining the soil.

Vegetation also interferes with water that is already moving through the landscape. It reduces the amount of energy the water has to carry its load of sediment, so rivers often deposit their sediment load in marshy parts of their course. Branches and other debris that fall into streams in forest areas produce temporary dams or logjams that trap sediments.

Some organisms have a direct influence on landforms. Mangrove swamps are areas of forest growing on flat shorelines between high- and low-water marks in tropical areas. The mangrove roots extend above the shore as a tangled mass that slows down the water passing through them, so its sediment is deposited. As the sediment accumulates the land becomes higher and is flooded less often.

Salt marshes are the equivalent of mangrove swamps in temperate areas. They are colonized by grasses and other salt-loving plants, and eventually become dry land. Also in temperate zones, plants in wet, acid peat bogs or (in Canada) muskeg decay only partially and can become solid land.

In some cases animals play a part in landscape development that is direct, distinctive and easy to recognize. Beavers dam rivers and create new lakes; termites build mounds of earth 2 m (6.6 ft) or more high that in savanna areas may be found every few meters. In seas that are shallow, warm and clean, coral reefs are formed from tiny sea creatures called polyps. They cluster together in colonies, and when they die the calcium carbonate of their skeletons provides the foundation for further colonies. The most spectacular of all reefs, the Great Barrier Reef, covers about 270,000 sq km (104,000 sq mi) and stretches for some 2,000 km (1,240 mi) along the northeast coast of Australia.

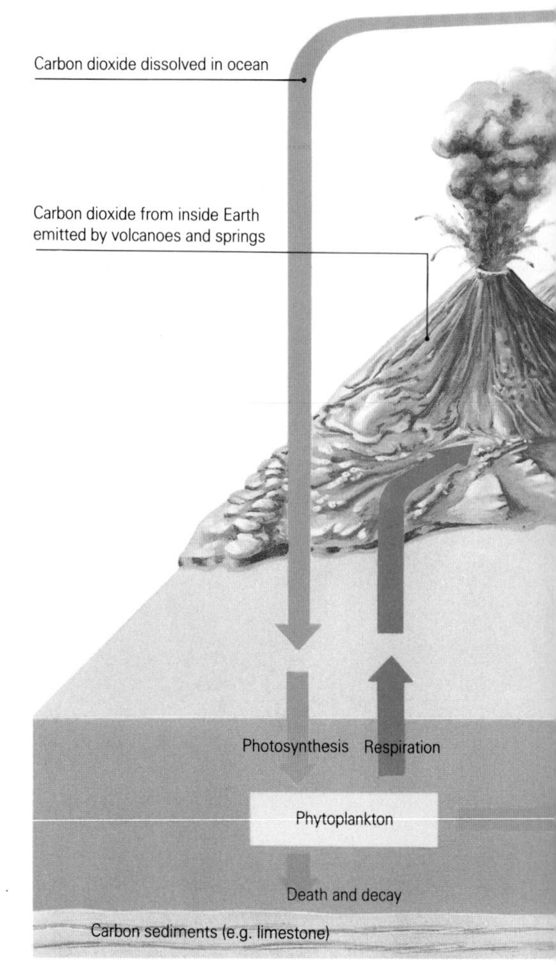

The carbon cycle Plants and animals play a vital part in the cycling of carbon between the rocks, soil, water and atmosphere. This natural cycle is increasingly being perturbed by the activities of humans, as we fell forests and burn fossil fuels.

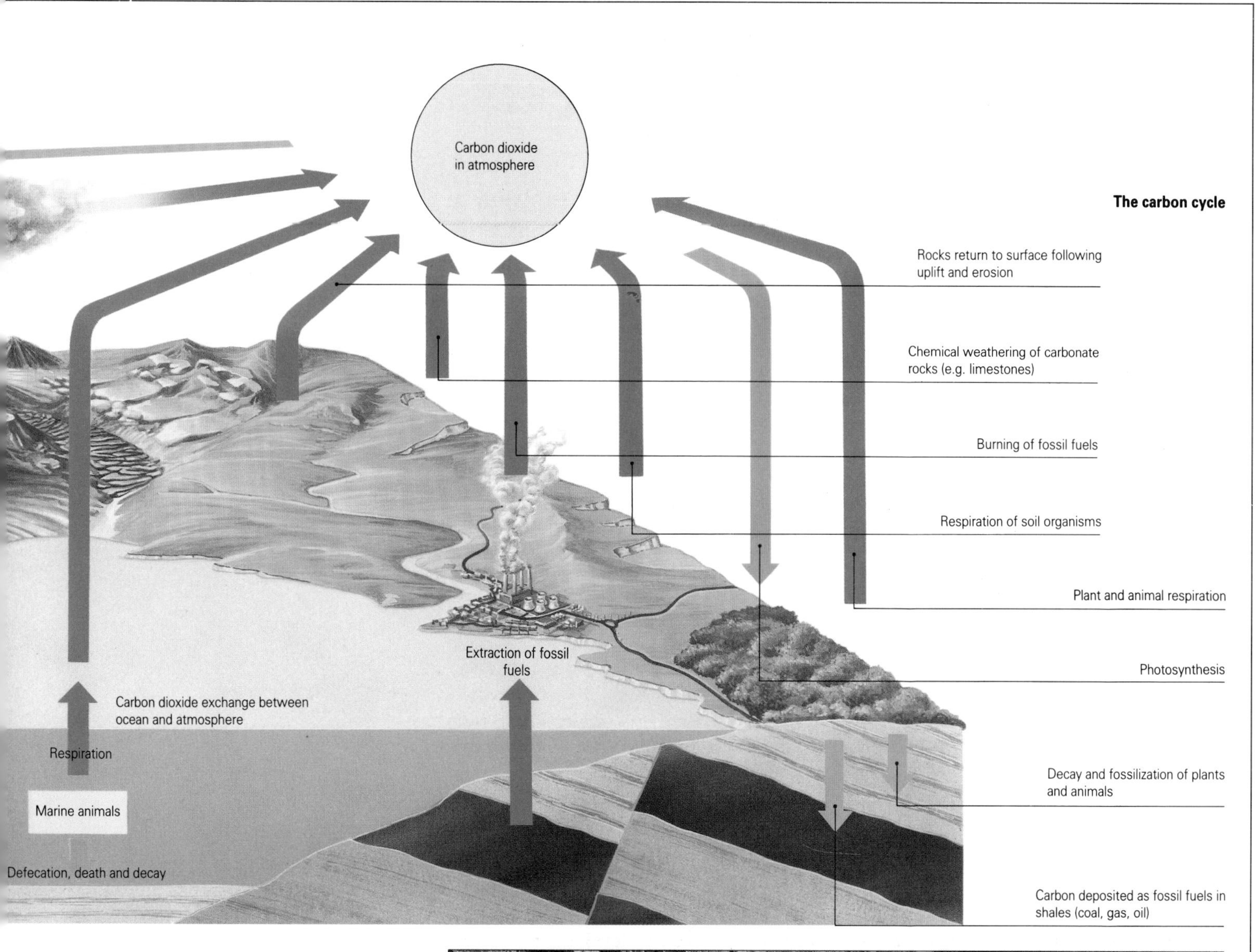

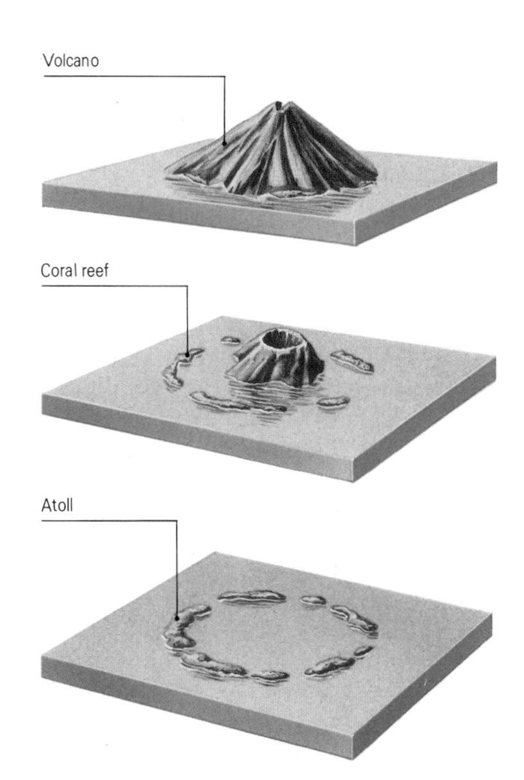

Formation of an atoll A volcanic oceanic island surrounded by a coral reef slowly subsides beneath the ocean, but reef growth keeps pace with subsidence.

Nutrient cycling in action When trees die and are left to rot on the forest floor, the organic material is broken down by animals such as woodpeckers, termites, wood-boring beetles, fungi and bacteria, returning minerals to the soil. The respiration of these decaying organisms returns carbon dioxide to the atmosphere.

The Formation of Soil

THE LAND SUSTAINS 99 PERCENT OF THE Earth's biomass, and without soil most of it would not exist. Soil is the material that plants grow in and that provides them with physical support and nutrients. It is a complex mixture of particles derived from rocks, living and dead organisms, water and gases.

The character of a soil changes with depth. A typical cross section (profile) has a number of distinct layers (horizons). Near the surface is an organic layer known as topsoil, which includes plant and animal matter at various stages of decomposition. Below this are paler, less fertile layers from which fine materials and minerals are removed in solution by percolating rainwater. These substances are deposited lower down in layers of infertile weathered rock (subsoil) and unweathered parent rock (bedrock).

Water generally moves downward in soils, but in deserts waters may draw salts upward through the soil. When it reaches the surface the water evaporates, leaving the salts as a white crust.

The principal factors affecting the development of soils are the rocks or deposits, the parent material; climate; the lie of the land, especially slopes; and plants and animals, which cycle nutrients and add organic matter to the soil. The length of time over which these environmental factors have interacted also affects both the characteristics of the soil and the amount of soil produced.

The parent rock affects the texture of the soil. It has most influence in the early stages of development, when the rock is being broken down by the processes of weathering. For example, soils formed on shale generally have a finer texture than those developed on sandstone. Certain rocks give very characteristic soil profiles. Limestone soils, including chalk, are often thin because the minerals of the parent material are removed in solution rather than being broken down by mechanical weathering.

The lie of the land affects soil development because weathered rock tends to move downslope, so soil does not have time to evolve to any depth. In hollows the reverse may be true, and deep soil profiles build up as sediment and organic materials accumulate. If water in the soil cannot drain away, the soil becomes waterlogged and dead plant material may not decompose. Under such conditions peat is formed.

Time is an important factor in the evolution of a soil. It takes thousands of years for a mature soil profile to develop. Eventually the topsoil may bear no resemblance to the parent material because of changes caused by chemical reactions, plants and animals. In these cases, climate and vegetation dominate soil formation.

SOIL TYPES

Soils defined by climatic zones and vegetation type

Name	Chief natural vegetation	Character and variations	US Department of Agriculture equivalent
Tundra soil, arctic browns	Tundra	Relatively shallow, leaching important, waterlogging frequent; slow organic decay gives peat; arctic brown is well drained; gley soils have alternate drained and waterlogged horizons	Entisols, histosols, inceptisols
Podzols	Moist coniferous forest	Thin acid humus, grayish upper horizon often with leached material and hard pan below	Spodosols, alfisols
Brown earths	Deciduous forest	Rich in humus, slightly acid, more so where rainfall and leaching greater	Alfisols, inceptisols
Chernozem (black earth), chestnut soils, prairie soils	Prairie or steppe grassland	Humus-rich upper layer; often with chalky horizon; thicker profile where rainfall greater	Mollisols
Terra rossa, brown soils, cinnamon soils	Mediterranean-type woodland	Organic decay rapid; soils dry out deep down, often with red horizons; *rendzina* on limestone bedrock	Inceptisols, ultisols
Red and gray desert soils, chernozems	Desert scrub or no vegetation	Thin, coarse with little humus or leaching; may have covering that is coarse (pavement) or fine (blown sand); some soils with white salty crust at surface; red mostly in tropics, gray in mid-latitudes	Aridisols
Red and yellow soils (latosols)	Tropical rainforest, savanna woodland and grassland	Deep, reddish leached soils; may have iron-rich oxide layer; dark gray grumusols in tropical grassland, often deeply cracked	Oxisols, vertisols, ultisols
Intrazonal soils, in which non-climatic influences predominate			
		Dominant influence parent material (bedrock), local topography (e.g. rich organic peat soils, histosols) or salt accumulation	Histosols, entisols
Azonal soils, too young for profile to develop			
		e.g. partly weathered fragments or where erosion removes weathered material quickly (in mountains) or shortly after retreat of a glacier; variations over very short distances	Entosols

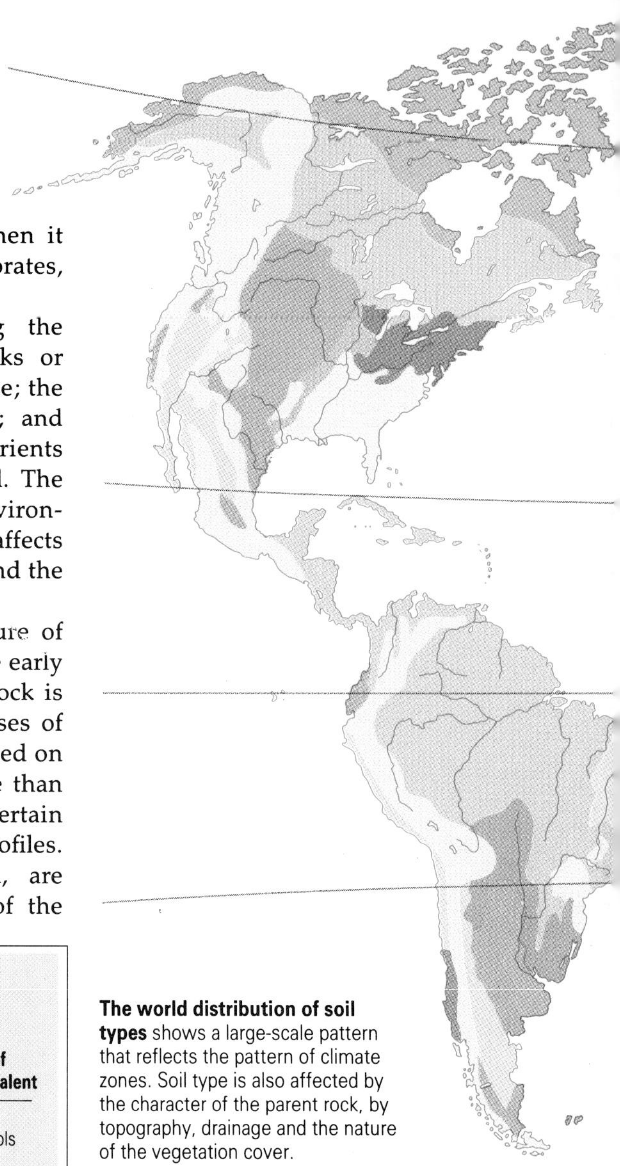

The world distribution of soil types shows a large-scale pattern that reflects the pattern of climate zones. Soil type is also affected by the character of the parent rock, by topography, drainage and the nature of the vegetation cover.

Soil types

The most common approach to describing soils is based on climatic zones. An alternative classification is based solely on the qualities of the soil itself. In tundra

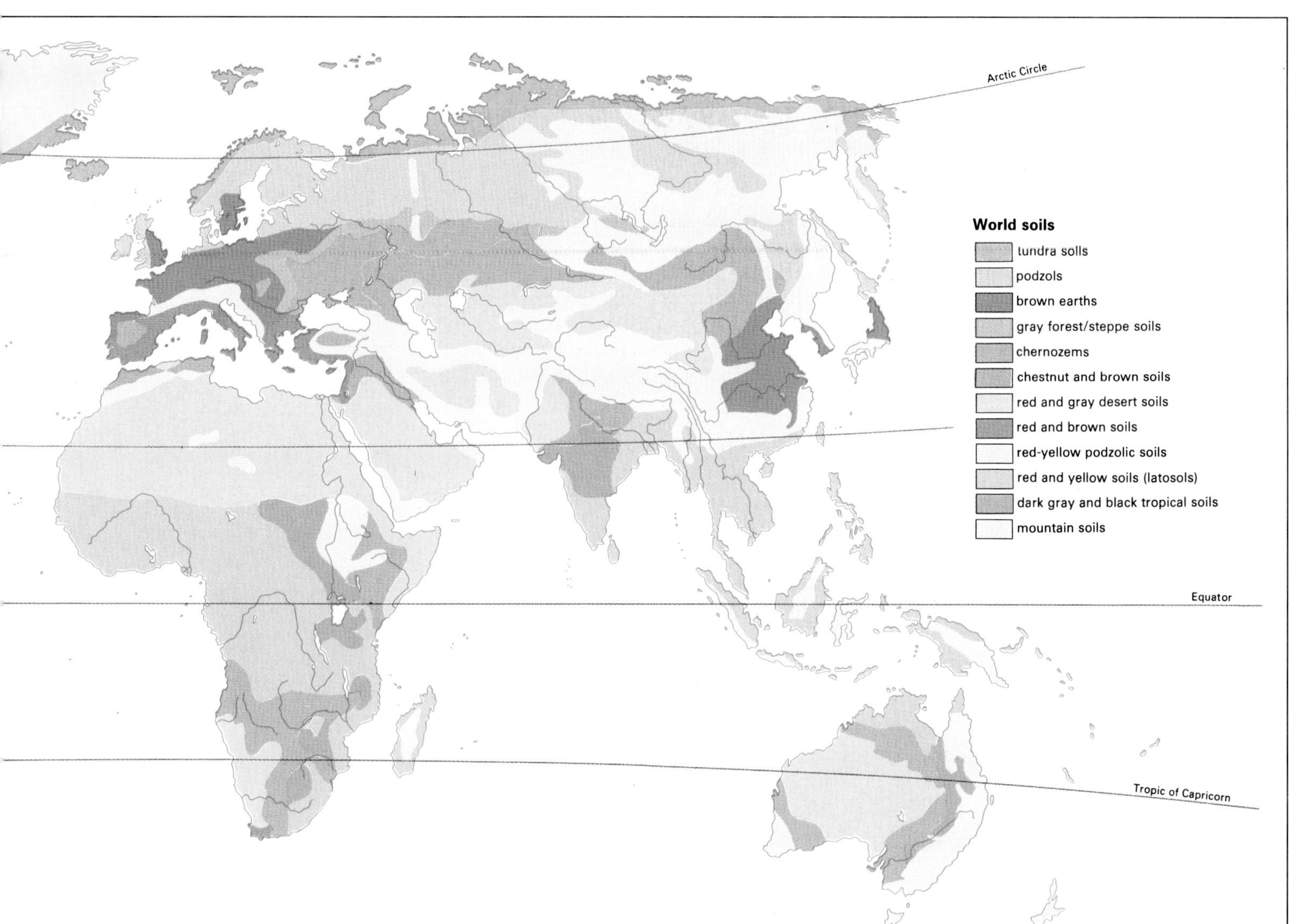

areas soils are often waterlogged or frozen, with a peaty layer on the surface and bluish mud below. Podzols are found in cool northern coniferous forest areas; they have a thin, acid layer of topsoil over a grayish, leached layer with a hard pan below. Brown forest soils rich in humus are found in temperate forest areas, and make good agricultural land. On the drier temperate grasslands the surface layer is also rich in organic material; dark-colored chernozem, prairie soil and chestnut-brown soil are found. Hot deserts have soils low in humus because of the lack of vegetation; salts often accumulate near or at the surface. Dark clay soils called grumusols develop in tropical grassland areas. Deep, reddish soils rich in iron underlie the humid tropics. Mountain soils are very thin, often consisting of little more than coarse rock fragments.

Soil and the landscape

Soils are sources of sediments and therefore a major influence on the appearance of the land. In humid temperate areas there is a substantial cover of vegetation, which protects the soils beneath and reduces erosion on slopes. Material moves downhill under the influence of gravity, and a slow soil creep can cause small terraces to form. Material can also move more dramatically, as an earth flow or landslide, especially when lubricated by water. It may also move over a great distance. Loess is a fine soil deposited by the wind, which in China has accumulated to depths of up to 300 m (1,000 ft). The Huang (Yellow) river is so called because of the amount of eroded loess that it transports.

Soil also affects drainage. While clay soils impede drainage and encourage water to flow over the surface, causing sheet erosion, coarser soils allow the water to drain through them quickly, so there may be little surface erosion.

A generalized soil profile, a vertical section from the surface to the bedrock. The number and relative sizes of the different layers (horizons) varies with the type of soil. There may be additional layers in waterlogged or mineral-impregnated soils. The thickness of the humus-rich layer of topsoil depends on the rate at which organic matter is broken down. In wet or acidic conditions decay is minimal and peat forms. The E horizon is present only in soils of cold, wet climates; rainwater leaches out minerals, which precipitate out of solution in lower horizons. In arid areas the upper horizons may be rich in mineral salts drawn to the surface as water rises and evaporates.

Organic layer (O horizon): plant and animal material in various stages of decomposition

Topsoil (A horizon): dark, fertile, rich in humus and minerals

Eluvial layer (E horizon): paler because fine particles and dissolved minerals are washed out (leached) by rainwater

Illuvial layer (B horizon): material leached from upper layers is deposited here

Subsoil (C horizon): infertile weathered rock

Bedrock (D horizon): unweathered parent material

Landscapes and Climates

To understand why scenery looks the way it does the processes responsible for shaping it need to be unraveled. Climate is a major factor; so is rock type, which determines how effective different processes can be. Patterns of temperature, rainfall, radiation, water, soil and vegetation also give clues to the evolution of different environments.

The world's landscapes fall into four main zones: the intertropical zone, which includes rainforests and savannas; the arid and semiarid zones, which include deserts and steppes; the forested zone of the middle latitudes; and the cold zone of high latitudes toward the poles.

Rainforest and savanna zone

Tropical rainforests straddle the Equator in Southeast Asia, Africa, South America and on some of the islands of the Pacific Ocean. They are hot and humid, with little seasonal change. Rainfall is over 2,000 mm (80 in) a year and temperatures average 25°C (77°F), allowing dense tropical vegetation to grow rapidly.

Rainforest environments are the product of erosion by chemical weathering as rainwater percolates through the soil. Because the climate is not seasonal, the weathering is continuous and can rot the underlying (parent) rock to depths of 20–30 m (66–87 ft) below the surface. Organic material accumulates on the surface and is rapidly broken down in the hot, humid climate, adding nutrients to the soil. In some areas chemical weathering proceeds faster than the rate at which the fine-grained eroded material is washed away by rivers, so over millions of years extensive plains form. The plains are scattered with island mountains (inselbergs), sometimes in the tall, conical form of a sugarloaf. These granite hills are resistant to the forces of erosion.

Savanna is a typical landscape of the seasonally wet tropics. As in the rainforest, chemical weathering is common, but because vegetation is sparser the weathered material is easily removed by water flowing over the ground surface (sheet flow). This can produce a concave slope called a pediment. Savannas often consist of a series of plains at different levels, each containing pediments.

In seasonal tropical zones, landscape processes operate at two levels. On the surface eroded rock and soil material falls or slips downslope or is moved by water action. Where the underlying rock is affected by chemical weathering, there is a level often described as the basal weathering front. The two levels are frequently separated by considerable depths of weathered material. This layer is thinner in more arid areas because surface processes have proportionately more influence on the landscape.

Deserts

One-seventh of the Earth is desert, defined as an area where less than 25 cm (10 in) of rain falls in an average year. There are hot and cool deserts in Africa, Asia, Australia, North and South America, and cold deserts in the polar regions.

Daytime temperatures in deserts often exceed 37.8°C (100°F), while at night they can fall below freezing point. In these extreme conditions rocks expand and contract, causing cracks and splinters. The resultant wind-blown rock particles and sand scour rocks and other obstacles in their path, producing beautiful and spectacular landforms. When the material is deposited dunes may form up to 200 m (660 ft) high and 50 km (30 mi) long. Blown by the wind, they can move as much as 50 m (160 ft) in a year.

Hot deserts are the driest places on Earth, but when rain does come it can be torrential. Because there is little vegetation to resist the flow of water, flash floods can carve out gorges and channels called wadis.

Forests and grasslands

Between the treeless Arctic and the hot equatorial regions, the middle latitudes were originally heavily forested. Much of the forest has been cleared by humans, but there are still distinct vegetation zones. Coniferous forests extend over great tracts of North America and Eurasia. They give way farther south to mixed and deciduous forest, and finally to the open evergreen forests of the Mediterranean type climates. Today the characteristic landscapes are formed by slope and river processes; fossils provide clues to tropical climates in the past.

The cold zones

Around the edge of the polar ice sheets are treeless periglacial tundra landscapes of low-growing cushion plants, mosses and lichens. Here the midwinter sun rises barely or not at all, and there is only a short summer growing season. Temperatures are often below −30°C (−22°F), with precipitation between 150 and 300 mm (6–12 in), so these are cold deserts. The subsoil is permanently frozen (permafrost). Above the permafrost is a thin layer of soil, which is frozen in winter but thaws in summer, washing out soil and other deposits over large areas. As the water cannot sink into the permafrost, the ground becomes boggy and series of lakes can form.

Wind action is also significant, with the strong winds in winter blowing snow into dune formations. Alternating freeze and thaw produce features such as ice-cored mounds (pingos) and patterned ground, and retreating ice sheets have exposed glacier-cut U-shaped valleys and sharp mountain peaks and crests, and have left behind mounds of material eroded and later deposited by the ice.

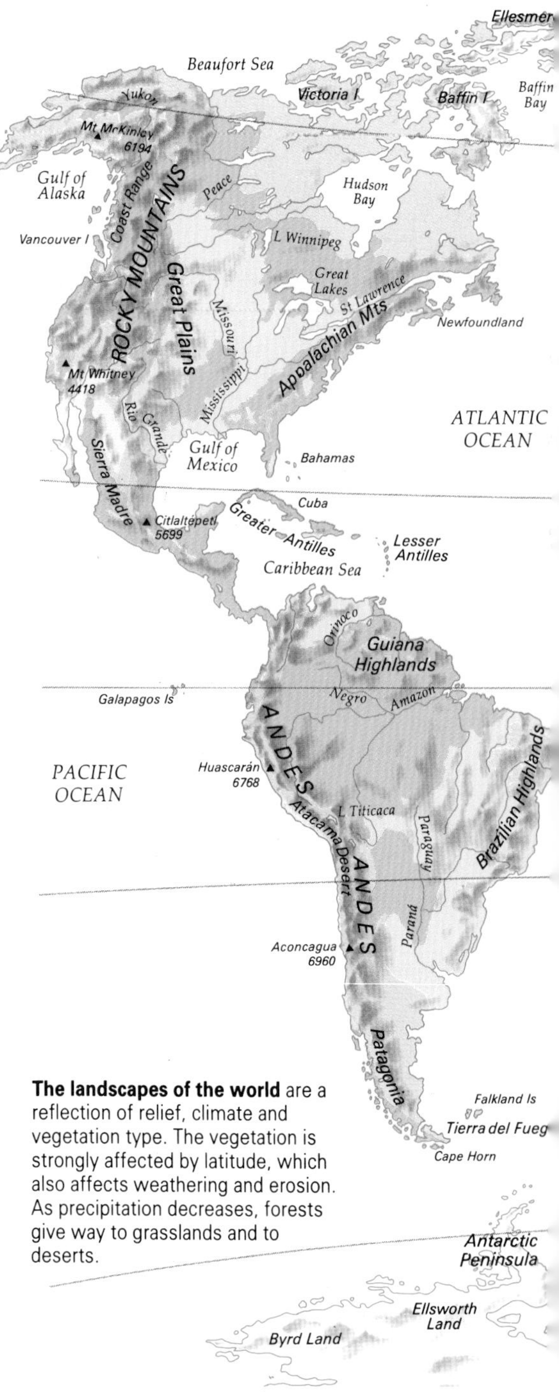

The landscapes of the world are a reflection of relief, climate and vegetation type. The vegetation is strongly affected by latitude, which also affects weathering and erosion. As precipitation decreases, forests give way to grasslands and to deserts.

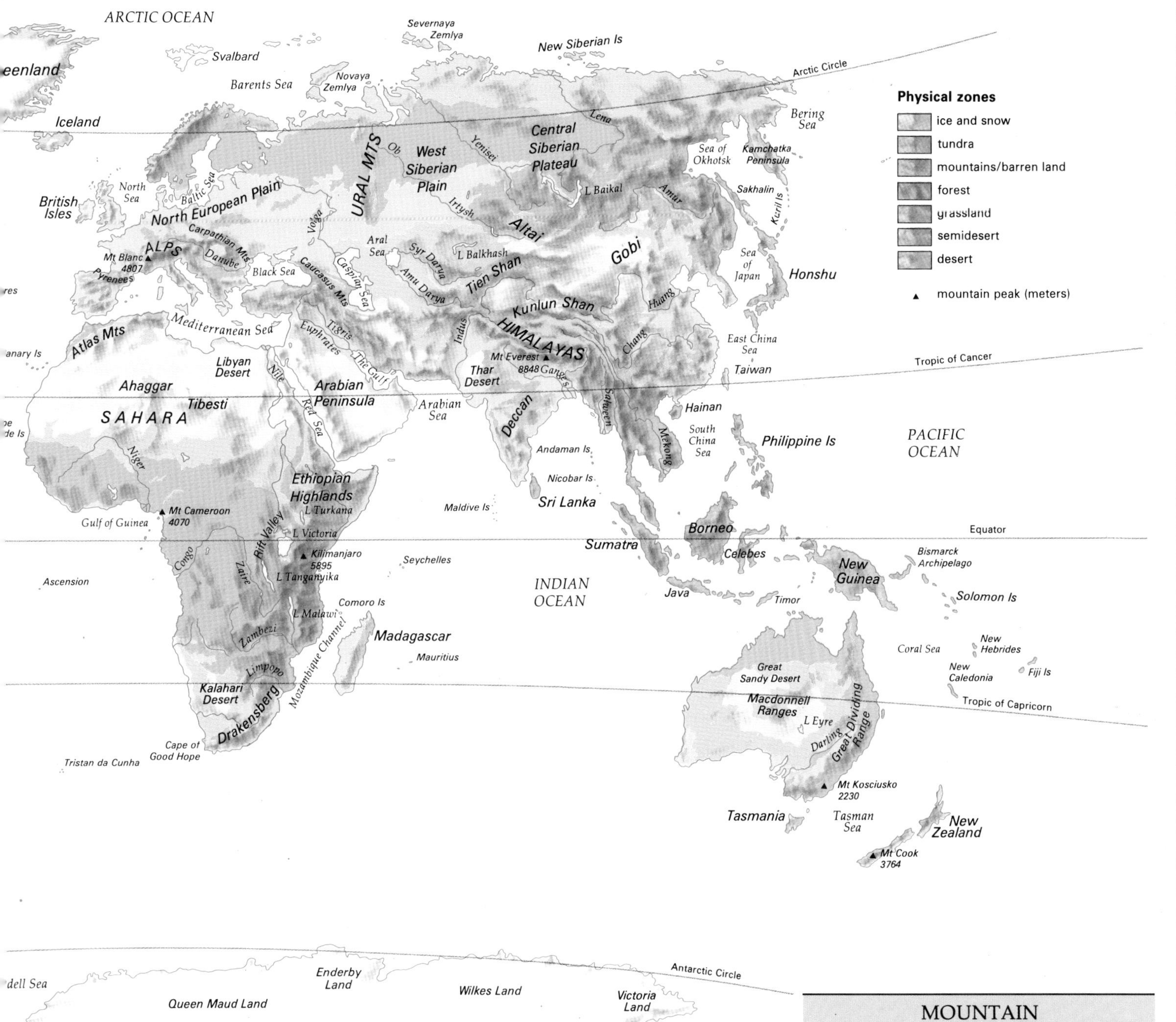

MOUNTAIN ENVIRONMENTS

In mountain areas the landscape, rather than distance from the Equator, can dominate the climate. With every 100 m (330 ft) of altitude temperatures drop about 0.5°C (1°F) and rainfall increases, creating a series of distinct climatic and vegetation zones. The pattern of these climatic zones resembles the pattern of world climatic zones from the Equator to the Arctic.

A climber of a tropical mountain such as Mount Kilimanjaro in eastern Africa would start by walking through tropical forest. Higher up this gives way to temperate deciduous forest, and higher still to coniferous forest. Above the treeline is low alpine scrub scenery similar to that of tundra landscapes, and toward the summit of the mountain is the permanent snowline. In drier tropical areas the first zone may be savanna or desert, but as the humidity increases with altitude the forest zones take over.

The tall pillars of Spider Rock in the Canyon de Chelly, Arizona, USA, were left standing when softer rocks around them were eroded. These dramatic landforms have resulted from the action of rivers on rocks of various types, reinforced by the flash floods that follow sporadic desert rainstorms. The dry river valley on the left may well become a raging torrent during such a storm.

Natural Hazards

EARTHQUAKES, VOLCANIC ERUPTIONS, HURRICANES, floods and other violent phenomena are part of the Earth's natural processes. These extreme events of nature can harm both people and the landscapes in which they live.

Certain parts of the world are prone to one or two particular types of hazard. Nearly all Bangladesh, in the northeast of the Indian subcontinent, occupies the floodplain and delta of the Ganges and Brahmaputra rivers and is subject to serious flooding. The same area is lashed by tropical cyclones. Many of the more hazardous parts of the Earth lie on or near the edges of tectonic plates where they traverse the tropics. Violent storms as well as earthquakes, volcanic eruptions and landslides are constant threats in these areas – as in the islands of Southeast Asia, for example.

People can themselves inadvertently exacerbate, trigger or accelerate the onset of natural hazards. Deforestation in the Himalayas strips away the natural vegetation that prevents heavy rainfall from washing the soil away, and increases the severity of floods. Felling trees in the European Alps to provide ski slopes has removed natural barriers, increasing the likelihood of avalanches.

Scientists divide natural hazards into four groups according to the sphere in which they take place: the atmosphere, the land or geosphere, water or the hydrosphere, and the living world or biosphere. Some disasters may result from a combination of natural hazards. They can also be distinguished as either pervasive or intensive. Pervasive hazards, such as droughts, have a large-scale and long-term effect, while intensive hazards, such as hurricanes, are shorter lived and often affect smaller areas.

NATURAL DISASTERS

Any list that aims to give an idea of the variety and different intensities of fatal natural hazards is bound to be arbitrary. The following is a selection of a hundred years or so of extreme geophysical events that caused immediate deaths (intensive hazards).

Pervasive hazards, those that are more extensive over time and space, such as drought or other climate change, and soil erosion, do not appear. Nor do those exacerbated or triggered by human activity, such as pollution, the greenhouse effect or acid rain, nor fatalities caused primarily by disease, or by poisonous or otherwise deadly plants and animals (hazards of the biosphere), such as the 30 million deaths in the 1959–61 famine in China and the 22 million who died in the 1918 influenza pandemic.

Year	Location	Nature of hazard	Deaths
1883	Krakatau, Indonesia	Island volcano eruption, tsunami	36,000
1887	Huang river, China	Flood leading to famine, disease	900,000
1888	Moredabad, India	Hailstorm	246
1889	Johnstown, United States	Failure of dam after rainstorms	2,200
1896	Honshu, Japan	Tsunami	26,000
1900	Galveston, United States	Hurricane, storm surge	6,000
1902	Mount Pelée, Martinique	Eruption, hot ash, volcanic gas	30,000
1908	Messina, Italy	Earthquake	85,000
1916	Italian-Austrian Alps	Snow avalanches along military front	10,000
1920	Gansu, China	Landslides triggered by earthquakes	200,000
1927	Tien Shan, China	Earthquake	200,000
1931	Huang river, China	Flooding, famine	20,000
1932	Gansu, China	Earthquake	70,000
1951	London, United Kingdom	Smog (smoke fumes and fog)	2,850
1953	Netherlands	Sea flooding in coastal areas	1,835
1959	North China	Floods	2,000,000
1963	Italy	Landslide caused flood over dam wall	2,500
1970	Huascarán, Peru	Mudslide started by eruption	21,000
1970	Bangladesh	Tropical cyclone, storm surge	300,000
1974	Honduras	Hurricane	7,000
1975	United States	Lightning, deaths in an average year	150
1976	Tangshan, China	Earthquake	240,000
1985	Ruiz, Colombia	Mudslide after eruption	25,000
1986	Lake Nyos, Cameroon	Poisonous volcanic gas cloud	1,600
1986	Bangladesh	Cyclone	2,000
1988	Armenia, Soviet Union	Earthquake	25,000

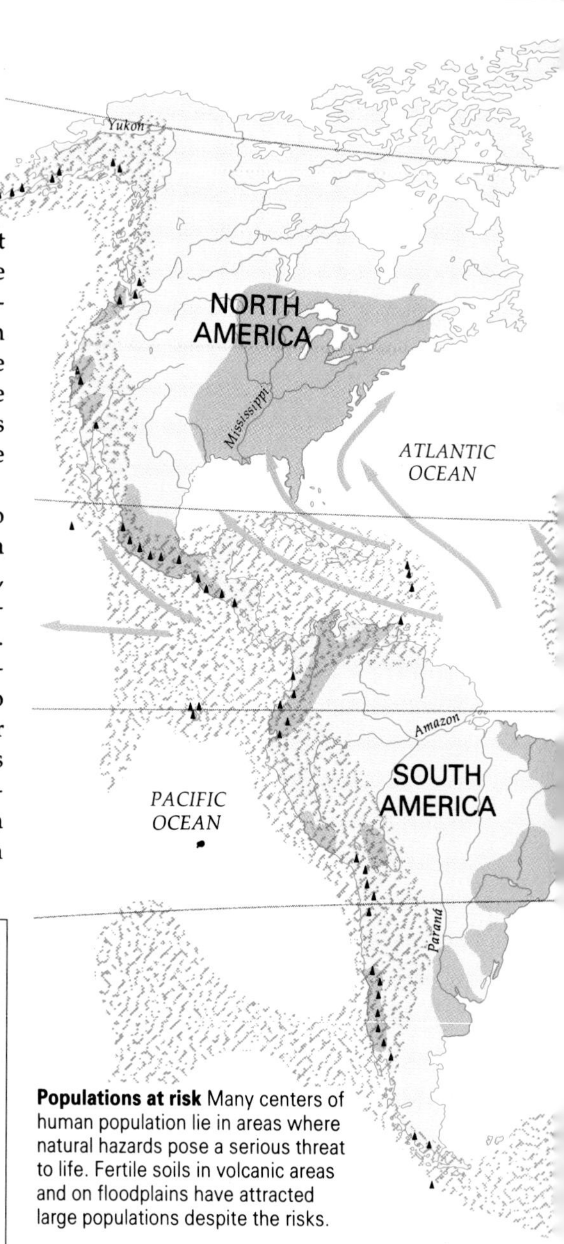

Populations at risk Many centers of human population lie in areas where natural hazards pose a serious threat to life. Fertile soils in volcanic areas and on floodplains have attracted large populations despite the risks.

Hazards of the atmosphere

Extremes of temperature are a major natural hazard of the atmosphere. At one end of the scale, high temperatures can encourage fires, cause heatstroke and lead to drought; at the other, low temperatures can be a hazard because of damaging frosts, snow and ice.

The wind also has the power to disrupt human settlement and activities. Violent winds can damage crops and result in massive soil losses. High winds combine with intense rainfall in hurricanes (known in Asia as typhoons or cyclones). A large hurricane can release as much energy in 24 hours as the United States uses in electricity in an entire year. This surge of power can increase the height of storm tides, causing severe damage in coastal areas.

Drought is defined as a prolonged period with rainfall markedly below average. Fifteen days with less than 0.2 mm (0.07 in) rain in any day is a drought in temperate areas, whereas in semiarid environments this level of rainfall could

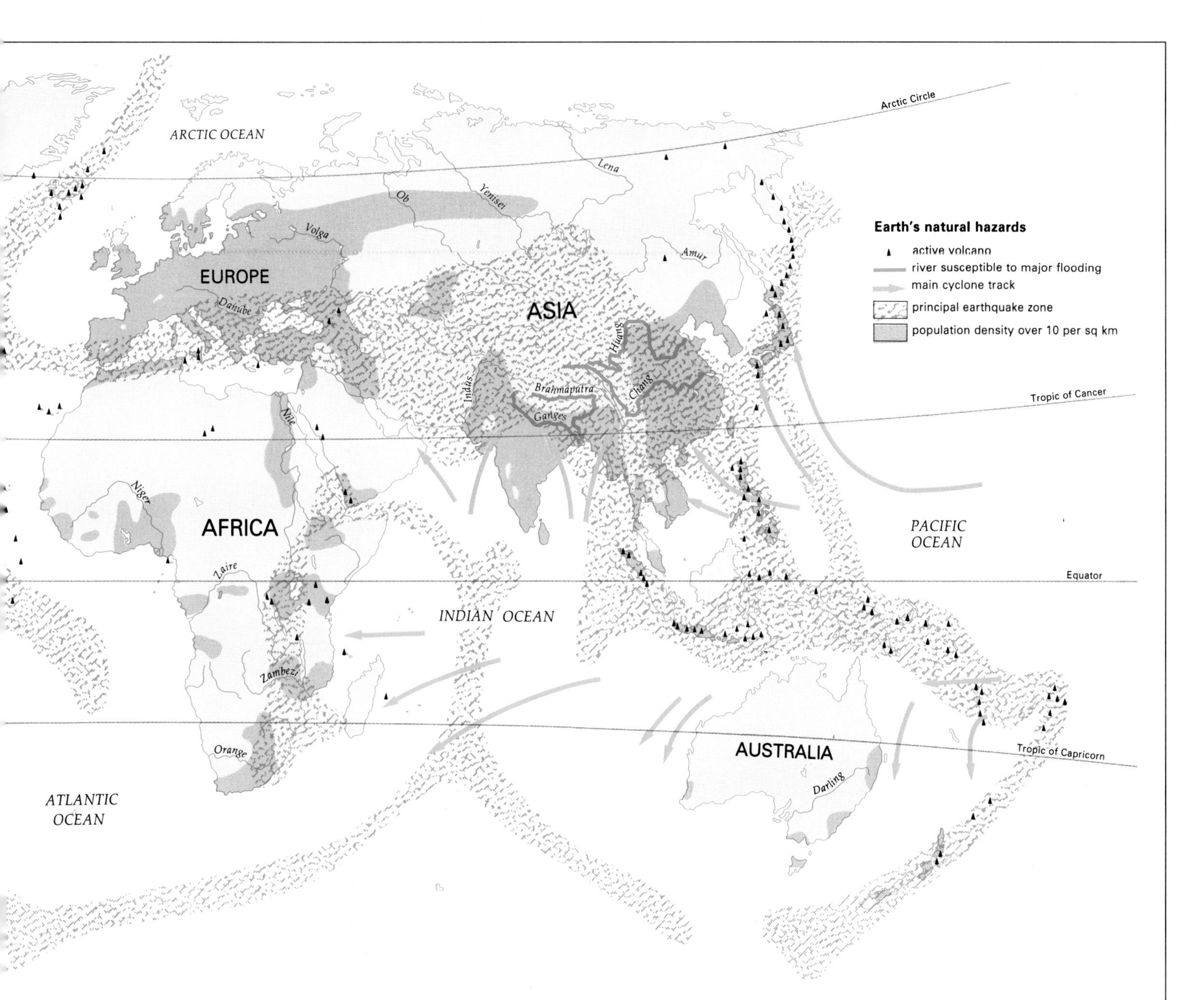

continue for months or even years before it was called a drought.

Hazards on the land

Hazards of the geosphere, such as earthquakes and volcanic eruptions, originate beneath the surface of the Earth, and most frequently take place along plate boundaries. There are about a million earthquakes a year, from small tremors to massive quakes; 455 volcanoes are currently active. Earthquakes and tremors can trigger landslides and mudslides.

Avalanches are a mixture of snow and ice, often with soil and rock. They can move at up to 320 k/h (200 mph) down a mountainside. One of the greatest avalanches on record was in 1962, when a vast mass of ice and rock fell from Peru's highest mountain, Huascarán, in the Andes. Snow and ice swept down the Ranrahirca valley at 100 kmh (60 mph), devastating nine villages and a town and killing 4,000 people and 10,000 animals. In 1970 an earthquake triggered an even more catastrophic avalanche on the same mountain; on this occasion a further 20,000 people lost their lives.

Earthquakes can also cause floods. Tsunami are huge sea waves generated by an earthquake or volcanic eruption beneath the ocean floor. The waves can travel across the sea at up to 800 k/h (500 mph), and measure up to 23 m (75 ft) high on reaching the coastline. One tsunami that hit the island of Honshu, Japan in June 1896 drowned 26,000 people.

Hazards of the hydrosphere

The greatest and most devastating floods are caused by rivers after prolonged or sudden torrential rainfall or a sudden thaw of snow. The condition of the ground receiving heavy rain is an important factor in determining how much water will run off the land into the rivers.

Hazards of the living world

Hazards encountered in the sphere of plant and animal life include fire, disease and infestations. Fires encouraged by drought or caused by lightning can destroy the habitats of plants and animals, and sweep over human settlements. Bacterial, fungal and viral diseases can damage plant and animal populations, including people. Particular environmental conditions can affect both disease and infestations: the entire living world is influenced by its physical environment.

THE MOST DEADLY NATURAL HAZARDS

A 25-year sample provided details of the frequency of intensive natural hazards and their effects.

Type of disaster	Index of frequency	Percentage of deaths
Floods	100	39
Typhoons, hurricanes	73	36
Earthquakes	41	13
Tornadoes	32	1
Gales, thunderstorms	15	5
Snowstorms	13	1
Heatwaves	8	1
Cold waves	6	1
Volcanic eruptions	6	2
Landslides	6	1

The Human Impact

HUMAN ACTIVITY HAS PLAYED AN INCREASingly significant part in the shaping of landscapes in the geologically very recent past. It is estimated that 55 percent of the Earth's land surface is now intensively used by people and 30 percent is partly modified; only 15 percent is untouched or only slightly modified.

People started to change the landscape some 10,000 years ago, when the Agricultural Revolution began. In the fertile river basins of the Nile, Euphrates and Tigris, Chang, Ganges and Brahmaputra, people who had been hunter–gatherers began to cultivate crops and domesticate animals. This system of food production was successful, and as it intensified it led to a radical change in the ways humans used the landscape. Land was plowed and irrigated, and as populations grew the first cities were founded.

With the Industrial Revolution, which began in Europe in the mid-18th century, the scale and intensity of human impact on the environment dramatically increased. The introduction of mechanization and spread of industrialization speeded up the demand for natural resources, and the use of land for settlement, industrial development and mines accelerated.

Advances in science, in medicine, hygiene and food production have led to a third revolution – the population explosion. It took from the beginning of time to the year 1830 for the world population to reach 1 billion, but only another hundred years to reach 2 billion. By 1975 it had doubled again to 4 billion, and in the late 1980s it stood at 5 billion. The projected total for the mid to late 1990s is 6 billion. This exponential population growth has created a huge demand for energy, land, food and resources.

Deforestation and erosion

One of the most dramatic ways in which humans have modified the appearance of the land is by cutting down trees. Deforestation has been taking place for thousands of years all over the world. In southern Asia 9,000 years of deforestation have removed two-thirds of the forest; areas of China that were once forested no longer have any trees. In the African tropics three-quarters of the moist forest has been cleared over the past 3,000 years. Today as much as 12 million hectares (30 million acres) of forest are cleared from the Earth every year.

Clearing forest has profound effects on the environment. A cultivator who clears a plot of land changes the appearance and character of the landscape directly, and he modifies it further by plowing, which alters the soil structure, and by introducing new plant species.

Bringing land under cultivation can have indirect effects that are just as striking and have an even greater impact on environments. Without plants to hold the soil in place, the amount of soil removed by erosion may increase by ten times or more. Gullies and underground pipes may be formed by the uncontrolled flow of water. During urban development the amount of sediment carried by a river may increase a hundred or even a thousandfold. The eroded soil adds to the sediment load carried by rivers, changing the landscape some distance away.

Creating new landscapes

In many wet tropical and subtropical parts of the world farmers have created a new landscapes of terraced hillsides to make flat areas for cultivation and irrigation and to reduce erosion. The terraces are just one example of manmade landscapes. In some parts of the world the land itself has been made, or at least reclaimed from the sea, by human activity – most notably in the Low Countries and parts of Japan. Dams and reservoirs form new lakes, such as the Kariba Lake in Zimbabwe, southern Africa. Channelization has affected the course and character of rivers: 1,000 km (620 mi) of the Nile, in northern Africa, and 4,500 km (2,800 mi) of rivers in North America's Mississippi basin have been affected in this way. Human conflict can also scar the land: between 1965 and 1971, 26 million craters were produced by bombing in the battlegrounds of Southeast Asia, displacing 2.6 billion cu m (3.5 billion cu yd) of soil.

Mining and its associated spoil tips, quarrying, and earth-moving to build railroads, highways and harbors are all major causes of landscape change.

Some individual mines have produced enormous holes in the Earth's surface. The Bingham Canyon coppermine in Utah, United States, affects an area of over 7 sq km (2.7 sq mi), and has been excavated to a depth of 774 m (2,539 ft). Mining for iron ore near Zheleznogorsk, north of Lake Baikal in the central Soviet Union, has created a hole 3 km (1.9 mi) long and 0.5 km (0.3 mi) wide that is still expanding as more ore is extracted.

Indirect impacts

The extraction of oil, water, coal and other minerals underground can cause the surface above to subside. Similarly, some activities carried out on the surface can have quite unforeseen effects, even far below: the construction of large dams and reservoirs can trigger small earthquakes by laying stress on the Earth's crust.

These disturbances take place in many different environments. In desert and semidesert zones the removal of vegetation, for example by overcultivation or through overgrazing by domesticated animals such as cattle or goats, make it possible for sand dunes to encroach, and this speeds up the process of desertification. This is happening in central Africa's Sahel region. In cold tundra regions, where human activity around settlements and oil pipelines melts the otherwise permanently frozen soil, hummocky surface features known as thermokarst result. On coasts, the construction of a sea wall to protect the coast in one area may lead to beach starvation, increased erosion and changes in the coastline elsewhere.

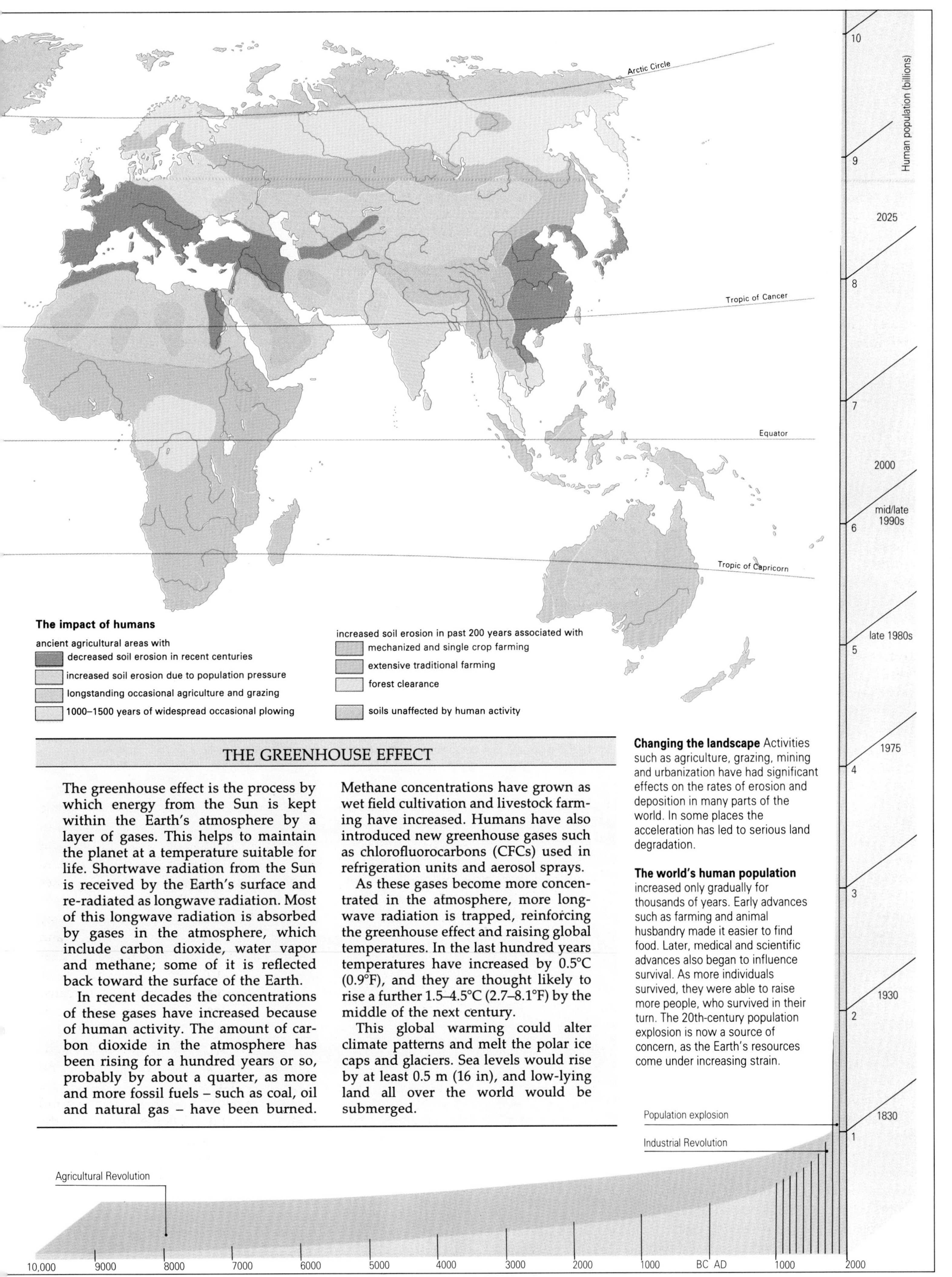

THE GREENHOUSE EFFECT

The greenhouse effect is the process by which energy from the Sun is kept within the Earth's atmosphere by a layer of gases. This helps to maintain the planet at a temperature suitable for life. Shortwave radiation from the Sun is received by the Earth's surface and re-radiated as longwave radiation. Most of this longwave radiation is absorbed by gases in the atmosphere, which include carbon dioxide, water vapor and methane; some of it is reflected back toward the surface of the Earth.

In recent decades the concentrations of these gases have increased because of human activity. The amount of carbon dioxide in the atmosphere has been rising for a hundred years or so, probably by about a quarter, as more and more fossil fuels – such as coal, oil and natural gas – have been burned. Methane concentrations have grown as wet field cultivation and livestock farming have increased. Humans have also introduced new greenhouse gases such as chlorofluorocarbons (CFCs) used in refrigeration units and aerosol sprays.

As these gases become more concentrated in the atmosphere, more longwave radiation is trapped, reinforcing the greenhouse effect and raising global temperatures. In the last hundred years temperatures have increased by 0.5°C (0.9°F), and they are thought likely to rise a further 1.5–4.5°C (2.7–8.1°F) by the middle of the next century.

This global warming could alter climate patterns and melt the polar ice caps and glaciers. Sea levels would rise by at least 0.5 m (16 in), and low-lying land all over the world would be submerged.

Changing the landscape Activities such as agriculture, grazing, mining and urbanization have had significant effects on the rates of erosion and deposition in many parts of the world. In some places the acceleration has led to serious land degradation.

The world's human population increased only gradually for thousands of years. Early advances such as farming and animal husbandry made it easier to find food. Later, medical and scientific advances also began to influence survival. As more individuals survived, they were able to raise more people, who survived in their turn. The 20th-century population explosion is now a source of concern, as the Earth's resources come under increasing strain.

Managing the Earth

People have long tried to manage the Earth and its processes, both to combat natural hazards and to use its natural resources. It is increasingly being understood that good environmental management is important. It demands a thorough knowledge of the landscape and the processes that operate in it. Research, measurements and experience all provide information about an area. Satellite monitoring provides instant information, and computers store and analyze the data. However, detailed records for the required area often do not exist, or they cover only the last few decades, providing inadequate information about the nature and frequency of landscape-forming events. Scientists have also said that existing records of climate are no longer valid, as the warming of the Earth as a result of the greenhouse effect is changing the world's climate patterns.

Many attempts at management have had unforeseen effects. In Egypt the Aswan High Dam was built to regulate the Nile, to control flooding and to provide hydroelectric power. The dam has benefited Egypt, but it has also created another set of problems. The silts and clays that used to flow downstream and fertilize the floodplain have been trapped behind the dam, silting up Lake Nasser. The disease schistosomiasis has been increasing among people living round the lake. Fertilizers now have to be imported for the floodplain areas downstream, and the brick-making industry that relied upon the silts and clays no longer has local raw materials. The dam has also altered the structure of the Nile delta and increased the salinity of agricultural land. Such experiences have led to a growing realization that human activity in the landscape can have undesirable consequences and that it is beneficial to work with nature rather than against it.

Strategies for management

There are four basic ways of approaching landscape management. The first is to do nothing, accepting the consequences that arise. Some events, such as landslides or tsunami, cause devastation and considerable loss of life but may just have to be accepted. Certain environments are known to be hazardous, but people still choose to live there, accepting that enjoyment of benefits has its attendant risks – perhaps, for example, the danger of a volcanic eruption in the vicinity of fertile volcanic soils. Hilo, the largest city on the island of Hawaii, lies 64 km (40 mi) from the active volcano Mauna Loa, whose lava flows seriously threatened the city in 1881, 1855, 1899, 1935 and 1942.

The second way is to try to reduce the effects of environmental hazards. Precautions such as weather forecasts or evacuation schemes can be developed to give the population time to prepare for a hazard. Immediately after a major disaster has taken place, emergency aid on a national or even an international scale can reduce losses, and in developed countries commercial insurance can offset the financial impact of disasters.

A third way of managing the environment is to use engineering or planning solutions to place vulnerable aspects of human activity out of range of environmental hazards. In parts of Southeast Asia and elsewhere, houses are built on stilts to avoid the effects of annual flooding. In North America developments on floodplain land have been zoned so that the most dispensable activities are sited closest to the river. The most expensive activities, involving industry and domestic housing, are placed in the safest zone farthest from the river.

A fourth type of environmental management is actually to modify the environment or its processes. There are many possibilities. Soil erosion can be prevented or reduced by terracing the land, grasses or trees can be planted to stabilize the soil or to form a windbreak. Flooding can be contained by building high walls along a river's course, and by constructing sea walls along the coast. Deserts can be irrigated, and rainfall induced by "seeding" clouds with dry ice.

Successful management schemes have been undertaken to ensure that areas

ENVIRONMENTAL IMPACT ASSESSMENT

The realization that human activity has an effect on the environment has led to the introduction of environmental impact assessment (EIA) legislation in many countries. EIA is defined by the United Nations Environment Program as "the examination, analysis and assessment of planned activities with a view to ensuring environmentally sound and sustainable development". EIA legislation requires planners and developers to evaluate and report on the short- and long-term impact of projects before permission is given for them to proceed. Another aim is ensuring that the effects of the projects on the ecosystems are monitored.

In the United States these assessments are called Environmental Impact Statements (EIS). They became compulsory in 1969, when the National Environmental Policy became law, and have to be prepared for all proposed major federal actions that could affect the quality of the human environment. The European Community (EC) and a number of other countries have since introduced similar legislation.

THE IMPACT OF DAM CONSTRUCTION

With increasing awareness of the environmental problems that may result from the construction of dams, an environmental impact assessment should be made. Many of the detrimental effects of dam construction take many years to show.

Expected benefits

Hydroelectric power generation

Improved control of seasonal floods on floodplain

Water available year-round for irrigation of farmland and for industrial and domestic users

Larger area of floodplain can be irrigated

Fishing industry centered on reservoir lake

Leisure facilities and tourism on reservoir lake

Possible costs

Relocation of settlements inundated by reservoir

Large amounts of water lost by evaporation from reservoir surface, or by seepage through rocks

Reservoir silts up (especially if watershed is deforested)

Enormous weight of water triggers earth tremors

Diseases such as bilharzia spread among people living around reservoir

River cuts a deeper channel because its sediment load is decreased; this upsets irrigation systems

Floodplain soils lose fertility because the river no longer deposits silt on the land; increased use of fertilizers is required to compensate

Wetland ecosystems are destroyed on the floodplain; some plant and animal life lost from river

Farmland in delta area becomes more saline due to reduced flushing with fresh water

Fertile delta eroded by longshore drift because of reduced deposition of silt; tourist beaches eroded

Sea fisheries decline with reduced flow of fresh water

retain their character and that the Earth's natural processes are not dramatically altered. One example is on Indian Bend Wash, a tributary of the Salt river in Phoenix, Arizona. A history of flooding led in the 1970s to a decision to implement a flood protection scheme that worked with the river rather than against it. As well as containing the floods and producing a channel that looked pleasing, the flood protection measures were deliberately integrated with other uses of the river, such as recreation.

It is now acknowledged that the Earth's environments can be managed effectively if the task is approached with respect for the land and with an understanding of its processes and living organisms. The careful and wise stewardship of the Earth in the future depends on the development of this understanding.

Contoured farmland in Pennsylvania, USA, has been plowed parallel to the contours of the slope. This reduces soil erosion, as rainwater cannot wash soil straight down the furrows.

The Hoover dam on the Colorado river in Nevada, USA, is used to generate hydroelectric power, to control flooding, for irrigation and to supply water to homes and industry. The removal of water by this and other dams has greatly reduced the river's flow.

REGIONS OF THE WORLD

CANADA AND THE ARCTIC
Canada, Greenland

THE UNITED STATES
United States of America

CENTRAL AMERICA AND THE CARIBBEAN
Antigua and Barbuda, Bahamas, Barbados, Belize, Bermuda, Costa Rica, Cuba, Dominica, Dominican Republic, El Salvador, Grenada, Guatemala, Haiti, Honduras, Jamaica, Mexico, Nicaragua, Panama, St Kitts-Nevis, St Lucia, St Vincent and the Grenadines, Trinidad and Tobago

SOUTH AMERICA
Argentina, Bolivia, Brazil, Chile, Colombia, Ecuador, Guyana, Paraguay, Peru, Uruguay, Surinam, Venezuela

THE NORDIC COUNTRIES
Denmark, Finland, Iceland, Norway, Sweden

THE BRITISH ISLES
Ireland, United Kingdom

FRANCE AND ITS NEIGHBORS
Andorra, France, Monaco

THE LOW COUNTRIES
Belgium, Luxembourg, Netherlands

SPAIN AND PORTUGAL
Portugal, Spain

ITALY AND GREECE
Cyprus, Greece, Italy, Malta, San Marino, Vatican City

CENTRAL EUROPE
Austria, Liechtenstein, Switzerland, West Germany

EASTERN EUROPE
Albania, Bulgaria, Czechoslovakia, East Germany, Hungary, Poland, Romania, Yugoslavia

THE SOVIET UNION
Mongolia, Union of Soviet Socialist Republics

THE MIDDLE EAST
Afghanistan, Bahrain, Iran, Iraq, Israel, Jordan, Kuwait, Lebanon, Oman, Qatar, Saudi Arabia, Syria, Turkey, United Arab Emirates, Yemen

NORTHERN AFRICA
Algeria, Chad, Djibouti, Egypt, Ethiopia, Libya, Mali, Mauritania, Morocco, Niger, Somalia, Sudan, Tunisia

CENTRAL AFRICA
Benin, Burkina, Burundi, Cameroon, Cape Verde, Central African Republic, Congo, Equatorial Guinea, Gabon, Gambia, Ghana, Guinea, Guinea-Bissau, Ivory Coast, Kenya, Liberia, Nigeria, Rwanda, São Tomé and Príncipe, Senegal, Seychelles, Sierra Leone, Tanzania, Togo, Uganda, Zaire

SOUTHERN AFRICA
Angola, Botswana, Comoros, Lesotho, Madagascar, Malawi, Mauritius, Mozambique, Namibia, South Africa, Swaziland, Zambia, Zimbabwe

THE INDIAN SUBCONTINENT
Bangladesh, Bhutan, India, Maldives, Nepal, Pakistan, Sri Lanka

CHINA AND ITS NEIGHBORS
China, Taiwan

SOUTHEAST ASIA
Brunei, Burma, Cambodia, Indonesia, Laos, Malaysia, Philippines, Singapore, Thailand, Vietnam

JAPAN AND KOREA
Japan, North Korea, South Korea

AUSTRALASIA, OCEANIA AND ANTARCTICA
Antarctica, Australia, Fiji, Kiribati, Nauru, New Zealand, Papua New Guinea, Solomon Islands, Tonga, Tuvalu, Vanuatu, Western Samoa

North America

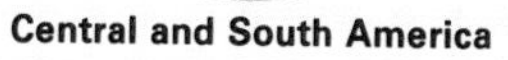

Central and South America

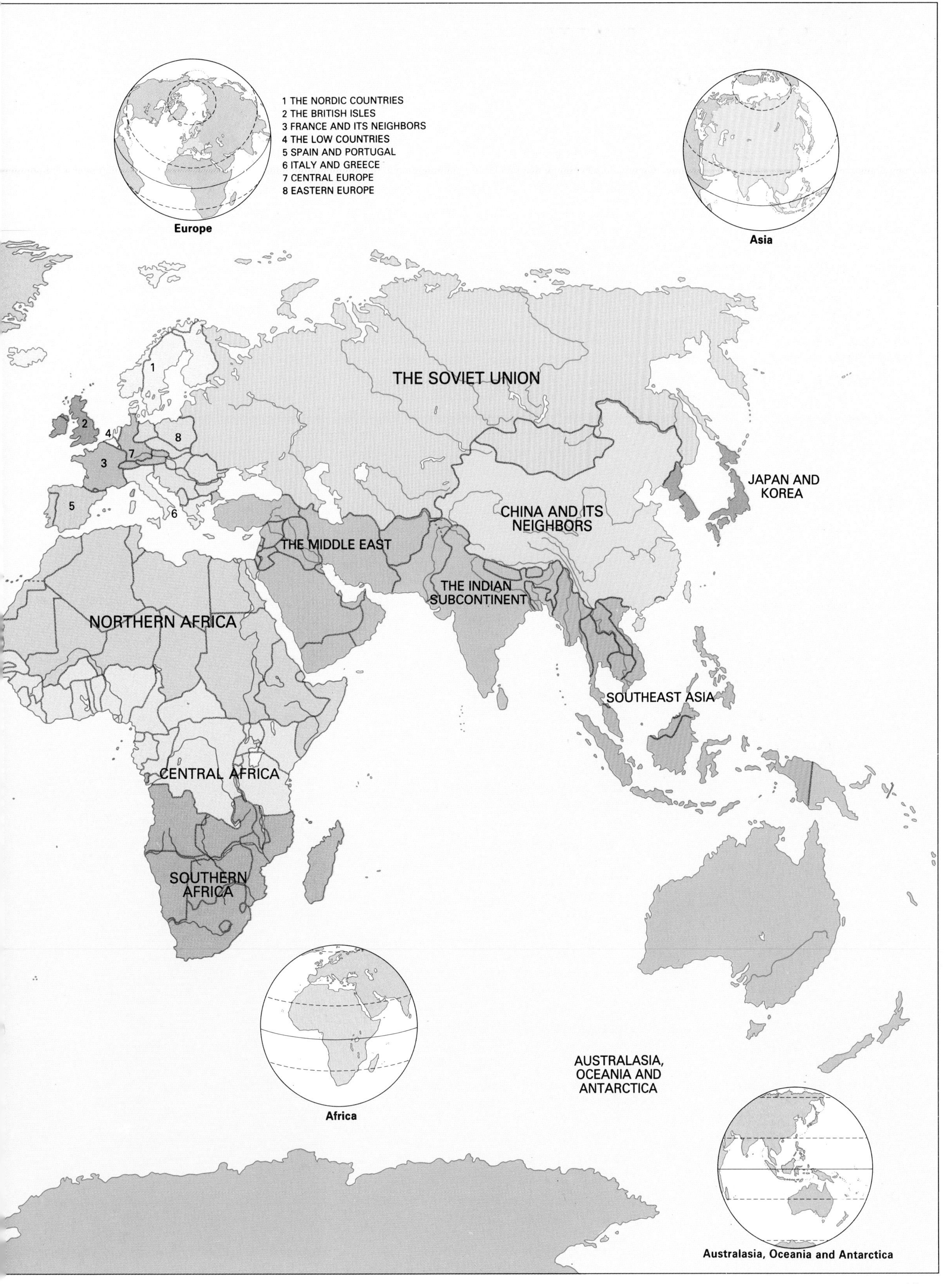

Europe
1 THE NORDIC COUNTRIES
2 THE BRITISH ISLES
3 FRANCE AND ITS NEIGHBORS
4 THE LOW COUNTRIES
5 SPAIN AND PORTUGAL
6 ITALY AND GREECE
7 CENTRAL EUROPE
8 EASTERN EUROPE
Asia
THE SOVIET UNION
JAPAN AND KOREA
CHINA AND ITS NEIGHBORS
THE MIDDLE EAST
THE INDIAN SUBCONTINENT
NORTHERN AFRICA
SOUTHEAST ASIA
CENTRAL AFRICA
SOUTHERN AFRICA
Africa
AUSTRALASIA, OCEANIA AND ANTARCTICA
Australasia, Oceania and Antarctica

FROM GREAT LAKES TO ARCTIC DESERT

MOUNTAINS, LAKES AND PLAINS · THE FROZEN NORTH · THE PAST IN TODAY'S LANDSCAPE

Some of the world's largest landscape features can be found in Canada and the Arctic islands to the north. Canada, second in size only to the Soviet Union, has more lakes than any other country and shares the world's greatest expanse of fresh water, Lake Superior, with the United States. The region contains four of the world's largest islands including Greenland, the world's largest, which has the biggest ice sheet in the northern hemisphere. There are also massive mountain ranges with forests, swift rivers and snowy peaks. The Rocky Mountains run parallel to the Pacific coast and the older Appalachian Mountains border the Atlantic. Between these ranges lie the farmlands of the vast prairies and, centered on Hudson Bay, the ancient and bleak Canadian Shield with its lakes, remnants of the retreating ice sheet.

COUNTRIES IN THE REGION

Canada, Greenland

LAND

Area 12,151,739 sq km (4,691,791 sq mi)
Highest point Mount Logan, 5,951 m (19,524 ft)
Lowest point sea level
Major features Rocky Mountains, Canadian Shield, Arctic islands, Greenland, world's largest island

WATER

Longest river Mackenzie, 4,240 km (2,635 mi)
Largest basin Mackenzie, 1,764,000 sq km (681,000 sq mi)
Highest average flow Saint Lawrence, 13,030 cu m/sec (460,000 cu ft/sec)
Largest lake Superior, 83,270 sq km (32,150 sq mi), world's largest freshwater lake

CLIMATE

	Temperature °C (°F) January	July	Altitude m (ft)
Resolute	−32 (−26)	4 (39)	64 (200)
Vancouver	2 (36)	17 (63)	0 (0)
Winnipeg	−18 (0)	20 (68)	248 (813)
Montreal	−9 (16)	22 (72)	30 (98)
Halifax	−4 (25)	18 (64)	30 (98)

	Precipitation mm (in) January	July	Year
Resolute	3 (0.1)	21 (0.8)	136 (5.3)
Vancouver	139 (5.5)	26 (1.0)	1,068 (42.0)
Winnipeg	26 (1.0)	69 (2.7)	535 (21.0)
Montreal	83 (3.3)	89 (3.5)	999 (39.3)
Halifax	137 (5.4)	96 (3.8)	1,381 (54.4)

World's highest recorded snowfall in 24 hours, 1,180 mm (46 in), Lake Lakelse, British Columbia

NATURAL HAZARDS

Cold and snowstorms, drought, gales, avalanches, rockfalls and landslides

MOUNTAINS, LAKES AND PLAINS

Canada can be divided into four broad physical areas: the western mountain system, the central lowlands, the Canadian Shield north of the five Great Lakes and the Appalachian Mountains in the east. Each of these main physical divisions has different geological origins, but climatic change was responsible for shaping the varied landscapes of today.

Changes in the climate affected the movement of ice sheets, which at their farthest advance covered the whole of Canada. The most recent advance, known in North America as the Wisconsin glaciation, lasted until 10,000 years ago. As the climate fluctuated, the ice sheets advanced and retreated. Glacial erosion flattened and scoured the landscape and redirected rivers; debris deposited by the ice formed landscape features such as moraines and provided the basis for the fertile soils of the central lowlands.

Rockies and prairies

The western mountain system or cordillera stretches some 800 km (500 mi) from the islands and the coast mountains along the Pacific seaboard to the eastern slopes of the Rocky Mountains. Between the Pacific ranges and the Rockies lies a high plateau region. The Rocky Mountains and neighboring areas are relatively recent. They were created some 80–40 million years ago when the Pacific Ocean crust was forced under the westward-moving North American plate. The area is still undergoing change, as the 1980 eruption of Mount Saint Helens in the northern United States showed. In the Rocky Mountains peaks 4,000–6,000 m (13,000–19,000 ft) high alternate with deep and frequently U-shaped valleys carved out by glaciers. The slopes are steep, and avalanches and rockfalls are common.

The central lowlands, also known as the Great Plains, offer a landscape of wide horizons. They are a continuation of the Great Plains of the United States, and taper northward from a width of about 1,500 km (900 mi) at the United States border, to the Mackenzie delta in the Arctic. The prairies in the south are Canada's main farmland area. Spread across the Great Plains are beds of rock of great thickness, evenly deposited on the floor of a former sea. Overlying this are softer materials, produced by erosion and deposited by rivers and glaciers. The relief is varied by some hilly areas and by long lines of steep slopes (escarpments), such as that between lowlands near Lake Winnipeg and the Great Plains to the west. These features are often concealed by glacial and postglacial lakes. Bitterly cold winters give way in the south to burning summers, which produce semi-arid or near-desert conditions in some places. By contrast, on the Pacific coast temperatures vary less, and moist air from the sea gives the Coast Mountains the highest rainfall in Canada.

The Shield, south and east

The vast semicircle of the Canadian Shield extends north from the Great Lakes, encompassing Hudson Bay. The shield is a plateau of much-eroded granite rocks, over 600 million years old. It has thin soils, yet the area is not barren but covered by forests. The mixed forest of the south gives way to a coniferous belt that thins to true tundra in the north.

South of the Canadian Shield there is an abrupt change. Here there is rolling

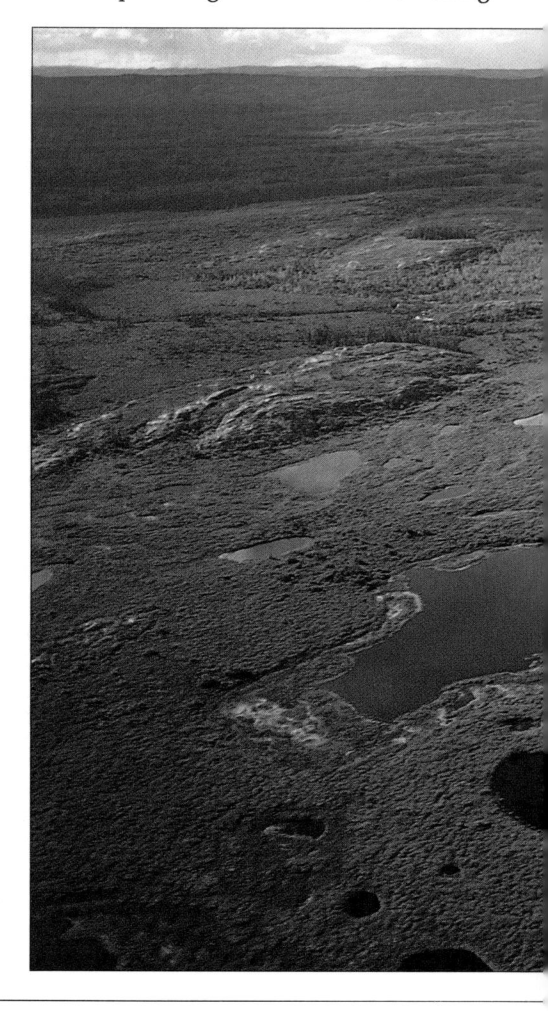

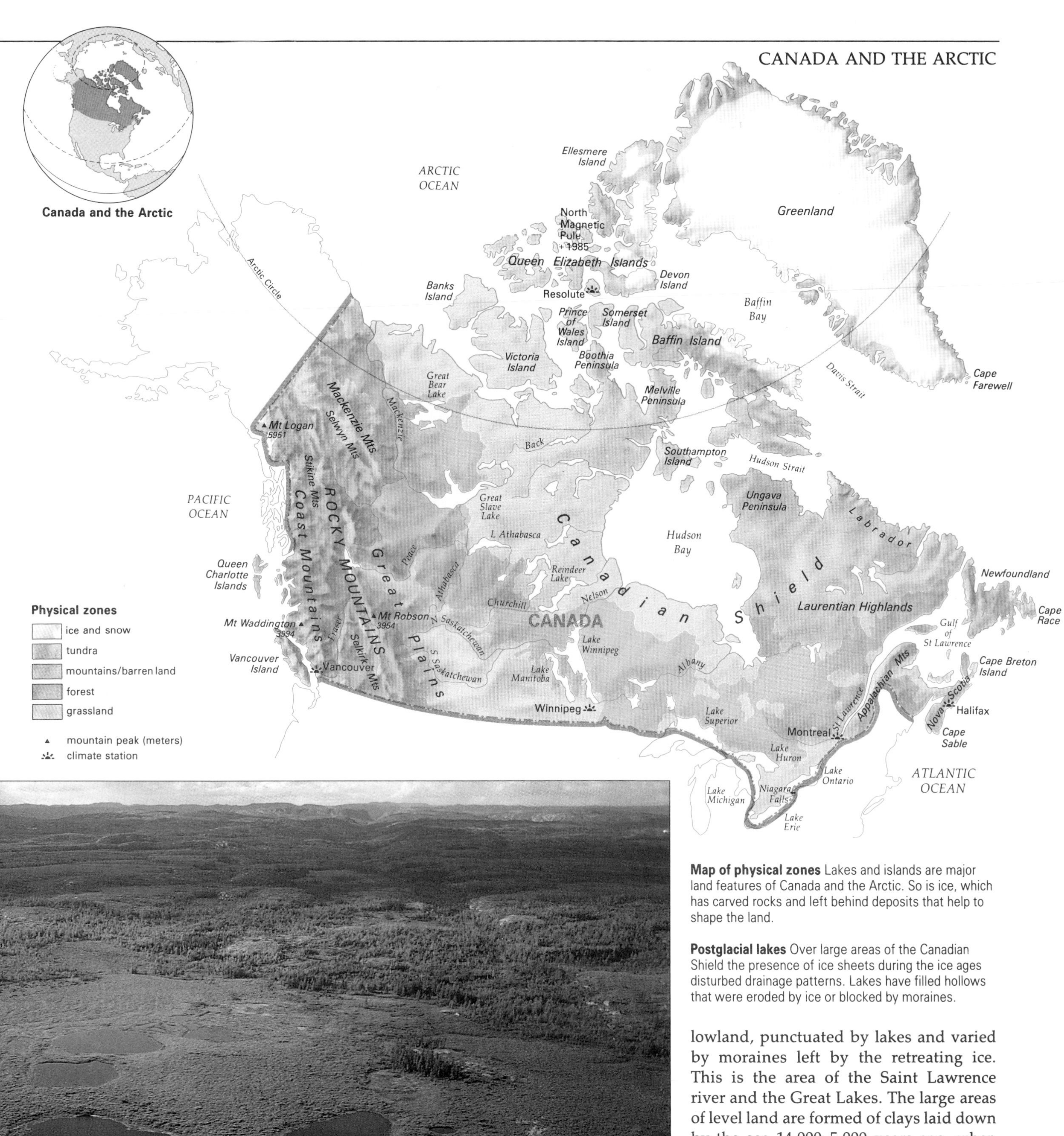

Map of physical zones Lakes and islands are major land features of Canada and the Arctic. So is ice, which has carved rocks and left behind deposits that help to shape the land.

Postglacial lakes Over large areas of the Canadian Shield the presence of ice sheets during the ice ages disturbed drainage patterns. Lakes have filled hollows that were eroded by ice or blocked by moraines.

lowland, punctuated by lakes and varied by moraines left by the retreating ice. This is the area of the Saint Lawrence river and the Great Lakes. The large areas of level land are formed of clays laid down by the sea 14,000–5,000 years ago, when the world's great ice sheets retreated. They make good agricultural land, but are subject to landslides and therefore form a rather unstable foundation for Canada's urban areas.

The Appalachians form a mountain system raised between about 400 and 225 million years ago as the North American plate met the African plate. The gently rolling uplands are separated by deep valleys, fringed by the rugged, indented coasts and islands of the Gulf of Saint Lawrence and the Atlantic Ocean.

The pingo is a striking feature in the Arctic tundra landscape. Pingos, named from the Inuit people's word for a mound, are formed by water under the ground expanding when it freezes. As the ice core grows it pushes up the surface of the land.

THE FROZEN NORTH

The two-fifths of Canada lying to the north of the 60th parallel are a continuation of the mountains, plains and shield to the south. However, the climate, vegetation and detailed landforms are all so overwhelmed by the extremes and duration of cold that Canadians regard this as a separate area. The frozen, barren and largely treeless landscapes extend from the precipitous peaks of the Yukon Territory and the delta of the Mackenzie river in the west to the arc of Baffin Island in the east. A narrow strait separates Ellesmere Island from the great ice sheet and bare coasts of Greenland.

The islands of the Arctic Archipelago cover an area of about 1,300,000 sq km (500,000 sq mi) and make up one-seventh of the land of Canada. They range over 2,200 km (1,400 mi) from the southern tip of Baffin Island to the northern tip of Ellesmere Island. Two-fifths of the Arctic is affected by ice, and on the islands the highland areas are covered by glaciers. The Arctic holds just one-eighth of the world's glacial ice; most of this is found in Greenland's huge ice cap.

Greenland is the world's largest island. At 2,175,600 sq km (840,000 sq mi) it is four times the size of France, and shares with Antarctica the distinction of preserving a major remnant of the ice sheets that covered much of the Earth in the high latitudes until about 10,000 years ago. The thick central ice dome is fringed by coastal mountain ranges rising to 3,700 m (12,000 ft). The ice sheet reaches the sea at about a hundred places.

A narrow and broken strip along the coasts, amounting to about a seventh of Greenland's area, is free of ice. The coasts have many raised beaches and terraces. They are very indented, with many deep fjords, so the coastline is some 40,000 km (25,000 mi) long. The climate is milder along the more populated south and west coasts, which are warmed by sea currents from the south. The west coast is enclosed by pack ice for fewer months of the year than the east coast.

To the west of Greenland, across Baffin Bay and the Davis Strait, is Baffin Island – the fifth largest island in the world. In the north is Ellesmere Island, another of the world's ten largest islands. At its closest it is only 25 km (15.5 mi) from Greenland. Ellesmere Island has an unusual climate: with only 65 mm (2.6 in) of precipitation each year, it is far drier than much of the Sahara desert. Most of the islands receive only 130–250 mm (5–10 in) each year.

Clues to a rising land: a series of old raised beaches in northern Baffin Island. The beaches are a legacy of the ice sheets of the last ice age. About 20,000 years ago ice extended as far south as the Great Lakes. The weight of the ice depressed the Earth's crust. The world's climate subsequently became warmer and the ice melted. As the ice sheet retreated the land began to rise again (isostasy). Pauses in this recovery are marked by landforms associated with coastlines, such as beaches,cliffs and wave-cut platforms. These shorelines have been abandoned by the sea as the land has risen still further. Raised beaches can also be formed by changes in sea level rather than land level, a process known as eustasy.

However, if the Arctic islands qualify as desert, much of the land is "wet desert" in the short summer as the surface of the land thaws.

Baffin, Ellesmere and the smaller Devon Island are all mountainous, like Greenland. The islands lying further west are much flatter, their tundra landscapes contrasting with those of their neighbors.

The tundra

The northern region of Canada has a tundra landscape. The key feature of this treeless zone is the permanently frozen ground beneath the land surface. This layer may be hundreds of meters thick and can represent tens of thousands of years of cooling. Across much of the Arctic tundra the top 20–100 cm (8–40 in) of the surface will thaw out each summer. The water is not able to flow into the underlying layer of permafrost; nor can it evaporate, because the air is usually too cold, so the landscape becomes waterlogged with many bogs and lakes. During the long winter the water refreezes and the land surface contracts, cracking into polygonal patterns 10 m (40 ft) or more across. The cracks fill with ice each year and in time grow to become buried wedges of ice.

In some places the process of expansion and contraction linked to freezing and thawing produces patterns of stones. The slow churning of the ground separates large stones from the surrounding soil, moving them into polygonal and striped arrangements.

Tundra mounds, or pingos, are another feature of this landscape. Water trapped underground that is frozen during the winter expands to produce mounds filled with ice. Where the underground water supply is abundant (as beneath an old lake bed) progressive growth of this ice core over long periods can produce a pingo up to 60 m (200 ft) high.

Not surprisingly, the tundra environment is one in which few plants and animals can survive. Much of the landscape is bare rock and ice. Vegetation is sparse, with lichens and mosses, tussock and sedge grass and dwarf shrubs and trees. Apart from polar bears, which feed on the abundant fish and seals, most large creatures forage over larger areas. They include Musk ox and caribou, the Arctic fox and hare, and game birds including grouse, partridge and duck.

THE ARCTIC GREENHOUSE

Some predictions of the greenhouse effect, which results from increased quantities of carbon dioxide and other compounds in the atmosphere, suggest a global warming of 3°C (5.4°F) in the next 50–100 years, with temperatures rising about 1°C (2°F) in the tropics, but by up to 12°C (22°F) in the Arctic.

In the Arctic the changes would be dramatic. The delicate balance of the tundra environment would be upset. When permafrost melts, the surface degenerates into a jumble of hummocks, hollows, lakes and mudflows. These features are called thermokarst, because the scenery resembles the karst features found in limestone country. At present the short tundra vegetation helps to insulate the soil from much of the warmth of the brief summers. However, if temperatures rise as predicted, thermokarst is likely to spread widely over the now frozen north, covering both empty wilderness and human settlement areas. Trees and shrubs would spread farther north, as would grazing animals. More snow would fall, so most of the Arctic Archipelago would cease to be the cold desert it is today.

THE PAST IN TODAY'S LANDSCAPE

Canada has only recently emerged from a tumultuous past, the geological history of which is imprinted on the landscape. The mountains, plains and coasts, which are subject to sea-level changes, all carry clear indications of having origins in conditions very different from those of today. However, there are also processes – such as glaciation and erosion – that have been at work for millions of years and are still going on.

Prairie badlands to the southwest

At first sight the prairies, or Great Plains, of southern Alberta give an impression of a barely undulating landscape clothed in wheat. On closer inspection, they reveal a dramatic past. Complex patterns of deposits left 10,000–5,000 years ago by retreating ice sheets show that this area had an unusually varied ice-age history, dominated by the ice sheets centered on the shield and the Rocky Mountains. Since then rivers flowing eastward from the Rocky Mountains and swollen by melt waters have carved deep, wide canyons (as much as 100 by 1,500 m/330 by 5,000 ft) through the nearly horizontal, multicolored sand stones and shales, producing a spectacular eroded and barren landscape often known as badlands.

Badlands are formed in arid and semiarid regions and are areas of deep winding gullies divided by steep ridges. The differences in rock type can produce stepped landforms, and the joining and separating of the gullies creates rock-spires, earth pillars (known as hoodoos) and flat-topped, steep-sided plateaus called mesas. The detail of the badlands has been formed as a result of recent intense summer rainstorms, but the canyon walls reveal rocks 80 million years old and are famous for the dinosaur fossils that can still be found there today.

Fantastic hoodoos on the Red Deer river. Here in the rain shadow of the Rocky Mountains the climate is semiarid; further west rainfall is heavy. The hoodoos are pillars formed by erosion caused by flash floods and wind-blown particles.

Banff National Park in the Canadian Rockes includes mountains, glaciers and lakes. Millions of years ago the walls of rock enclosing this ice-sculpted valley were deposits beneath the seas. Debris falls from the frost-shattered peaks, forming scree slopes.

A changing landscape

Present landscapes in many parts of the world owe much to the last period of glaciation and its aftermath. Canada is one of the few places where the evolving environment can still be found by those who visit the mountains of the north.

The glacial landscapes of today include the Columbia icefield in the Rocky Mountains. It is small by global standards, only covering an area of 310 sq km (120 sq mi)

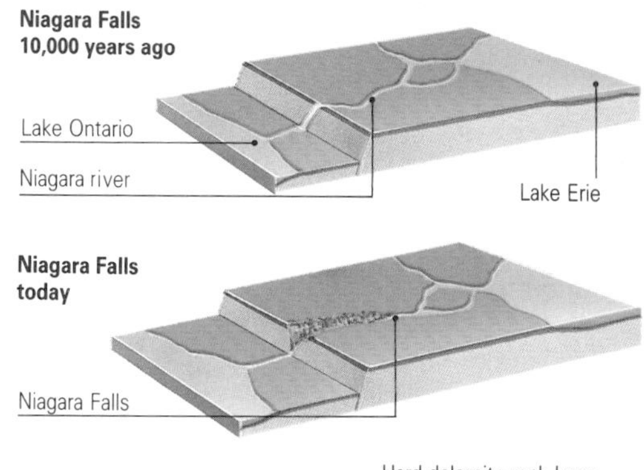

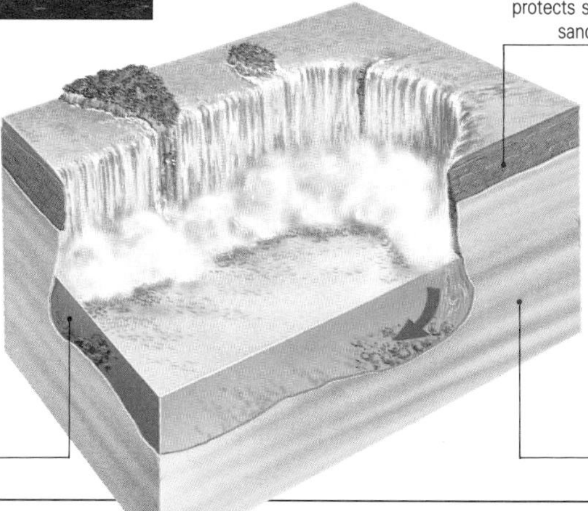

The dramatic landscape of the Niagara Falls and gorge on Canada's southern border owes its origins to the meltwaters of the retreating ice sheet some 10,000 years ago. Water from this part of North America once flowed to the sea through the Mississippi river. With the land to the north sagging under the weight of the ice sheet, water overflowed from Lake Erie to Lake Ontario and then into the North Atlantic. On the way the Niagara river crossed an escarpment whose capping of dolomite (a hard limestone) gave some protection to the softer sandstone and shale beneath. The plunging waters of the river gradually undermined the dolomite, and the waterfall slowly retreated upstream at approximately 90 cm (3 ft) a year, cutting a gorge some 60 m (200 ft) deep and 11 km (7 mi) long. At this rate the falls will reach Lake Erie in 25,000 years' time.

and 1,000 m (3,300 ft) thick, but unique in its accessibility to thousands of visitors to the Rocky Mountains' national parks. Although only 10 km (6 mi) long, the Athabasca glacier tongue in the Columbia icefield perfectly demonstrates the change that has dominated Canada's landscapes. Parallel to its tip, or snout, lie concentric rubble ridges (moraines) produced every year or two by debris dumped by the retreating ice margin. So regular has this landscape change been that the path leading to the snout is marked with signboards that date the positions occupied by the ice front as it has retreated over the last few decades.

THE HILL THAT DISAPPEARED

Late in the evening of 4 May 1971, in the village of Saint-Jean-Vianney just north of the Saint Lawrence, a Mme Laforge telephoned a friend to say she could see the lights of the neighboring town, Chicoutimi. Her surprised call was prompted by the fact that this view was usually obscured by a hill. In fact the hill and its surrounds had turned to liquid clay and started to flow downhill.

Clay was laid down here when the sea inundated areas around the retreating ice sheet some 14,000 to 5,000 years ago. These marine clays can become fluid if the amount of water that is retained in the soil increases.

On a slope, the increased weight of the waterlogged clay overcomes the cohesion between soil particles, and the earth starts to move. Such flows often come after heavy rains, and in some parts of the world are triggered by earthquake shocks.

At Saint-Jean-Vianney a towering wall of liquid clay 20 m (66 ft) high flowed 3 km (1.9 mi) down the valley at 25 km/h (15.5 mph) to the Saguenay river, carrying with it a bridge, a bus, 40 houses and 31 people. Mme Laforge was among the victims.

Icesheets, glaciers and icebergs

The Greenland ice sheet is the largest in the northern hemisphere and the second largest in the world, after Antarctica. Measuring 2,600 by 1,200 km (1,600 by 750 mi) at its greatest extent, the ice sheet is on average some 2,000–3,000 m (6,500–10,000 ft) thick. This volume adds up to about a tenth of the world's ice.

Ice in the form of glaciers and ice sheets now covers some 10 percent of the world's land area. During the last glacial period, between about 110,000 and 10,000 years ago, ice sheets covered up to 30 percent of the land. On the North American continent the Laurentide ice sheet extended over an area that was nearly as large as Antarctica's ice cap today and as far south as parts of the United States. Further ice sheets developed in the Rocky Mountains and on Greenland.

Since this last (Wisconsin) glaciation the ice has been in retreat, though there have been occasional readvances, such as the so-called Little Ice Age, which lasted several hundred years and ended about a hundred years ago. Mounds of rocky debris (moraines) that were left by the retreating glaciers can still be seen, for example in Glacier National Park in the Selkirk Mountains in the Rockies.

The ice sheet is at present very stable. The central dome thickens by 3–10 cm (1–4 in) a year because of snowfall, while the margins become 20–50 cm (8–20 in) thinner. Every year snow accumulates to an equivalent of an estimated 500 cu km (120 cu mi) of water. This is balanced by the melting of ice equivalent to 295 cu km (71 cu mi) of water, together with the separation of icebergs equivalent to a further 205 cu km (49 cu mi) as they break off the ice sheet into the sea.

Icebergs and sea ice

Continued pressure from the great weight of ice and snow inland forces the ice sheet to flow outward to the sea. In the northern hemisphere both ice sheets and glaciers usually meet the sea at deep inlets or fjords. Here they produce large numbers of tall, jagged icebergs. Most Antarctic icebergs, by comparison, are broad and flat, as they break away from ice shelves, floating sheets of ice, rather than glaciers.

Sea ice is formed when the sea freezes. Most of the Arctic Ocean, with the North Pole at its center, is permanently frozen. During the summer some of the ice melts, releasing large blocks of ice known as pack ice that drift into the northern oceans. About a quarter of the world's oceans at any one time are affected by pack ice, the area remaining more or less constant because of the alternating Arctic and Antarctic winters.

The future of glaciers and ice sheets is closely linked to that of the global climate. We are most probably living in a period between glaciations: the ice could return. However, predictions of rising temperatures as a result of the greenhouse effect open up the possibility of major and irreversible change to the normal pattern of ice ages. If global temperatures rise, more ice would probably melt, tending to raise world sea levels. At the same time there might also be increased snowfall, as the climate is likely to become generally wetter. Would the balance be one of growth or shrinkage in the world's great concentrations of ice, and for how long? Such questions are of particular interest to the many people who live on low-lying land in the world's coastal cities, which may become flooded.

ICE SHEETS AND GLACIERS

Ice sheets and glaciers account for three-quarters of the world's fresh water. Most permanent ice lies in the polar regions, but altitude is also important. There are several small glaciers in mountains near the Equator.

Landmass	Ice area	
	sq km	sq mi
Antarctic including ice sheets, glaciers and islands	12,588,000	4,860,250
North America including Greenland ice sheet and glaciers, Canadian Arctic islands and Alaska	2,032,649	784,810
South America	26,500	10,232
Europe including Svalbard, Iceland, Scandinavia, Alps, Caucasus	79,465	30,682
Asia including Arctic islands, Himalayas, Kunlun and Karakoram mountains	170,679	65,899
Africa	12	5
Oceania including New Zealand	1,015	392
World total	14,898,320	5,752,270

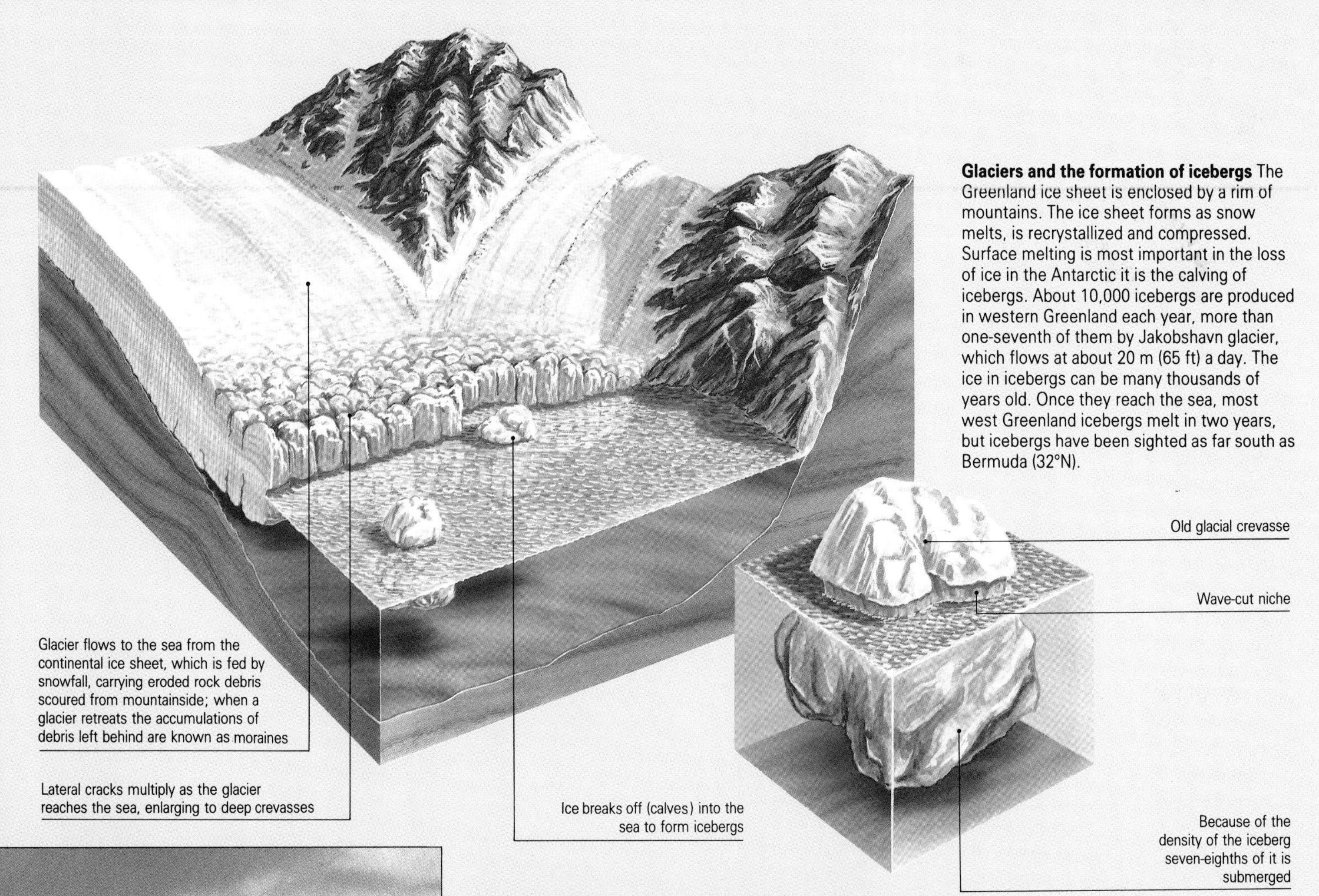

Glaciers and the formation of icebergs The Greenland ice sheet is enclosed by a rim of mountains. The ice sheet forms as snow melts, is recrystallized and compressed. Surface melting is most important in the loss of ice in the Antarctic it is the calving of icebergs. About 10,000 icebergs are produced in western Greenland each year, more than one-seventh of them by Jakobshavn glacier, which flows at about 20 m (65 ft) a day. The ice in icebergs can be many thousands of years old. Once they reach the sea, most west Greenland icebergs melt in two years, but icebergs have been sighted as far south as Bermuda (32°N).

A tongue of ice marks the end of a glacier fed by Greenland's ice sheet. The barren glacial valley in the foreground shows that the glacier extended along the valley until quite recently. The end or snout of the glacier has receded because rising temperatures since the last glaciation caused glaciers like this to melt faster than they can move down the valley. This broad U-shaped valley is in Stauning's Alps, east Greenland.

Icebergs in the midnight sun On the west coast of Greenland the Jakobshavn glacier receives about a tenth of the Greenland ice sheet's total flow. When the glacier meets the sea, huge chunks of ice break off, forming icebergs. The glacier produces many of the 15,000 icebergs that Greenland releases into the ocean each year. It was one of these that sank the passenger liner *Titanic* on 15 April 1912.

The ultimate light show

The greenish-blue glow on the horizon forms an arc across the sky. The bottom of the arc brightens and streamers are sent out toward the zenith, sometimes converging. The arc slowly transforms into shining curtains of white, green, blue and occasionally pink light, the brilliant display of patterns slowly changing as if blown by a heavenly wind. After perhaps an hour the curtains dissolve into luminous patches, then fade away altogether.

Such light shows take place in the high latitudes. In the northern hemisphere they are known as the aurora borealis or northern lights, in the south as the aurora australis or southern lights. They are named after Aurora, the Roman goddess of dawn. The show can recur on many nights in succession. Within the atmosphere the lights stretch from several hundred kilometers down to about 100 km (60 mi) above the surface and are seen from points on the ground several thousand kilometers apart. Yet they are less than 1 km (0.6 mi) in depth and the stars can shine through to add to the splendor.

Auroras are most brilliant when there is intense activity in the Sun, indicated by the presence of sunspots. Electrically charged particles emitted by the Sun (the solar wind) are drawn to the Earth's magnetic poles, where they bombard molecules and atoms in the ionosphere. As the atoms become excited they emit radiation: oxygen atoms emit greenish light, nitrogen atoms pink light.

The aurora borealis or northern lights over Canada
The North Magnetic Pole, focus in the northern hemisphere of the magnetic activity that produces the aurora, is situated in the islands of the Canadian Arctic.

GREAT PLAINS BETWEEN MOUNTAINS

A CONTINENT OF CONTRASTS · LEGACIES OF THE ICE · CLIMATES OF THE UNITED STATES · THE APPALACHIANS AND THE COASTAL PLAIN · PLAINS AND LOWLANDS OF THE INTERIOR · MOUNTAINS AND BASINS OF THE WEST

Bounded by two great oceans east and west, the United States is characterized by open spaces. The huge expanses of the Great Plains, the heartland of the country, rise to meet the long mountain chains that span it from north to south. In the east the old, eroded Appalachian Mountains follow the Atlantic coast. In the west are the younger, rugged Rocky Mountains and other ranges along the Pacific coast, much of which is subject to earthquakes and volcanic activity. Famous landforms include the deep gorge of the Grand Canyon, the badlands of South Dakota, Niagara Falls, and the great American rivers such as the Colorado and the Mississippi, named "Father of Waters" by the Indians. The climates span a range from polar in Alaska to tropical in southern Florida; the most rain falls in the Pacific island of Hawaii.

COUNTRIES IN THE REGION

United States of America

LAND

Area 9,371,786 sq km (3,618,467 sq mi)
Highest point Mount McKinley, 6,194 m (20,320 ft)
Lowest point Death Valley, −86 m (−282 ft)
Major features Rocky Mountains, Great Plains and central lowlands, Appalachian Mountains

WATER

Longest river Mississippi–Missouri, 6,020 km (3,740 mi)
Largest basin Mississippi–Missouri, 3,222,000 sq km (1,224,000 sq mi)
Highest average flow Mississippi, 17,500 cu m/sec (620,000 cu ft/sec)
Largest lake Superior, 83,270 sq km (32,150 sq mi)

CLIMATE

	Temperature °C (°F) January	July	Altitude m (ft)
Barrow	−27 (−17)	4 (39)	4 (13)
Portland	4 (39)	20 (68)	12 (39)
San Francisco	9 (48)	17 (63)	5 (16)
New Orleans	12 (54)	27 (81)	9 (30)
Chicago	−3 (27)	24 (75)	190 (623)

	Precipitation mm (in) January	July	Year
Barrow	5 (0.2)	20 (0.8)	110 (4.3)
Portland	136 (5.4)	10 (0.4)	944 (37.2)
San Francisco	102 (4.0)	0 (0)	475 (18.7)
New Orleans	98 (3.9)	171 (6.7)	1,369 (53.9)
Chicago	47 (1.9)	86 (3.4)	843 (33.2)

NATURAL HAZARDS

Hurricanes and tornadoes in south, earthquakes and volcanic eruptions in west, drought and blizzards in Midwest

A CONTINENT OF CONTRASTS

The United States, with its vast plains, deep canyons and rugged mountains has some of the most spectacular natural scenery in the world. The center is one enormous plain that stretches from the Appalachians to the Rocky Mountains, and from the Canadian border south to the Gulf of Mexico.

Along the eastern seaboard lies the Atlantic coastal plain. In Alaska to the far north, where the ground is almost permanently frozen, are the lowlands of the Arctic. In the west between the Coast, Cascade and Sierra Nevada ranges of the Western Cordillera, there are other smaller plains, and also in the basin-and-range area of the southwest.

Plateaus occupy about one-seventh of the continent. They include those of Appalachia and those of the Yukon, Columbia and Colorado rivers. The eastern plateaus are very old; those in the west are much younger. Here great rivers such as the Colorado and the Columbia have cut down through the plateau rocks to form deep canyons.

Mountains account for another fifth of the continent, and again there are sharp contrasts between the east and the west. In the east lie the Appalachians, old mountains worn down by erosion; and in the west the Rockies and the mountains of the Pacific rim, which are younger, higher and more rugged.

The coasts are also very varied. On the Atlantic coast the land slopes gently into the sea as a continental shelf. This shallow seabed zone fringing the continent is more than 160 km (100 mi) wide; it is much narrower on the Pacific coast.

A stable mountain platform

The pattern of the plains, plateaus, mountains and coasts of the United States is best understood by looking at how the continent was formed. The North American plate, one of the large lithospheric plates on which the Earth's continents rest, was once part of the ancient supercontinent of Laurasia. The continent itself is formed from some of the oldest rocks in the world, Precambrian granites. These outcrop in the Canadian Shield, or Laurentian Highlands, which extend into the United States to the south and west of Lake Superior and northeast of Lake Ontario. Centered on Canada's Hudson Bay, this ancient shield forms an upland plain of great stability. To the south lie other platforms, such as those underlying the Great Plains between the Rockies and the Mississippi valley, and the lowlands that extend east to the Appalachians.

A faulted landscape The Crystal Spring reservoir near San Francisco, California lies in a valley formed by the San Andreas fault. Over 1,050 km (650 mi) long, the fault is still active today, and the risk of earthquakes in this area is high.

It is along the margins of these platforms that tectonic activity – as the movement of the plates is called – such as folding, faulting and the formation of volcanoes takes place. Down folding and subsidence were followed by rocks being lifted up to form the mountain ranges that can still be seen. Chief among these are the Appalachians, which were formed over 225 million years ago, the Rockies, which are less than 65 million years old, and the ranges of the Western Cordillera, which are younger still.

North America, like other continents, has moved and is still moving. As it has drifted westward, the North American plate has overridden the smaller, adjacent plate on the edge of the Pacific and has almost consumed it. The Coast and Cascade ranges and the Sierra Nevada were all formed by earth movements as the plates collided and the rocks were pushed up. Inland, the crust has been stretched by the great volume of basaltic oceanic crust being pushed underneath it. This has produced the parallel block-faulted mountains and valleys that make up the basin-and-range area of the southwest. Farther north, plate movements are still generating earthquakes and volcanic activity, as the 1980 eruption of Mount St Helens showed. The powerful forces of earth movements in the west contrast with the more stable conditions on the eastern side of the continent, where there is no plate boundary.

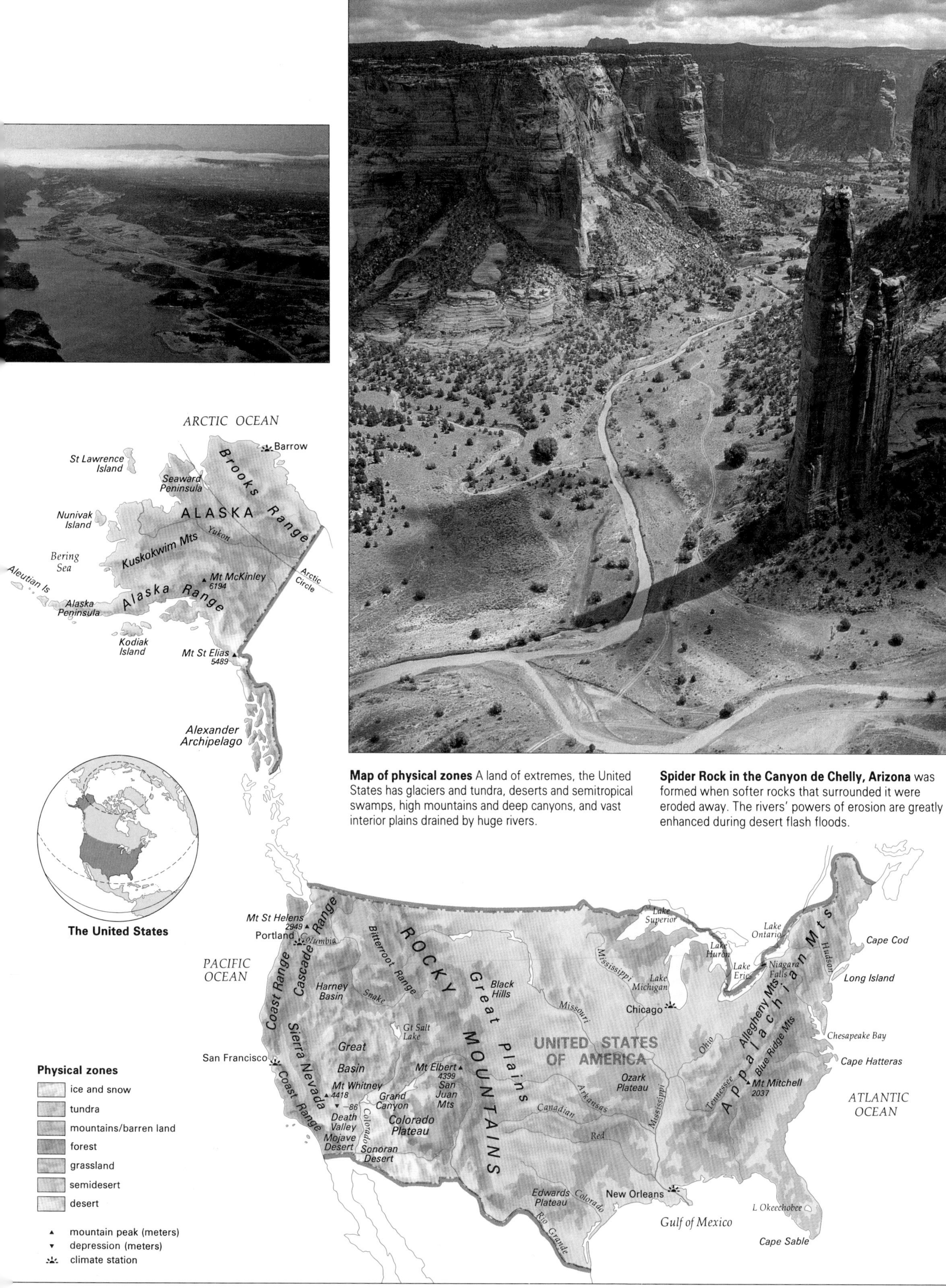

Map of physical zones A land of extremes, the United States has glaciers and tundra, deserts and semitropical swamps, high mountains and deep canyons, and vast interior plains drained by huge rivers.

Spider Rock in the Canyon de Chelly, Arizona was formed when softer rocks that surrounded it were eroded away. The rivers' powers of erosion are greatly enhanced during desert flash floods.

LEGACIES OF THE ICE

There have been several ice ages in the Earth's history. During the most recent period of glaciation, which began about 2 million years ago, the polar ice sheet advanced deep into North America at least four times. In the sequence in which they took place, these four glaciations are named the Nebraskan, Kansan, Illinoian and Wisconsin, after the states that were most affected by the advancing ice sheet. The advance and retreat of the ice sheet has determined the shape of the land surfaces that lay under it.

The most extensive glaciation took place during the Illinoian advance, when nearly two-thirds of North America was covered by ice. The ice sheet blocked all the rivers draining to the north; huge tongues of ice carved out the courses of the great rivers. Two rivers flowed along the edge of the ice sheet during the Illinoian advance. In the west the Missouri flowed southeast, and in the east what is now the Ohio flowed southwest. They joined to form the Mississippi river, which drains to the south.

Each glaciation has left its mark on the land, but the impact of the most recent episode of glaciation, the Wisconsin, is most evident today; it began between 125,000 and 100,000 years ago. The ice sheet reached its maximum size about 18,000 years ago. It began to shrink 4,000 years later, and by 10,000 years ago had retreated north of the Great Lakes.

Features formed by ice

The great pressure of the ice sheet scraped and scoured the land surface underneath it as it spread. Soils were eroded and rock was ground up; the mixture of cobbles, gravel and powder was incorporated into the moving ice and carried southward. When the ice melted, the load of soil, sediment and rock (known as glacial till) was deposited on the land.

A huge blanket of till covers the once-glaciated areas of the United States. Much of the landscape is flat; elsewhere it is punctuated by hummocks and holes. The hummocks are streamlined hills of till called drumlins. The holes, or kettles, were left when buried pockets of ice melted. Kettle holes often fill with water; most of the lakes to the west of Lake Superior are in kettle holes.

Farther east the retreat of the ice sheet was sporadic, and the margin remained stationary for long periods. Great thicknesses of till built up parallel to the ice front. These ridges, called terminal moraines, enclosed the meltwater of the ice sheet to form the Great Lakes – from west to east, Lakes Superior, Michigan, Huron, Erie and Ontario. Ridges up to 100 m (330 ft) high and hundreds of kilometers long still curve around the southern end of Lake Michigan.

The melting of the ice sheet produced enormous volumes of water, which also formed many other lakes. These lakes were at their largest toward the end of the Wisconsin glaciation, when melting was at its peak. Their shrunken remnants can be seen in Lake Champlain and other finger lakes southeast of Lake Ontario.

Where the sediment-laden meltwaters were free to flow, the streams issuing from the ice spread deposits around the former ice margins and also carried them to the south. These landforms include the channeled scablands between the Columbia and Snake rivers of the northwest, and the sand and gravel terraces (braided bars) on the valley floors of the Ohio and Missouri rivers.

Vegetation and soils

To the south of the glaciated area, the Great Plains and the central lowlands were carpeted in dust that was produced by glacial grinding and blown down from the ice margins. These wind-blown sediments, called loess, produce some of the most fertile soil on Earth; and in the Midwest the rich loess soils provide a basis for intensive agriculture. But the forces of wind and water that delivered these bountiful soils are also quite capable of taking them away again. Stripped of their protective cover of vegetation, the soils are highly susceptible to erosion, as America discovered in the 1930s when the Dust Bowl was formed.

The climate changes that lead to ice ages also affect plants and animals. Species that are unable to adapt to changing conditions are simply doomed to extinction. The vegetation changes of the early Holocene epoch, which began 10,000 years ago, show a succession of plants. Vegetation patterns have continued to fluctuate with changes of climate in the last few thousand years. The boundaries between the prairies and the forests, for example, have moved to and fro by hundred of kilometers.

The Nutzotin mountains of Alaska show the steep-sided valleys and angular ridges and peaks typical of a glaciated landscape. Glaciers are still active in the higher land to the south.

The young Mississippi river at Bemidji, Minnesota, not far from its source in Lake Itasca, meanders lazily across low marshy ground.

THE LAKE MISSOULA FLOOD

Between the valleys of the Columbia and Snake rivers in the northwest, the plateau has been scoured clean of soil, and has been dissected by deep, sharp-rimmed channels called coulees. One of the largest is the Grand Coulee, home of waters impounded by the Grand Coulee Dam. Other coulees are still dry and, if viewed from the air, reveal a braided pattern.

The origin of this landscape was a mystery for many years. In the 1920s a geologist called J. Harlen Bretz suggested that the channeled scablands were all formed at one time, by a deluge of biblical proportions. Today Bretz's theory, which had been condemned as preposterous when he put it forward, is generally accepted as true.

The Clark Fork river drains through a narrow gap around the northern end of the Bitterroot Range into the Columbia river valley. During the Wisconsin glaciation, a tongue of advancing ice blocked this gap, trapping Clark Fork river and forming a huge body of water, Lake Missoula.

The ice dam eventually broke under the pressure of the lake water, releasing a huge flood toward the meeting point of the Columbia and Spokane rivers and beyond. The water swept across the Columbia plateau, stripping away soil and cutting deep channels before draining through the Columbia river gorge to the Pacific Ocean.

At its peak the flow of water in the flood was greater than that of all the world's rivers combined. Within weeks Lake Missoula drained and Clark Fork river resumed its course, but the landscape had been changed forever.

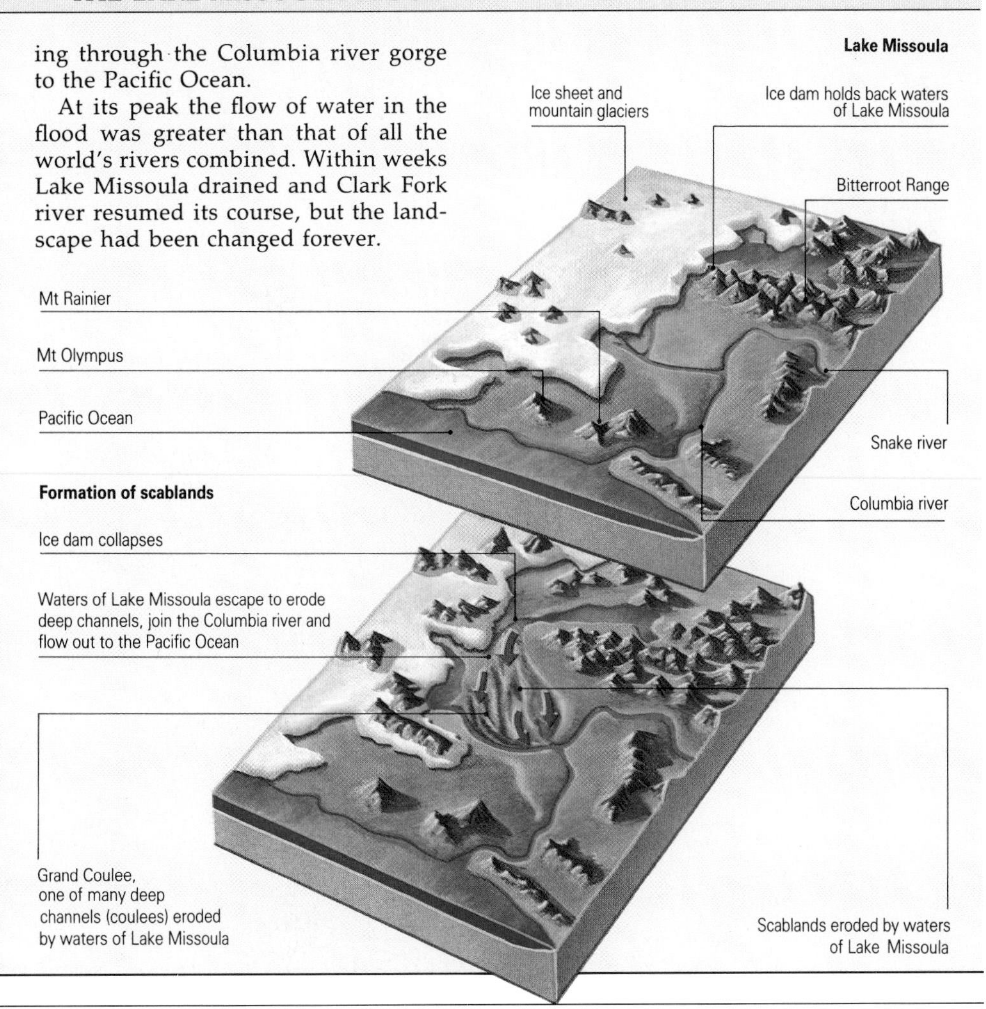

CLIMATES OF THE UNITED STATES

The United States lies between two climatic extremes. In the far north the Arctic coast of Alaska is truly polar, while in the far southeast the peninsula of Florida extends into the tropics. The lands in between experience the full range of climate types.

Cold and wet climate zones

Polar climates are characterized by low temperatures throughout the year and a very short growing season. Milder polar climates such as that of northern Alaska do have average temperatures above freezing during the warmest months but they are still below 10°C (50°F).

A moderate polar climate is also found at high altitudes farther south because the temperature falls with increasing height. A polar climate can be found at just 900 m (3,000 ft) in the Alaska Range at about latitude 62°N. In the northern Cascades Range, at 47°N, there would be polar conditions at 2,000 m (6,500 ft); in the Sierra Nevada, a polar climate can still be found at altitudes over 3,500 m (11,500 ft).

South of the polar zone the climate becomes wet, and winters are still severe. This is the climate of most of Alaska and the eastern United States north of 40°N. The coldest months have temperatures well below freezing, but the warmest months are above 10°C (50°F).

Along the southern coast of Alaska several weather stations record more than 3,800 mm (150 in) rainfall each year, making this one of the wettest areas in the United States. The wettest point of all is Kauai island, Hawaii, where 11,500 mm (450 in) of rain a year falls.

The Pacific coast as far south as San Francisco also has a rainy climate, but winters are milder with temperatures mostly above freezing even in the coldest months. The climate here is similar to that of the southeastern United States. Both areas receive over 1,000 mm (40 in) of rain a year, and the figure rises in some places to 2,500 mm (nearly 100 in).

Dry climate zones

From the western ranges east to 100°W, the climate is dry. This area includes the Mojave and Sonoran deserts, which receive less than 200 mm (8 in) rainfall each year, the Great Basin and Death Valley, where rainfall averages only 40 mm (1.6 in) a year. The rest of this area is semiarid, with between 200 and 500 mm (8–20 in) of rain.

Watercourses flow only during and immediately following a bout of rain. If less than about 500 mm (20 in) of rain falls during the year it is unlikely that streams will flow at all. Rivers that originate in the mountains and cross this area actually shrink as they flow downstream, because they lose more water through evaporation and seepage than they gain from tributaries and rain. This is true of both the Colorado and the Rio Grande.

Another feature of this area is the local wind known as the chinook. Air warms as it descends from mountainous areas, increasing its capacity to hold moisture. The result is the parching winds that occasionally sweep across the Great Plains from the Rocky Mountains. The heat from a chinook wind can melt and evaporate 15 cm (6 in) of snow in a day.

Humid and continental climates

East of the 100°W line of longitude the climate becomes more humid. Annual rainfall is between 500 and 1,250 mm (20–50 in), most of it falling in spring and summer. Average temperatures at the center of the continent are similar to those found at the coasts, but these averages can be deceptive. Land both heats and cools more rapidly than water, so the interiors of large continents, which are far from the sea, experience far greater extremes of temperature than coastal areas. This characterizes what is known as a continental climate. In the center of the continent the temperature in winter on the Canadian border averages −40°C (−40°F), and in summer soars to over 38°C (100°F). On the Pacific coast at more or

less the same latitude the equivalent averages are 4°C (39°F) and 20°C (68°F).

It is in areas of continental climate that the greatest extremes are found. Death Valley holds the record high at 56°C (134°F), while Prospect Creek, Alaska has the record low at −62°C (−80°F).

The southern tip of Florida in the extreme southeast is the only area with a humid tropical climate. Even the coolest month averages above 18°C (64°F), and 1,500 mm (60 in) or more of rainfall is spread throughout the year.

The Koolau Mountains on the northeast coast of Oahu, Hawaii, intercept moisture-bearing winds from the Pacific Ocean. The heavy rainfall has eroded deep gullies in the young volcanic rocks.

The hottest, driest place in the United States, Death Valley in California sometimes experiences summer temperatures in excess of 55°C (131°F). The high evaporation rate draws salts to the surface, forming a white crust on the valley floor.

CLIMATIC HAZARDS

The whole of the United States is vulnerable to hazardous extremes of climate and exceptional local weather patterns. Drought is one example. There have been three severe droughts in the Midwest during this century – in the 1930s, the early 1970s and the late 1980s. In the 1930s the terrible effects of high winds in a drought-stricken area were vividly demonstrated in the Great Plains, where much of the exposed topsoil of the farmland was blown away and the Dust Bowl was formed. In other years so much rain has fallen that there has been severe flooding in the major river valleys.

Hurricanes are another, and more frequent, problem. They develop over the Caribbean Sea to the south in late summer and early fall, gaining strength from the warm ocean and humid air masses before moving northward. These great tropical storms average 650 km (400 mi) in diameter, and have very low barometric pressures at their centers. Winds rotate anticlockwise around the still air at the center – the eye of the hurricane – and may gust at 160–260 km/h (100–160 mph).

Hurricanes strike the coast of the Gulf of Mexico and the Atlantic seaboard every few years. Those that reached the United States in 1969, 1972, 1980 and 1985 caused a total of some 800 deaths. In Hurricane Agnes (1972) winds were measured at 135 km/h (85 mph), with storm tides 1.8 m (6 ft) above normal; at Washington DC, near the east coast, 290 mm (11.5 in) of rain fell in 24 hours.

In the continental interior there are also severe local storms. Like the most local of storms, tornadoes, they usually develop when different air masses mix. In September 1977 over Kansas City on the lower Missouri, for example, a warm, moist air mass from the south met a cold mass coming in from the west. The resulting storm dumped over 150 mm (6 in) rain in six hours, and a further 150–180 mm (6–7 in) the following day. Floods wrecked homes and claimed many lives.

Farther north, winter storms on the Great Plains produce blizzards that threaten the lives of both humans and livestock. Hailstorms are not as dangerous, but they can nevertheless cause serious damage.

Tornadoes and waterspouts

Tornadoes, or twisters, are the most violent and dangerous disturbances of the Earth's atmosphere. About 200 tornadoes a year are recorded across the United States. They are a major hazard in central and eastern areas, and particularly in the valley of the Mississippi, which is sometimes known as "Tornado Alley". Between 1920 and 1950 every county in the Mississippi valley and adjoining coastal plains of the Gulf of Mexico was struck by at least a dozen tornadoes. Damage to property cost millions of dollars and 2,000 people were killed.

A tornado is a sinuous, funnel-shaped cloud that descends from storm clouds to reach the ground. A typical tornado is a few hundred meters across at its base; it is much smaller, and also lasts less long, than a hurricane.

Most tornadoes form in spring and early summer. The deep south is often hit in early spring; the tornadoes strike farther north as the season progresses. They can develop at any time of day or night, but most form in the late afternoon or early evening. Most tornadoes move from the southwest toward the northeast, at a speed of some 50 km/h (30 mph). They travel a distance of about 26 km (16 mi).

Precisely what causes a tornado is not fully understood, though they are known to be the result of great instability in the atmosphere when cold, dense air aloft traps a layer of warm, buoyant air below. Windshear, a marked difference in wind direction between the two layers, also seems to play a part.

Tornadoes are formed when the warm lower air breaks through the cold upper air and drains upward, rotating anticlockwise (clockwise in the southern hemisphere). This is a tornado cyclone. As the base of the cyclone extends toward the ground, sucking up the air in the lower layer, a white, funnel-shaped cloud reaches down. The whiteness is due to the formation of water droplets in a region of low pressure. After the funnel strikes the ground the intensely low pressure at its center sucks up dust and debris, giving the tornado its grim color.

In the American Midwest, and particularly over the Mississippi valley, conditions are often right for tornadoes to develop. Cold air moving east from the Pacific Ocean overlies warm, moist air moving north from the Gulf of Mexico, particularly in spring. The lower air may be farther warmed locally by the sunbaked ground.

Air in a tornado rotates rapidly. Its speed has not been measured, but analysis of damage suggests that the wind

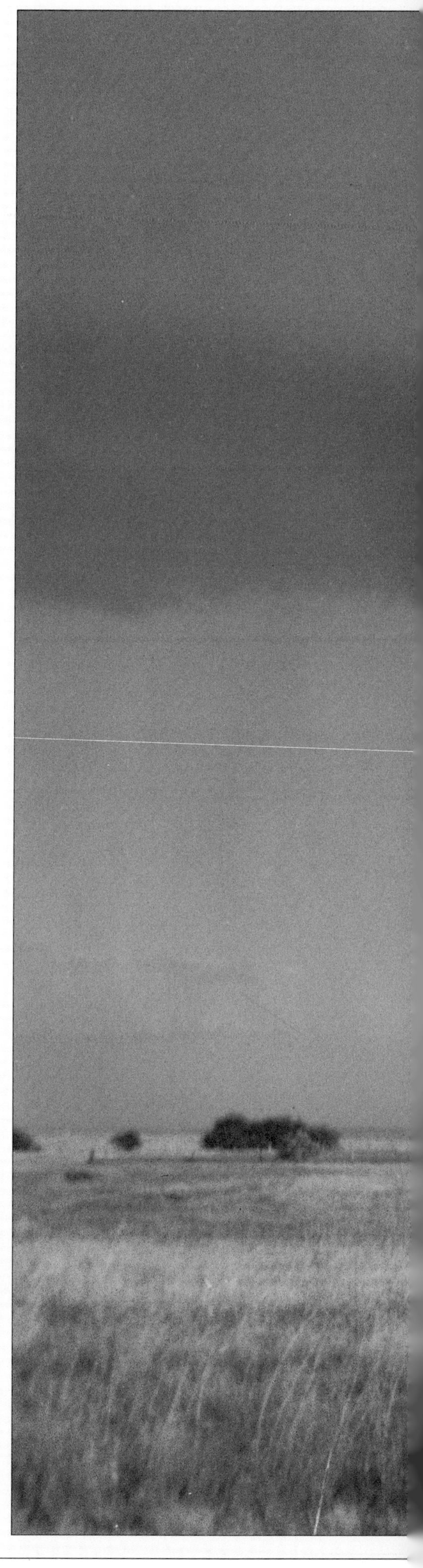

A tornado over Oklahoma Dirt and debris are sucked skyward as a twisting funnel of air reaches down from the cloud.

Devastated by a twister Whirling winds of up to 800 km/h (500 mph) have wreaked havoc on this Texan town. The base of a tornado is seldom more than a few hundred meters across, and houses a few streets away remain unscathed.

A giant waterspout in the Bermuda Triangle, in the Atlantic Ocean. A spiraling column of water is swelling the ominous black cloud above.

may reach as much as 800 km/h (500mph). The destructive power of the tornado is increased as debris is sucked up and whirled away by the rotating winds. Tornadoes cut through the landscape, demolishing everything in their path. Trees are simply screwed up out of the ground by the combination of suction and spinning winds. Many buildings are blown flat; others literally explode, as the pressure inside them cannot fall quickly enough to match the rapid fall in outside pressure. No barometer has survived the passage of a tornado to record the fall, but it is thought that a drop in atmospheric pressure of 100 millibars is typical.

Tornadoes can be spotted on weather radar as they form, and warnings are issued to people living in their path. They continue both to take life and to damage property in the United States but fortunately, unlike other climatic hazards, the impact of their damage is local, as most tornadoes touch down only for a short time.

Waterspouts

Waterspouts are intense columns of rising water formed when a tornado descends over the sea or, less spectacularly, where strongly converging surface winds ascend. They are often seen off the coast of the Gulf of Mexico and over the Atlantic Ocean near Florida. Waterspouts can suck up considerable volumes of sea water. For example, after a large waterspout had formed off the Atlantic coast near Cape Cod in 1896, the area was deluged some hours later by a salty downpour.

"One hell of a place to lose a cow"

In late summer, brief torrential rainstorms produce flash floods in the ravines of Bryce Canyon. They are moments punctuating a millions-of-years-old process of erosion by water, frost, sun and wind that has carved out the canyon's towering pinnacles from many-hued sedimentary rocks. In the hundred years since rancher Ebenezer Bryce passed judgment on this arid, fantastic and beautiful landscape, the cliffs have receded some 60 cm (2 ft).

Bryce Canyon is on the eastern edge of Paunsaugunt Plateau, southwestern Utah. The rocks comprise limestones, sandstones and shales derived from sediments that were deposited some 60 million years ago. As the deposits were compressed and hardened (consolidated) under the weight of further layers, metals tinged the rocks with color: fiery reds and delicate pinks from iron compounds, green from copper, purple from manganese.

Erosion began some 13 million years ago, when enormous movements in the Earth's crust slowly lifted up the block that is now the Paunsaugunt Plateau, cracking the component rocks in the process. The plateau edge is raised some 650 m (2,000 ft).

The maze-like ravines follow the courses of streams and torrents that have enlarged cracks, scoured the softer rocks and swept away the debris eroded by sun, wind and frost. Harder limestone layers have in places shielded softer rocks below, so the running water has cut down deeply to produce a unique array of freestanding pillars and spires, colonnades, walls and amphitheaters.

The colorful vertical faces reveal a geological past extending over hundreds of millions of years, in a rainbow of hues that is constantly changing with the hour of the day, the weather and the season.

Many-hued pinnacles of Bryce Canyon in the southern United States tell a story of deposition and erosion of rocks going back 60 million years.

THE APPALACHIANS AND THE COASTAL PLAIN

The Appalachian Mountains of the eastern United States extend northeast from the coastal plains of the Gulf of Mexico past the Great Lakes and up to the Canadian border. The ancient sandstone, limestone, slate and other rocks have been folded, eroded, lifted and again eroded. These are some of the oldest mountains in the world, formed some 350–250 million years ago. Between the foothills of the Appalachians and the Atlantic Ocean lies the coastal plain, which narrows to the northeast.

The coastal plain

The Atlantic coastal plain extends from Cape Cod to the Rio Grande and stretches inland for 160–320 km (100–200 mi). It continues into the Atlantic Ocean as the continental shelf.

Along the coast there are sandy beaches, mud flats and marshes, offshore islands, and sand bars sheltering shallow lagoons. Coral reefs are found off the coast of the Florida peninsula in the southeast. Inland the area is level and low-lying, with large areas of swamps, mud flats and wetlands crossed by meandering watercourses called bayous. There are some low ridges running parallel to the coast, but these are nowhere higher than 150 m (500 ft).

During the last ice age the land in the north was depressed by the great weight of the ice that covered it. Since the ice melted the land has slowly been readjusting, but it has not risen fast enough to avoid being invaded by the rising sea. Flooded river valleys have formed bays and inlets, including Chesapeake Bay and Long Island Sound.

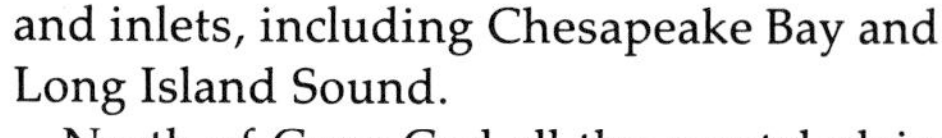

North of Cape Cod all the coastal plain is submerged and the Appalachian Mountains reach the coast. Rocky headlands are separated by flooded glacial valleys to provide some of the most scenic coastline in the United States. The flooded coastal plain extends for nearly 1,600 km (1,000 mi) under the sea as a series of shallow banks that support a thriving fishing industry.

Cadillac Mountain, on the Maine coast at the northeast tip of the USA, was formed some 395 million years ago when sediments that had collected in a great dip of the Earth's crust, the Appalachian geosyncline, were uplifted, folded and metamorphosed.

The northern Appalachians

The northern Appalachians were affected by the last glaciation, when ice eroded the valleys and transported huge quantities of rock and soil. These glacial deposits cover much of the area to depths of 8–15 m (25–50 ft); at the finger lakes southeast of Lake Ontario they are up to 300 m (1,000 ft) deep.

To the north, between Lake Ontario and the upper Hudson river, lie the Adirondack Mountains. These have developed from a structural dome formed when Precambrian rocks at the core, 600 million years old, were pushed up by over 3 km (2 mi). The highest peak is Mount Marcy at 1,629 m (5,344 ft). Away from this core, the hills get progressively lower and the rocks younger. The area in the northeastern corner of the United States known as New England is an extension of the piedmont plateau, but here the rocks have been folded to form the Green Mountains. They are easily eroded; only isolated mountains remain where there are more resistant rocks.

The central and southern Appalachians

The coastal plain is separated from the forest hills of the Appalachian mountains by the fall line. It takes its name from the line of rapids and waterfalls formed where rivers fall from the resistant rocks on the eastern edge of the plateau down onto the coastal plain. The fall line marked the upstream limit of navigation in the 18th and 19th centuries, and the lowest crossing point of the larger rivers. As a result, many of the great cities of the east have been built here. The eastern foothills or piedmont plateau of the Appalachians extend east from the fall line to the Blue Ridge. More rugged than ranges farther west, the Blue Ridge contains Mount Mitchell, at 2,037 m (6,684 ft), the highest peak in the Appalachians.

Beyond the Blue Ridge to the west is a

The "Grand Canyon of the East", formed by the Genesee river in New York State, has been carved into the ancient sediments of the Appalachians. As the land has risen the river has cut down through some 240 m (800 ft) of rock.

belt of long, parallel limestone ridges separated by narrow valleys. The limestone has been sculpted into underground caverns, among them the Mammoth Cave system, the world's longest, with a total length of 530 km (329 mi). Beyond lies the Appalachian plateau, a great tract of nearly horizontal rock that was raised up by tectonic activity to between 300 and 900 m (1,000–3,000 ft), then incised by meandering rivers. The landscape now consists of jumbled hills and low mountains separated by narrow stream valleys. Among areas of higher relief are the Allegheny Mountains.

THE NIAGARA FALLS

The Great Lakes were formed at the end of the last glaciation. The enormous quantities of meltwater that were released drained to the south via the Mississippi and Hudson rivers, because routes to the north were still blocked by ice. About 11,000 years ago the ice margin had retreated north of the St Lawrence river, and the Great Lakes started to drain to the northeast. Relieved of the great weight of the ice, the land was also rising at this time.

The level of Lake Ontario, the easternmost lake, fell sharply when its waters drained into the St Lawrence river. Lake Erie overflowed into Lake Ontario. One by one each of the Great Lakes began to overflow to the east into its neighbor, setting up the drainage pattern that exists today.

The swollen waters of the Niagara river carried the entire drainage of Lakes Superior, Huron, Michigan and Erie steeply down into the much lower Lake Ontario. The river eroded the land quickly at first but then met a hard layer of limestone. A waterfall formed as erosion by undercutting lowered the surface on the downstream side.

Today the American Falls are 51 m (167 ft) high and some 300 m (1,000 ft) wide; the Canadian, or Horseshoe, Falls are over twice as wide but not quite so tall. The limestone is slowly eroding, and the falls have retreated upstream by about 11 km (7 mi). They have maintained their height because the layer of limestone is almost horizontal.

PLAINS AND LOWLANDS OF THE INTERIOR

The landscapes of the heart of the North American continent are the result of both the most ancient and the most recent geological periods. Their great extent, their stability, and their generally low-lying and gentle terrain are a legacy from the Precambrian period, over 600 million years ago. Glacial deposits (drift or till) laid on top of this stable platform since the Quaternary period began 2 million years ago are responsible for the drainage pattern, surface landforms and rich soils of the area.

The interior plains and lowlands are dominated by the Mississippi, Missouri and Ohio rivers. The present drainage pattern of this enormous river system was established during the Illinoian glaciation, when drainage to the north was blocked by ice. Stretching west and south of Lake Superior, to the headwaters of the Mississippi and beyond, there are outcrops of granite. These ancient rocks, part of the Canadian Shield or Laurentian Highlands, were buried by sedimentary rocks that were subsequently eroded by wind, water and ice. Deep valleys have been cut along lines of weakness; these are separated by ridges of more resistant rock to form the peninsulas and bays of Lake Superior.

The central lowland

South of the Great Lakes is a vast area of almost level land that covers 1.7 million sq km (650,000 sq mi). This is the central lowland area, which slopes gently from 300 m (1,000 ft) in the east, where it meets the Allegheny Mountains on the edge of the Appalachians, down to 150 m (500 ft) at the Mississippi river, before rising again to about 600 m (2,000 ft) at its western boundary with the Great Plains. The uniformity of the landscape is caused by the great thickness of the glacial deposits that cover the underlying geological formations.

The deposits around the Great Lakes date from the last glaciation. The surface is roughened by hummocks, drumlins and kettle holes, but the main features of the landscape are terminal moraines, which form concentric ridges around the southern shores of the lakes. South of the moraines the glacial drift is older, and the humps and hollows have been smoothed out by erosion and deposition. The river valleys are incised, broad and steep-sided. West of the Mississippi the drift is older still, the plains are more dissected and the uplands more rounded.

In contrast to this postglacial landscape, to the west of Lake Michigan the area adjoining the Mississippi valley escaped glaciation, as the ice was contained by the deep valleys now occupied by Lakes Superior and Michigan. There are ancient landforms such as natural bridges, arches and buttes, which could not have withstood being covered by ice.

The plain that extends south from the

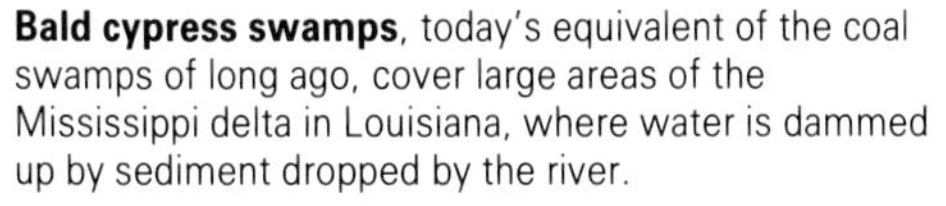

Bald cypress swamps, today's equivalent of the coal swamps of long ago, cover large areas of the Mississippi delta in Louisiana, where water is dammed up by sediment dropped by the river.

Missouri river in the north to between the Red and Colorado rivers, including the Ozark Plateau, also lay beyond the limit of the ice, and the landscape reflects the underlying structure. Ridges called cuestas have formed where there are bands of resistant limestone or sandstone. They are separated by long valleys cut through shales, which are more easily eroded.

The Great Plains

Only slightly smaller than the central lowland, the Great Plains run parallel to the Rocky Mountains. They cover about 1.47 million sq km (575,000 sq mi) of the United States and Canada. The land slopes gently eastward from a height of 1,675 m (5,500 ft) at the foot of the Rocky Mountains to about 600 m (2,000 ft) where the plains meet the central lowland. Changes in climate and vegetation rather than underlying structure are responsible for the appearance of the landscape. The eastern limit of the Great Plains coincides with the 500 mm (20 in) line of annual rainfall, which separates the short-grass prairie on the drier side from the longer grass prairie on the wetter eastern side.

A notable feature of the Great Plains in the north are the Black Hills of South Dakota, formed from Precambrian granite that was pushed up through the overlying sedimentary rocks 70–40 million years ago. They are famous because of the colossal heads of Presidents Washington, Jefferson, Lincoln and Theodore Roosevelt that have been carved out of the slopes of Mount Rushmore. The younger and weaker rocks that surround the Black Hills have been eroded down, but where a more resistant band is found steep ridges called hogsbacks have formed. East of the Black Hills lie the barren and deeply eroded Badlands, which have given their name to similar landscapes in other parts of the world.

Badlands The splashing raindrops and concentrated runoff from sporadic desert storms combine to turn areas of relatively soft rock into badlands.

AMERICA'S GREATEST EARTHQUAKE

North America's greatest known earthquake, at New Madrid on the Mississippi not far south of where the river is joined by the Ohio, shook the heart of the continent. Eyewitnesses described how "the ground rose and fell as earth waves, like the long slow swell of the sea, passed across its surface, trees tilting until their branches interlocked and opening the soil in deep cracks." The effects on the Mississippi were equally remarkable: "Great waves overwhelmed many boats . . . the return currents breaking down thousands of trees . . . sand bars and whole islands disappearing beneath the surge."

The earthquake consisted of three separate shocks between 16 December 1811 and 7 February 1812. Each shock reached magnitude 12, the highest on the modified Mercalli scale of intensity. It was reported that the quakes woke sleepers in Washington DC and rang church bells in Boston, Massachusetts, respectively over 1,000 km (620 mi) and 1,600 km (1,000 mi) northeast of New Madrid.

Most of the earthquakes that take place not only in the United States but throughout the world are associated with active plate margins, but the New Madrid earthquake was an example of a mid-plate quake. Deep beneath the Mississippi there are ancient faults created by continental movements long ago. These are buried by a great thickness of sediments. The continental crust occasionally sags under their immense weight, generating "deep-centered" earthquakes.

Following the quakes the flow of the Mississippi was temporarily reversed because the land was tilted in a different direction. Nearby, the 29 km (18 mi) long Lake Reelfoot was formed; it still exists today. Fortunately the area affected by the earthquake was sparsely populated. If the same quake were to strike today, cities such as St Louis and Memphis would be devastated.

MOUNTAINS AND BASINS OF THE WEST

The Rocky Mountains stretch from the Arctic Ocean to the Gulf of Mexico in an almost unbroken chain. About 80 million years ago the North American lithospheric plate began to move westward; in time it forced the oceanic crust of the Pacific plate deep under the continent. This caused lifting, faulting, folding and tilting of the continental granites and the overlying sedimentary rocks, and the Rocky Mountains were formed. In the Rocky Mountains proper (ranges close to the Pacific coast have a younger history) the highest peak is Mount Elbert, 4,399 m (14,433 ft). Uplift is continuing, and together with weathering and erosion has produced rugged mountains interspersed with basins filled by sediment.

Plateaus in the mountains
The Colorado Plateau is an area of sand deserts, badlands and canyons lying between the Rocky Mountains and the basin-and-range areas to the east. It is part of the ancient continental platform. The Colorado river flowed through the plateau, cutting down through the rock at the same rate as the land rose. In the Grand Canyon, which demonstrates superbly the erosive power of the river, rocks have been exposed that were formed over 2 billion years ago.

Elsewhere, mountains of igneous and volcanic rock, such as the San Juan Mountains, punctuate the landscape. To the southeast of the Grand Canyon a crater 1.2 km (0.75 mi) across marks where a meteor collided with the Earth 25,000 years ago. Where alternate layers of resistant and weaker rocks lie horizontally the protective cap produces a mesa or, if the mountain's height is greater than its width, a butte such as those to be seen in Monument Valley, between the Grand Canyon and the San Juan Mountains.

Huge overhangs are created where erosion by a river undercuts a resistant layer of rock. Natural bridges are formed when an incised meander loops round on itself and bores through the narrow neck of land separating the loops. In San Luis Valley at the head of the Rio Grande, east of the San Juan Mountains, huge sand waves over 180 m (600 ft) high have been deposited where the wind has been funneled through the mountains.

Disappearing hills in New Mexico Hard-capped rocks resist erosion, but are gradually undermined and cut back from the sides to form flat tablelands or mesas. Eventually the hills will be completely eroded, and only a flat desert will remain.

The southwest
Mountain ranges running north–south, and the basins that lying between them, cover much of southern California and all the state of Nevada. Over a distance of only some 160 km (100 mi) altitudes vary from 86 m (282 ft) below sea level in Death Valley to 4,418 m (14,494 ft) above sea level at the top of Mount Whitney in the Sierra Nevada, the highest point in the United States outside Alaska. The mountains and basins are the result of volcanic activity and faulting.

Death Valley is a dried-up lake bed that is sinking faster than sediments can fill it. A typical basin such as this has two parts, a central alluvial plain, or playa, and surrounding gravel fans deposited by intermittent streams issuing from the canyons. At the base of the mountains there are often gently sloping plains of solid rock strewn with gravel, called pediments. They have been eroded by streams leaving the mountains.

Along the Pacific coast the landscape is dominated by the effects of the colliding plates. The mountains of the Western Cordillera (the Coast, Cascade, Sierra Nevada and other local ranges) are the most obvious result. They are arranged in a north–south chain. The Sierra Nevada is formed from granite intrusions that pushed up the overlying rocks in great domes and arches. The Cascade Range to the north has volcanic peaks of which some are still active, like Mount St Helens.

LOS ANGELES SMOG

Los Angeles, on the Pacific coast in the far south of the United States, is notorious for its smog, a stifling smoky fog that is largely the result of exhaust fumes from the millions of automobiles used there. Los Angeles suffers more than most cities because it has a sunny climate and sits on a coastal plain backed by mountains, which hinder the dispersal of polluted air.

Engine exhausts emit polluting gases into the atmosphere, including carbon dioxide, carbon monoxide, sulfur dioxide, nitrogen compounds and hydrocarbons. The ultraviolet in strong sunlight encourages the photochemical production of ozone and other toxic substances by oxidation of the hydrocarbon gases.

The pollution becomes trapped by temperature inversions. These are of two types. In the first, air is chilled at ground level at night and is overlaid by a layer of warmer air that acts like a lid, trapping the polluted air in the basin. Los Angeles is also prone to stable anticyclonic conditions, with sinking air preventing the warm polluted air near the ground from rising. Light onshore winds prevent the air from moving out to sea, and the mountains prevent it from moving inland.

The most immediate damage is to people who breathe the polluted air, and may suffer painful eye and skin irritation caused by ozone and oxidized hydrocarbons. There is also damage to buildings, plant life is affected, and visibility is obscured, causing a hazard to air traffic. The California state authorities have the most stringent exhaust emission controls in the United States, but the problem remains.

A sleeping volcano? Mount Adams towers above the mists over the Cascade Range in Washington. Some of America's youngest and highest mountains, the Cascades are still a center of volcanic activity, as the eruption of Mount St Helens in 1980 showed.

Complex folds and faults where the continental crust has been crumpled have produced the Coast Range, which borders the Pacific Ocean.

The movement of the plates still continues. In the San Andreas Fault, which extends along the coast through the peninsula of San Francisco toward the southeast, movement is horizontal. The Pacific and North American plates are locked together for long periods while stresses gradually build up deep within the Earth's interior; in due course the fault suddenly gives, and there is an earthquake. The last great movement caused the San Francisco earthquake of 1906; another major shock was felt as recently as 1989.

Alaska – the last frontier

Only a narrow coastal strip links Alaska to the rest of the United States. The mountains of this Arctic land form the northern ranges of the Rockies. The Brooks Range in the far north was formed when part of the Pacific plate collided with the continental plate and rode over it. Erosion and weathering have since produced its jagged peaks and steep valleys.

The Yukon plateau, which lies between the Brooks Range and the Alaska Range to the south, has been cut into deeply by the Yukon river. The lower Yukon crosses an alluvial fan about 240 km (150 mi) in diameter, similar to the Mississippi delta but twice as big. The coastal plain extends for about 160 km (100 mi) to the Arctic Ocean. Here the ground remains permanently frozen, though the surface thaws for just a few weeks each year.

The Alaska Range is a continuation of the mountains of the Pacific coast, and they rise precipitously from the sea. Mount McKinley, at 6,194 m (20,320 ft), is the United States' highest peak. Some valleys excavated by glaciers have been flooded to become fjords, while others are still occupied by glaciers. The Alaska Range extends to the southwest to form the volcanic mountains of the Alaska Peninsula and the Aleutian Islands. Offshore, the Aleutian Trench follows the arc of the islands, and marks the active zone at the boundary between the continental and oceanic plates.

The mighty Mississippi

The Mississippi–Missouri river system extends for 6,020 km (3,740 mi) from the source of the Missouri river high in the Rocky Mountains to the Gulf of Mexico. It drains an area of about 3.2 million sq km (1.25 million sq mi). The average discharge at the mouth of the Mississippi is about 17,500 cu m (620,000 cu ft) a second; in flood this has increased to almost 85,000 cu m (3 million cu ft) a second.

In its upper course to the west of the Great Lakes the Mississippi wanders through lakes and between hummocks of glacial deposits for about 250 km (155 mi) before its course becomes better defined. About 800 km (500 mi) from its source the river enters a 1,600 km (1,000 mi) stretch with a broad floodplain carved out from the rock of the valley walls.

South of St Louis the Mississippi is joined by the 3,946 km (2,466 mi) long river Missouri from the west. Its dark brown waters can be traced for many kilometers after the two rivers have joined. Farther south the waters of the 1,560 km (975 mi) Ohio flow in from the east, of which adds greatly to the volume of water.

During the last 60 million years the sediments washed down from the Appalachian Mountains and the Rockies have accumulated to a great thickness in the Mississippi valley and its delta. The Missouri, which contributes a relatively small volume of water, nevertheless provides most of the sediment today.

GREAT RIVERS

Counting the Missouri, its longest tributary, the Mississippi is the world's third longest river. The Mississippi–Missouri river basin is also the third largest in the world. The quantity of water and the amount of sediment carried are other measures of a river's influence on the landscape.

River and countries passed through	Length km	mi
Nile, Uganda/Sudan/Egypt	6,690	4,160
Amazon, Peru/Brazil	6,570	4,080
Mississippi–Missouri, United States	6,020	3,740
Chang, China	5,980	3,720
Yenisei, Soviet Union	5,870	3,650
Ob–Irtysh, Soviet Union	5,570	3,460
Huang, China	5,464	3,395
Congo (Zaire), Zambia/Zaire/Congo	4,630	2,880
Paraná, Brazil/Paraguay/Argentina	4,500	2,800
Amur, Soviet Union	4,444	2,761

Technicolor Mississippi An infrared photograph of the Mississippi river shows former meanders, now truncated to form oxbow lakes. These sites support lush vegetation, which shows up in red.

Loaded with silt, the Mississippi carries some 1,000 million tonnes of sediment into the Gulf of Mexico every year. As the river wanders across its floodplain the sediment tends to block its passage, slowing it down still further and increasing deposition.

Deposition of these sediments has formed the alluvial plain of the lower Mississippi. It stretches for 1,000 km (625 mi) to the Gulf of Mexico. The plain is 40 km (25 mi) wide at its northern end, and broadens out to 325 km (203 mi) where the river reaches the coastal plains. The weight of the sediments is so great that the underlying continental crust has been depressed. Seismic activity has been increased as a result.

Movements of sediment

When the Mississippi floods huge quantities of water overflow onto the floodplain. Beyond the main channel, where the water is shallower, the current slows down. The slower the current, the more sediment is deposited, particularly where there is vegetation to trap it.

The thickest layer of sediment builds up along the river itself, where it forms banks called levees parallel to the main channel. This vertical accretion of sediments seems to have taken place on average about twice every three years.

Sediments in the river channel are constantly moved from one side of the river to the other by meanders in the river. As the river follows its winding path, sweeping back and forth across the floodplain the outer bank, where the speed of flow is greatest, is eroded; on the inner bank sediment is deposited. The width of the river remains the same, but the meanders gradually migrate across the floodplain.

Great floods are a natural feature of the

lower Mississippi. In 1539 Garcilaso de la Vega noted: "The Indians build their houses on the high land, and where there is none they raise mounds by hand and here they take refuge from the great flood." Major floods have been recorded about twice every 10 years over the past 400 years, despite everything that has been done to contain the river by heaping soil on top of the natural levees.

Since the floods of 1927, when 650,000 people were driven from their homes, dams have been built on the major tributaries to hold back the floodwater and release it gradually. Floodways have been constructed to allow controlled flooding in certain areas. The course has been straightened and sediment dredged out to speed up the passage of water downstream. However, the faster the water flows the more power it has to erode, so to protect the new course hundreds of kilometers of embankments, made from concrete blocks wired together, have been built beside the river, particularly around the outside bends of meanders where they prevent the bank from being undercut. But floods still take place, and it is too early to say whether the Mississippi has been successfully controlled.

Sparkling rock terraces

Thermal springs in many parts of the world produce remarkable landscapes. Most thermal springs are found in volcanic areas but some, like these in Yellowstone National Park in Wyoming, are caused by water sinking down through the folded and faulted rocks to be heated by molten rock (magma). The superheated water then rises back up to the surface.

While underground, the water dissolves calcium carbonate from the limestone. The water becomes rich in minerals in solution; when it reaches the surface it cools, and the minerals are deposited as a solid rock known as travertine or calcareous sinter.

Steps form around the pools of water. The edges of the pool cool faster than the center, so the travertine builds up faster here. As the deposition continues the edges grow, creating columns of rock. Springwater spills over the edges, depositing layers of travertine onto the sides of the column and extending the rock terrace.

Yellowstone National Park has about seventy hot springs, and the calcareous terraces cover about 80 ha (200 acres). There are other hydrothermal features such as geysers, the most famous of which, Old Faithful, ejects steam and water many meters into the air.

Cascades of white rock form the Minerva Terrace at Mammoth Hot Springs in Yellowstone National Park, Wyoming, in the western United States. The giant steps are filled with steaming pools of mineral-rich water from thermal springs deep underground.

LAND BRIDGE BY A TROPICAL SEA

MOUNTAINS, VOLCANOES AND ISLANDS · PATTERNS OF CLIMATE · EVOLVING LANDSCAPES

The funnel–shaped isthmus of Central America joins the continents of North and South America and separates the Pacific from the Atlantic Ocean. With the island chain of the Antilles or West Indies it encircles the tropical Caribbean Sea. Most of this mountainous land lies within the tropics. Tropical rainforest and torrential downpours are characteristic of the northeastern coasts and islands, contrasting with the hot deserts of the northwest. Climates are cooler in the high plateaus and mountains. A chain of volcanoes follows the mainland's western coasts on the Pacific Ocean's "Ring of Fire". The region lies on an unstable part of the Earth's crust, and is particularly subject to earthquakes, seismic sea waves (tsunami) and volcanic eruptions; it also suffers from landslides, hurricanes and storm surges.

COUNTRIES IN THE REGION

Antigua and Barbuda, Bahamas, Barbados, Belize, Bermuda, Costa Rica, Cuba, Dominica, Dominican Republic, El Salvador, Grenada, Guatemala, Haiti, Honduras, Jamaica, Mexico, Nicaragua, Panama, St Kitts–Nevis, St Lucia, St Vincent and the Grenadines, Trinidad and Tobago

LAND

Area 2,735,515 sq km (1,056,183 sq mi)
Highest point Citlaltépetl, 5,699 m (18,700 ft)
Lowest point Lake Enriquillo, Dominican Republic, –44 m (–144 ft)
Major features volcanic mountain chain and Mexican plateau on isthmus, island chain of the West Indies

WATER

Longest river Conchos–Grande, 2,100 km (1,300 mi)
Largest basin Grande (part), 445,000 sq km (172,000 sq mi)
Highest average flow Colorado, 104 cu m/sec (3,700 cu ft/sec), at head of Gulf of California
Largest lake Nicaragua, 8,029 sq km (3,100 sq mi)

CLIMATE

	Temperature °C (°F) January	July	Altitude m (ft)
Guayamas	18 (64)	31 (88)	4 (13)
Zacatecas	10 (50)	14 (57)	2,612 (8,567)
Mexico City	12 (54)	16 (61)	2,306 (7,564)
Havana	21 (70)	27 (81)	49 (161)
Bluefields	25 (77)	26 (79)	12 (39)
Seawell	25 (77)	27 (81)	56 (184)

	Precipitation mm (in) January	July	Year
Guayamas	8 (0.3)	47 (1.9)	252 (9.9)
Zacatecas	7 (0.3)	69 (2.7)	313 (12.3)
Mexico City	8 (0.3)	160 (6.3)	726 (28.6)
Havana	51 (2.0)	93 (3.7)	1,167 (45.9)
Bluefields	264 (10.4)	746 (29.4)	4,370 (172.0)
Seawell	68 (2.7)	141 (5.6)	1,273 (59.1)

World's greatest recorded rainfall in 5 minutes, 305 mm (12 in) at Portobello, northern Panama

NATURAL HAZARDS

Earthquakes, landslides, volcanic eruptions, hurricanes

MOUNTAINS, VOLCANOES AND ISLANDS

Central America forms a land bridge, or isthmus, between North and South America, which became joined toward the end of the Pliocene epoch (5–2 million years ago). The isthmus and the Caribbean islands can be divided into nine areas based on their geological history and topography. These areas are defined by the effects of earth movements – folding and lifting up of rocks, earthquakes and the eruptions of volcanoes – and by quieter episodes when the land surface was eroded and sediments were laid down.

The Central American land bridge

The region is dominated by the Mexican plateau; it rises from an average 1,200 m (4,000 ft) in the north to 2,400 m (8,000 ft) in the south. The limestone mountains of the Sierra Madre Oriental in the east and the volcanic peaks of the Sierra Madre Occidental in the west form a high rim to the lower "hill-and-basin" relief of the plateau. Many of the rivers drain into interior basins; some of the westward-flowing rivers have cut deep canyons (*barrancas*), some of them rivaling the Grand Canyon in the United States in both size and splendor.

To the south the low, hot and dry basin of the river Balsas separates the Mexican plateau from the rugged crystalline rock terrain of the highlands of the Sierra Madre del Sur. This area has numerous peaks of 2,300–3,050 m (7,500–10,000 ft), and is heavily dissected by deep V-shaped valleys.

Bermuda's coral-fringed islands are situated in the western Atlantic away to the north of the islands of the Caribbean. The offshore reefs are one of the most northerly coral formations in the world.

West of the Mexican plateau lie the desert landscapes of the Pacific coastal hills and the peninsula of Baja California or Lower California, a raised block of granite and volcanic rocks separated from Mexico proper by a rift valley drowned by the Gulf of California.

East of the Mexican plateau lies the coastal plain around the Bay of Campeche in the Gulf of Mexico. This is a geologically young area of marl, shale and sandstone marine sediments containing major oil and gas deposits, and more recent river-borne sediments. Barrier beaches enclosing the swamps and lagoons that fringe the coast reflect the softness of the rocks and the still-active process of sedimentation.

The oldest and geologically most complex part of the region encompasses Belize, Guatemala, Honduras and eastern Nicaragua on the isthmus, and also the

Map of physical zones The land bridge of Central America extends from the high Mexican plateau south to Panama, the gateway to South America. Both the mountainous mainland and the islands of the Caribbean are shaped by tectonic disturbance and climatic extremes.

islands of the Greater Antilles in the Caribbean, which include Cuba, Jamaica, Hispaniola and Puerto Rico. In the Cretaceous period, 100 million years ago, this "Old Antillia" formed a single landmass. It has since been broken up by numerous episodes of lifting, folding and erosion. Old Antillia is composed of a series of mountain ranges and depressions that run from east to west. In places the limestone and sandstone rocks have been completely eroded to expose the underlying granite. Where limestone still predominates, the combination of wet tropical conditions and high mountain relief has produced spectacular weathered terrain with deep enclosed hollows and tall limestone towers.

To the west and north of Old Antillia lie the Yucután peninsula and the Bahamas. Both are low-lying plains of limestone resting on crystalline rocks and have emerged relatively recently from the sea. Here the landscape is dominated by steep-sided hollows, and by underground streams and expanses of fluted and etched limestone pavement. Shorelines are marked by low, notched limestone cliffs, coral reefs and mangrove swamps.

The active volcanic belts of Central America and the Lesser Antilles to the east are the result of movements in the Earth's crust. The Cocos plate of the tropical Pacific has undercut the North American and Caribbean plates and the

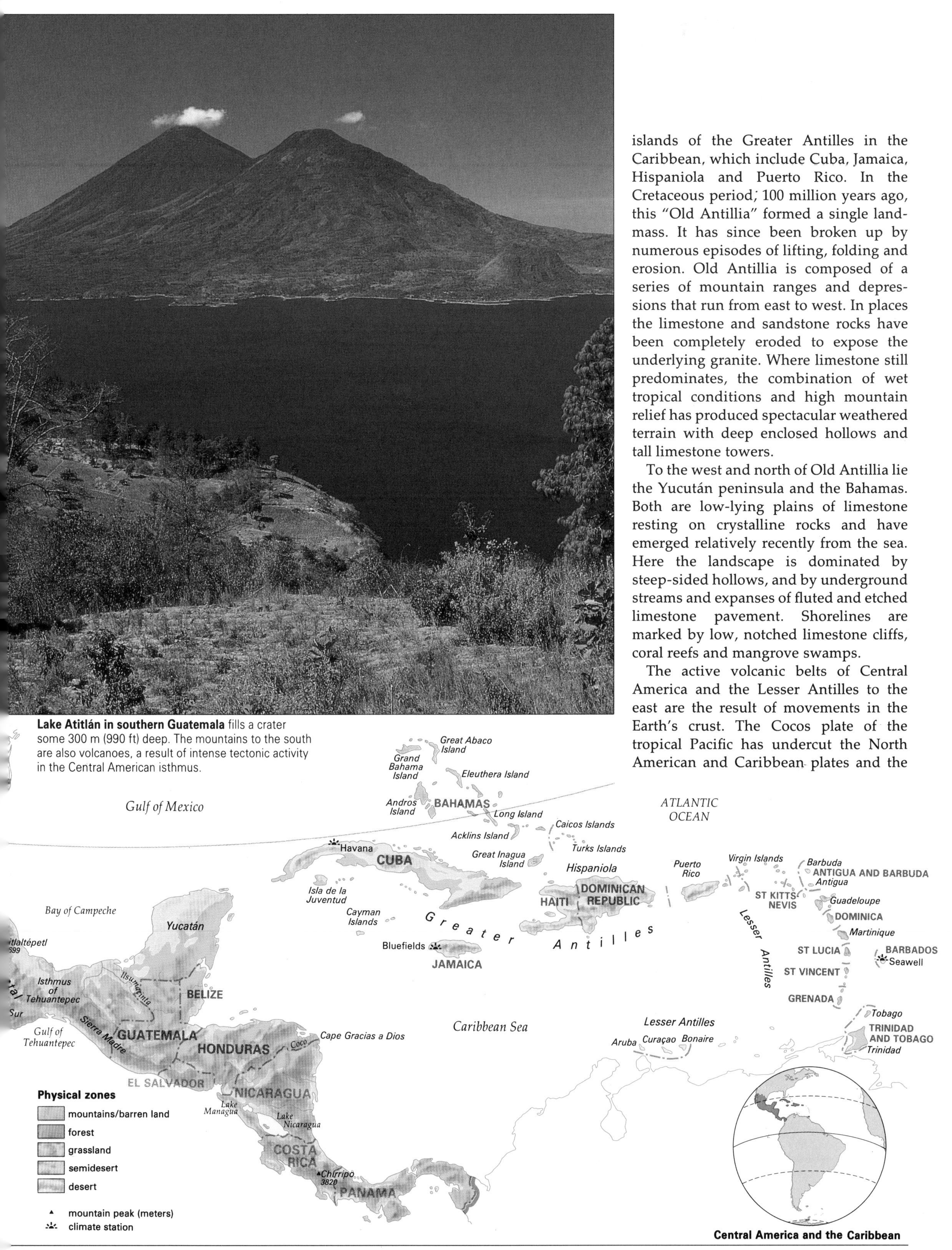

Lake Atitlán in southern Guatemala fills a crater some 300 m (990 ft) deep. The mountains to the south are also volcanoes, a result of intense tectonic activity in the Central American isthmus.

Central America and the Caribbean

Central American landmass to produce a line of volcanoes from western Guatemala to Panama. Volcanic deposits cover the older sedimentary rocks beneath.

The Lesser Antilles chain of volcanic islands was formed when the North American plate thrust under the Caribbean plate. The double arc of islands is the outcome of two distinct phases of volcanic activity. The low-lying eastern arc represents one phase, which lasted from 40 to 10 million years ago. These volcanoes were thoroughly eroded, then covered by limestones laid down during times of raised sea levels. In the second volcanic phase, which began about 7.5 million years ago, the arc of the volcanic belt moved westward and built new islands, such as St Christopher (Kitts) and Dominica, and added westerly extensions to older islands such as Guadeloupe, Martinique and St Lucia.

LIVING IN AN EARTHQUAKE ZONE

There is hardly a town or city in western and southern Mexico, the Pacific side of Central America or the West Indies that has not been devastated by earthquake or by volcanic disaster at some time in its history. San Salvador, capital of El Salvador, has been partly or completely destroyed nine times since 1528.

The impact of earthquakes is often increased by the landslides, fires and, on the coastline, giant sea waves (tsunami) that they trigger. In 1692, when an earthquake hit Port Royal near present-day Kingston, Jamaica, two-thirds of the town slipped into the sea and what remained was inundated by a tsunami. Nearly all the 2,500 inhabitants perished.

In 1902 the eruption of Santa Maria in the Sierra Madre in Guatemala cost 6,000 lives. In the same year all but two of the 30,000 inhabitants of St Pierre in Martinique were killed by a glowing cloud of volcanic dust and poisonous gas that sped down from the erupting Mont Pelée to the north of the town. Only the day before, 130 km (90 mi) to the south, St Vincent's Soufrière had erupted, killing 2,000 people. The 1979 eruption of the same volcano destroyed about a third of the island's banana crop.

Paradoxically, the unstable land attracts people to it: the benefits of cultivating the fertile volcanic soils outweigh the risk of damage from earthquake or eruption.

PATTERNS OF CLIMATE

The climates of Central America and the Caribbean are dominated by the northeast trade winds, which blow almost incessantly all year. In winter a zone of high pressure (anticyclone) lies over the Atlantic to the east, and the northeast trades are cool and mostly dry. Rainfall is low except over northeast-facing coasts and mountain ranges, where airflows are forced to rise, cool and release moisture.

In summer the belt of low atmospheric pressure over the Equator, with its rainy weather, advances north with the overhead sun and by July lies over Panama and Costa Rica. At the same time the Bermuda–Azores "high" weakens and retreats northward. Although the northeast trades still prevail the air is much less stable, and there are frequent rains as waves of convection currents and tropical cyclones track westward across the region. A few cyclones also develop off the Pacific coast.

The region's mountainous nature and latitudinal extent result in a wide range of climatic types. In general it is driest in northwest Mexico and wettest along northeast-facing coasts toward the Equator. In the extensive mountain and plateau areas of both Central America and the Caribbean the pattern of tropical climates is modified by altitude.

From rainforest to desert

In the tropical wet zone climates are hot and humid throughout the year, with an annual rainfall of at least 2,000 mm (80 in) and little or no dry season. Along the Atlantic coast of Central America from southern Mexico to Panama, and on the windward sides of the higher Caribbean islands, exposure to the northeast trades ensures substantial winter as well as summer rainfall. Most of the rain is concentrated in heavy falls of a few hours each day. Small areas such as the central mountains of Dominica, Guadeloupe and Martinique are "super-wet", with annual rainfalls of 5,000–9,000 mm (200–360 in), and never experience a dry month. The small annual range of temperature increases with latitude. In these hot, wet conditions tropical rainforest is the natural vegetation.

A tropical wet-dry climate characterizes areas in the rain shadow southwest of mountain ranges, including much of the Pacific side of Central America, and lower islands and peninsulas of the Caribbean. Temperatures are similar to those of the tropical wet zone, but there is a 4–7 month winter dry season. Annual rainfall is generally in the range 750–2,000 mm (30–80 in). Deciduous forests form the natural vegetation.

Much of the northern interior of Mexico is semiarid. Most of the 250–500 mm (10–20 in) of rain each year falls in just a few summer thunderstorms. Temperatures vary greatly with season: at Sabinas at the northern end of the Sierra Madre Oriental they range from 10°C (50°F) in January to 33°C (91°F) in July, with some frosts in winter. In the south, the Lesser Antilles islands of Curaçao, Bonaire and Aruba also have a semiarid climate. Their low rainfall is the result of cool temperatures above the sea surface where there is an upwelling of cold water. The natural vegetation of these semiarid areas is low thorn woodland, but human activity has usually modified this to a grass and scrub landscape.

Parts of the extreme northwest and north of Mexico lie under the influence of a subtropical anticyclone throughout the year and have a hot desert climate. Only 75–250 mm (3–10 in) of rain falls each year. There is a comparatively wide average temperature range, for example from 18°C (64.4°F) in January to 31°C (87.8°F) in August at Guayamas in the desert of northwest Mexico. Maximum daily temperatures frequently reach 45–50°C (112–122°F) away from the cool and often foggy Pacific coast. The scanty vegetation is dominated by cacti and grasses.

Climate variations

Climates have been very different at times in the past. Studies of old lake levels have revealed that 18,000 years ago areas of northern Mexico that are now semiarid and desert were much wetter, while tropical Mexico was considerably drier. Even in the last hundred years climates have varied significantly. Records show that since 1959 rainfall in the eastern Caribbean has been only about three-quarters what it was in 1871–1901 and 1928–58.

Mount Gimie on St Lucia in the Lesser Antilles is an old and highly eroded volcanic mountain. Its sharp ridges and steep slopes, and the humid climate, ensure luxuriant tropical rainforest and a high rate of chemical weathering, making the slopes prone to landslides.

EVOLVING LANDSCAPES

The diversity of climate, relief and rocks in Central America and the Caribbean is reflected in a great variety of landscapes. The low-lying limestone coastlands of Yucatán and the Cayman Islands northwest of Jamaica are fringed by mosquito-ridden mangroves, while on the high plateau of northern Mexico the land is cold, bleak and dry. The rock desert of Baja California contrasts with the steamy, forest-clad mountains of Panama, Costa Rica, Nicaragua and Guatemala.

Landslides and weathering

Landslides are a major factor in landscape formation. The steep mountainous slopes, frequent earthquakes, the mainly humid tropical climate, the character of the soils and vegetation all contribute to landslides. In these warm, wet conditions weathering rapidly breaks down the rocks, forming layers of soil. Soil erosion by sheetwash (water moving over the surface) and rainsplash is restricted by the dense tropical forest. As a result deep soils build up. Plant nutrients are concentrated in the top layer of tropical soils, so rainforest trees typically have shallow roots and the soil beneath is liable to slip. On the slopes intense tropical rainstorms can saturate the soil, which may then lose cohesion, triggering a landslide.

Another important process of erosion is chemical weathering. Areas of limestone and volcanic rock are particularly vulnerable, and the process also reflects the high rainfall and temperatures of most of the region. Weathering is caused by rocks being dissolved by rainwater that is acidic because of carbon dioxide absorbed from the atmosphere. In the very wet volcanic island of Dominica, in the Lesser Antilles, the rock surface is being eroded

by some 85–220 mm (3.3–8.7 in) every thousand years, one of the fastest rates of chemical erosion in the world. In places with active sulfur springs the rate rises to 775 mm (30.5 in) a thousand years.

The history of volcanic landscapes

The volcanic landscapes that cover much of the region vary greatly with the age of the volcano. The little-eroded cone or dome of a volcano that has erupted relatively recently, such as Soufrière on St Vincent in the Lesser Antilles or El Chichón in southeast Mexico, is shallowly incised by a radiating pattern of long hills and gullies. Older volcanoes, such as the rugged Mount St Catherine in Grenada and Mount Gimie in St Lucia, are deeply dissected, with knife-edge ridges and steep slopes vulnerable to landslides. The pattern of the drainage system is still basically radial, but more powerful streams have captured weaker flows. In long-eroded landscapes the original volcanoes can no longer be discerned. The drainage pattern has become branched and the slopes more gentle; north and central St Lucia provide an example of this type of landscape.

The impact of Europeans

People have had a considerable effect on the landscape of the region since the Europeans arrived. Before the Spanish Conquest in the early 16th century the Indians practiced mainly slash-and-burn farming. With long fallow periods and little long-term distrubance, only minor damage was caused to the soils and natural forests of the area. In some areas the Spanish Conquest may actually have protected the forests – but only because the Indian population was destroyed. The 3–4 million Arawak Indians of Hispaniola (Haiti and the Dominican Republic) were reduced to 60,000 by 1510, and virtually eradicated by smallpox in 1518.

In the low-lying islands of the West Indies, such as Barbados and Antigua, the natural seasonal forest was almost totally cleared for sugar planting from the 1640s onward. This had a serious impact, damaging the soil and disturbing plants and animals. The land surface was often rapidly eroded by heavy rain falling on the bare soil after clearance and later after cropping. Most of the sugar lands were so eroded and exhausted of soil nutrients by the time the sugar boom ended in the 19th century that the abandoned lands have reverted only to low scrub woodland rather than the tall forests of the pre-sugar era. In Cuba the same cycle was initiated when the forest was cleared for sugar planting in the late 19th century; today the impact of soil erosion is greatest in the southeast of the island.

Soil erosion and loss of natural habitats have accelerated dramatically in the 20th century. In southern Mexico population pressure, use of the plow, overcropping, and a shortening of the fallow period have all caused land degradation. Over wide areas of once-fertile volcanic soils sheet erosion has exposed underlying limey hardpan layers that are useless for cultivation. In the highlands north of the eastern end of the Sierra Madre del Sur disastrous gullying has transformed former cornfields and pasturelands into a bare, arid area of badlands that can no longer be used for agriculture. The coniferous highland forests of central north Mexico and the tropical rainforests of Central America farther south are today being cleared for agriculture and timber production. In the West Indies surviving rainforests on the higher islands have increasingly been cleared for banana and grapefruit cultivation as farming switches from the ruined and abandoned sugar-growing areas near the coasts.

FOREST CLEARANCE AND EROSION

The way rainforest clearance triggers soil erosion is clearly shown by measurements taken in the very wet interior of the Lesser Antilles island of Dominica in 1972–75. An area was cleared by hand for agricultural use and planted with bananas and citrus trees. During the clearance and planting phase, intense erosion removed half the original 20 cm (8 in) of rainforest topsoil in the area planted with bananas and virtually the entire topsoil in the citrus plot, which had required more thorough removal of tree roots and ground clearance.

Once the crops were established erosion rates changed. A grass-and-weed cover between the trees (and the fact that much of the topsoil had already disappeared) reduced erosion rates there to less than 2 mm (0.08 in), whereas the rates in the muddy, labor-intensive banana plantation, some 6 mm (0.24) a year, were approximately five times as high as in the rainforest next to it. In areas cleared by bulldozers, and on steeper slopes, the rate of erosion was even more severe.

An arid landscape on the southern Mexican plateau to the west of the Bay of Campeche. This is part of the hill-and-basin relief of the plateau, where eroded volcanic mountains separate large basins whose rivers often drain inland. Vegetation is sparse, a tropical scrub of low trees, shrubs and cacti.

Gray Whale Lagoon on the Baja California coast in Mexico. An offshore coastal bar has been created by ocean currents depositing sand and shingle. The lagoon has become stranded behind the bar, and sediments are slowly being built up in it. It will eventually become dry land.

Where hurricanes strike

The weaponry of a hurricane is fearsome – violent winds, very low pressure and extreme rainfall. High winds can destroy crops and light buildings, and severe hurricanes can defoliate and even flatten swathes of rainforest.

Tropical cyclones are classified by wind speed as storms (62–119 k/h, 39–72 mph) or hurricanes (winds upward of 120 k/h, 75 mph). In the west Pacific these are known as typhoons, in the Bay of Bengal east of India as cyclones. They are revolving systems of low atmospheric pressure accompanied by violent winds circulating in an anticlockwise direction (clockwise in similar latitudes in the southern hemisphere).

The combination of very low pressure and high winds can also generate storm waves of awesome power, which play a major part in coastal erosion and the building of barrier ridges and reef islands in the region. The accompanying floods in coastal areas are often the main cause of loss of life. Farther inland extremely high rainfall can lead to river floods and massive erosion of slopes, frequently triggering landslides.

Hurricanes in the Caribbean

Hurricane David in August 1979 in Dominica and Hurricane Gilbert in September 1988 in Jamaica had devastating effects on housing, property and human life. This was so partly because most of the islands' mainly young populations had never witnessed a hurricane before. Where cyclones are frequent, human preparedness and building standards are better. The forest vegetation also adapts: it tends to be shorter, more "streamlined" and hence less susceptible to hurricane damage.

A hurricane normally brings exceptionally heavy rain, between 200 and 500 mm (8–20 in) in under twelve hours. In 1909 a hurricane moving slowly over the Blue Mountains of Jamaica deposited 3,428 mm (135 in) of rain in a week, including 728 mm (29 in) in one day!

In the Caribbean cyclones have traditionally been given girls' names, but since 1979 boys' names have been included too. On average seven cyclones (including four hurricanes) hit the Caribbean each year; within the region their frequency declines toward the south and west. They develop in the northeast trade winds in the tropical Atlantic and in the Caribbean Sea itself. Cyclones generally pass through the Caribbean in a west-northwest direction, but at latitudes of around 20–25°N they curve to the northeast. Over 90 percent develop between June and October, when the sea temperature is at its warmest, the air above it humid, and the atmosphere in the region least stable.

The tracks and frequencies of cyclones in the Caribbean have varied greatly over the past few hundred years. Over the Caribbean as a whole cyclones were comparatively frequent in 1871–1901 and 1928–58, whereas 1902–27 and the years since 1959 have been periods with fewer cyclones. The recent decline has been particularly dramatic, with less than half the expected number of cyclones in the Bahamas, Jamaica, Hispaniola and the northern Lesser Antilles.

Periods of very few cyclones, such as 1650–1760 and 1840–70, are associated with sea-surface temperatures some 1.5°C (2.7°F) below average. During the cyclone peak this century, sea temperatures were 1°C (1.8°F) above normal. Some climatologists predict a significant increase in the frequency of cyclones if the expected global warming takes place as a result of the greenhouse effect.

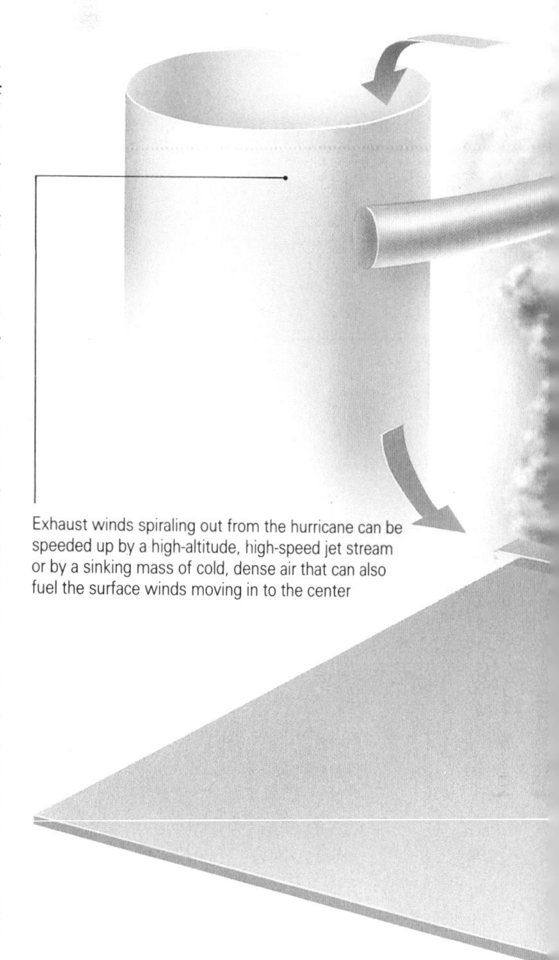

Exhaust winds spiraling out from the hurricane can be speeded up by a high-altitude, high-speed jet stream or by a sinking mass of cold, dense air that can also fuel the surface winds moving in to the center

Devastation at Kingston, Jamaica When Hurricane Gilbert passed through Jamaica it had wind speeds of up to 200 km (125 mi) an hour, and deposited 200–500 mm (8–20 in) of rain in 8 hours. Gilbert caused damage worth an estimated $7 billion, and three-quarters of the island's houses were seriously damaged.

The path of Hurricane Gilbert The most intense western hemisphere tropical cyclone on record, the hurricane gathered its strength over the eastern Caribbean. It struck Jamaica, the Yucatán peninsula and Monterrey, Mexico; its zone of rain stretched 1,000 km (600 mi) north and south of its center.

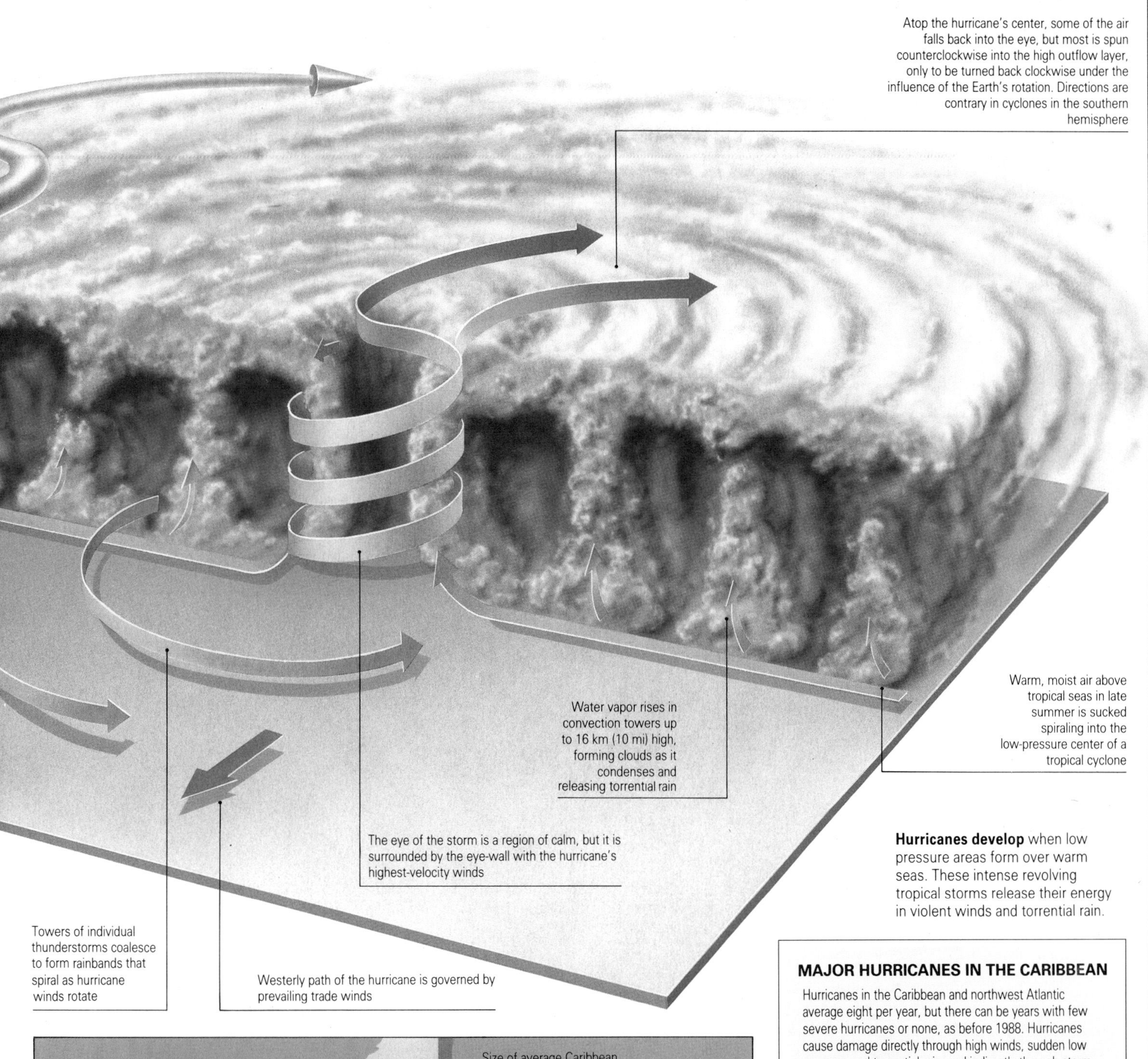

Hurricanes develop when low pressure areas form over warm seas. These intense revolving tropical storms release their energy in violent winds and torrential rain.

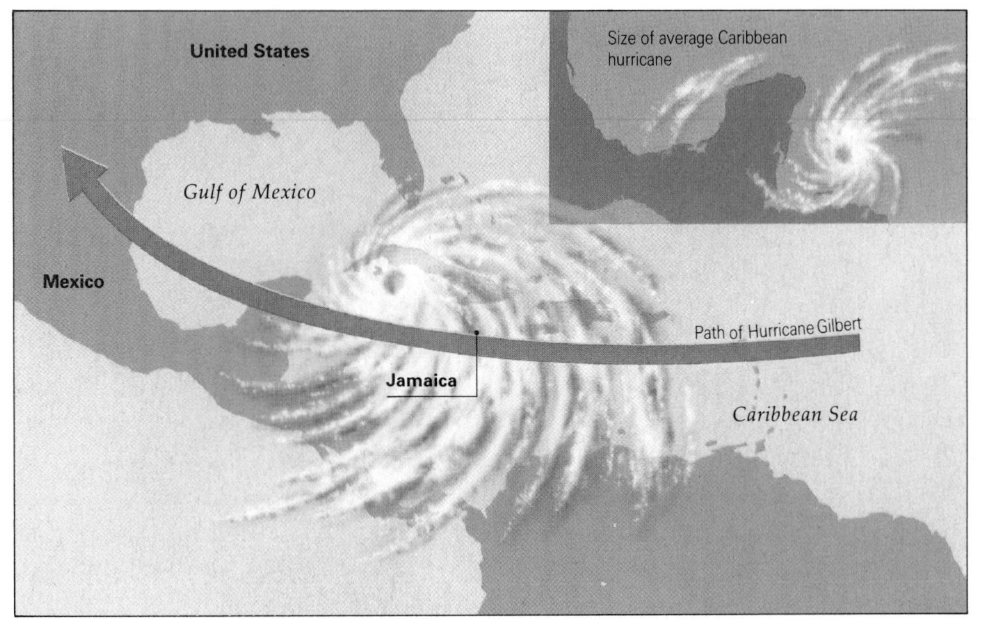

MAJOR HURRICANES IN THE CARIBBEAN

Hurricanes in the Caribbean and northwest Atlantic average eight per year, but there can be years with few severe hurricanes or none, as before 1988. Hurricanes cause damage directly through high winds, sudden low pressure and torrential rain, and indirectly through storm surges at sea and floods. The rainstorms can erode soils, start landslides and burst river banks.

Year	Location	Casualties
1966	Guadeloupe, Hispaniola, Honduras, Mexico	over 1,200 dead, 30,000 homeless
1969	United States	over 400 dead, 200,000 homeless
1974	Honduras	7,000 dead 600,000 homeless
1976	Mexico	400 dead
1979	West Indies	over 1,200 dead, 500,000 homeless
1980	United States	272 dead
1982	Guatemala, El Salvador	over 1,000 dead
1988	Jamaica, Mexico	260 dead 1,000,000 homeless
1988	Nicaragua, Costa Rica to Venezuela	110 dead 300,000 homeless

FROM THE EQUATOR TO THE ANTARCTIC

FOREST, GRASSLAND AND DESERT · MOUNTAINS, BASINS AND PLAINS · WATER, EARTH AND MUD

Tapering from north to south, South America extends from the tropical Caribbean to the stormy seas of Cape Horn, just 800 km (500 mi) from Antarctica. The continent covers almost one-seventh of the world's land surface. South America's pattern of mountains, high plateaus *(altiplanos)*, tropical grasslands *(llanos)* and temperate grasslands *(pampas)*, rainforests *(selvas)* and rivers provides landscapes of striking diversity. The snow- and cloud-capped Andes are the world's longest unbroken mountain system. Forming part of the Pacific "Ring of Fire", the Andes have many active volcanoes and include Aconcagua, the highest peak in the Americas. The Amazon is the greatest river on Earth, decanting one-fifth of the world's flow of fresh water into the sea, and draining its greatest area of rainforest.

COUNTRIES IN THE REGION

Argentina, Bolivia, Brazil, Chile, Colombia, Ecuador, Guyana, Paraguay, Peru, Uruguay, Surinam, Venezuela

LAND

Area 17,084,526 sq km (6,874,600 sq mi)
Highest point Aconcagua, 6,960 m (22, 834 ft)
Lowest point Salinas Grande, Argentina, −40 m (−131 ft)
Major features Andes, world's longest mountain chain, Guiana Highlands and Plateau of Brazil, Amazon basin

WATER

Longest river Amazon, 6,570 km (4,080 mi)
Largest basin Amazon, 6,150,000 sq km (2,375,000 sq mi)
Highest average flow Amazon, 180,000 cu m/sec (6,350,000 cu ft/sec)
Largest lake Titicaca, 8,340 sq km (3,220 sq mi)
Amazon has world's largest drainage basin and greatest flow
Angel Falls, Venezuela, 979 m (3,212 ft) are world's highest, Iguaçu Falls, Brazil–Argentina, one of the widest, 4 km (2.5 mi)

CLIMATE

	Temperature °C (°F) January	July	Altitude m (ft)
Maracaibo	26 (79)	29 (84)	48 (157)
Manaus	26 (79)	27 (81)	83 (272)
La Paz	9 (48)	8 (46)	4,103 (13,461)
Buenos Aires	24 (75)	11 (52)	25 (82)
Ushuaia	9 (48)	2 (36)	6 (20)

	Precipitation mm (in) January	July	Year
Maracaibo	3 (0.1)	28 (1.1)	387 (15.2)
Manaus	276 (10.9)	61 (2.4)	2,102 (82.7)
La Paz	139 (5.5)	4 (0.2)	555 (21.9)
Buenos Aires	104 (4.1)	61 (2.4)	1,027 (40.4)
Ushuaia	58 (2.3)	47 (1.9)	574 (22.6)

Atacama Desert has recorded no rain in 400 years

NATURAL HAZARDS

Volcanic eruptions, earthquakes, landslides and mudslides

FOREST, GRASSLAND AND DESERT

The Pacific rim of the Andes dominates the physiography of South America, and separates most of the continent from the Pacific Ocean. West of the Andes environments are dominated by the ocean. The air above the cold waters welling up along the coast is stable and dry, and winds blow along the shore. As a result the narrow central coastal strip is largely desert, though in the north and south onshore winds bring rain: the narrow Pacific coast of Colombia has more than 5,000 mm (200 in) a year.

To the east of the Andes the climate is dominated by the influence of the Atlantic Ocean, with its warm onshore currents and winds. Rainfall in general decreases toward the south away from the Equator and westward away from the coast. The interior basins of the eastern Andes in Argentina have less than 500 mm (20 in) of rain a year, while large areas of the equatorial forest *(selvas)* of the Amazon basin receive over 2,000 mm (80 in). An important exception is the northern coast along the continent's eastern edge, which receives only 650 mm (26 in), mainly from November to April.

This broad pattern hides sharp contrasts, especially in the Andes, where rainfall can decline from 2,000 mm (80 in) a year to 300 mm (12 in) a year in as little as 20 km (12 mi). Here relief is the main factor controlling climate.

There are also strong seasonal contrasts, which become more marked with distance from the Equator. The Amazonian lowlands are hot and wet all year round, but winters are temperate and dry to the south (July), and hot and dry to the north (January). On the interior plains June, July and August tend to be the driest months; in the Andean foothills of Argentina winters are very dry. In the south of the continent rainfall is evenly spread; July temperatures become progressively lower toward the extreme south.

Climates and vegetation

During several periods in the last 2 million years the Andes have contained extensive glaciers, and the continent's southernmost coasts have been surrounded by nearly permanent sea ice. There are also fossil dunes in areas that are now covered with rainforest, suggesting that

Map of physical zones South America stretches from north of the Equator almost to Antarctica. The western edge is divided from the east by the great mountain range of the Andes, which run almost the full length of the continent. To the west the land drops sharply down to the sea; there is only a narrow coastal plain. The sea bed plunges to the depths of the Peru–Chile trench, where the Nazca plate meets the South American plate. The bulk of the continent, to the east of the Andes, is a more stable land, less subject to earthquakes and volcanic activity. The great Amazon river dominates much of the region.

Forces at work in the Andes The Valley of the Moon in north-central Chile demonstrates the strength of the forces that created the mountain chain. The rocks have been warped to form an inverted arch known as a syncline; the crest of fold (the anticline) has been worn away. Piles of rock debris (scree) build up on the slopes as a result of frost action, though the valley lies within tropical latitudes.

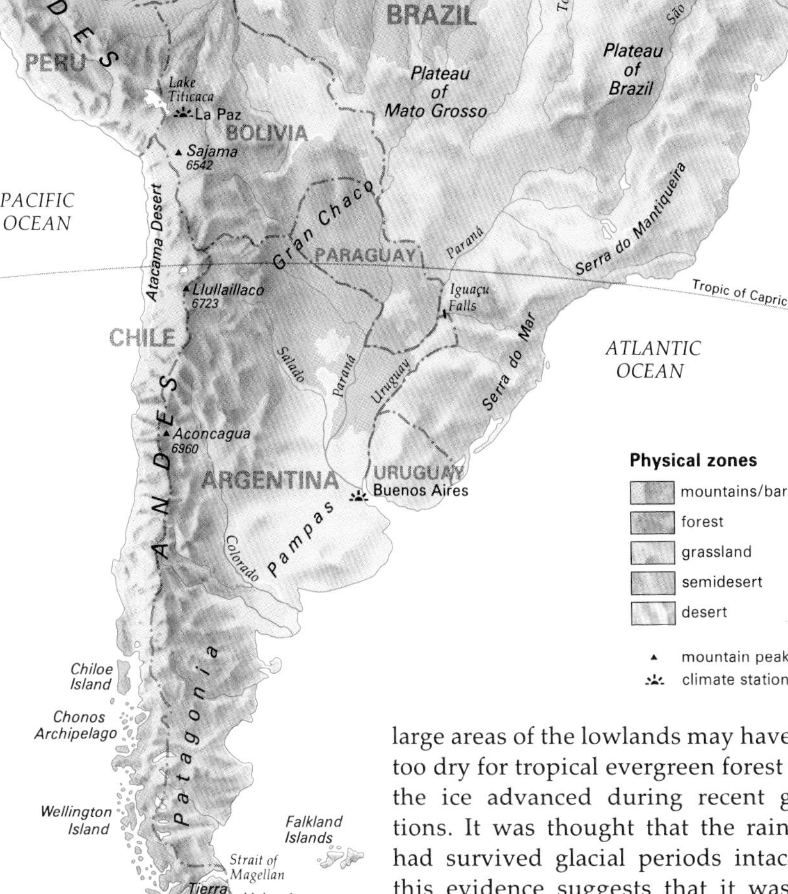

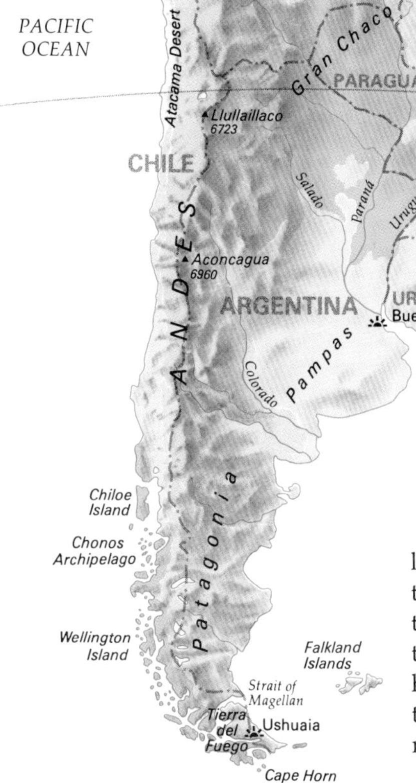

Where rain never falls The sandy desert that runs along the coast of Peru is almost totally devoid of vegetation because rain hardly ever falls. The cold current off the west coast is the principal reason for the extreme aridity. Cold water from the south moves northward along the coast until it is forced by the bulge in the land and by easterly winds to move westward out into the Pacific. These dry offshore winds blow for much of the year; even when air moves east from the sea over the coast there is no rain, because the air is cool and contains little moisture. When the air heats up over the land it becomes even more dry. The coastal desert is narrow only because the Andes, whose foothills can be seen in the distance, are never far from the coast.

large areas of the lowlands may have been too dry for tropical evergreen forest when the ice advanced during recent glaciations. It was thought that the rainforest had survived glacial periods intact, but this evidence suggests that it was restricted to small isolated patches (refugia).

The vegetation today largely reflects the influences of heat and moisture. Most of the Amazon basin is covered by the world's greatest rainforest, though large areas of it are destroyed each year. Lianas climb beneath the canopies of hardwood trees of enormous size. The Amazon has both wet and very wet seasons, causing seasonal flooding along the great rivers. Toward the south in the Plateau of Mato Grosso (the name means great forest), and toward the north in the plains, vegetation responds to seasonal shortages of moisture with a mixture of trees and grasses in the tropical grasslands. In the *pampas* of Argentina in the southwest the grasses take over completely.

MOUNTAINS, BASINS AND PLAINS

The major landform regions of South America include the Andean cordillera (the parallel mountain ranges and the land between them), the ancient shield lands of Brazil and Guiana, the sedimentary basins of the great rivers, and the Atacama Desert.

The continent of South America sits on one of the crustal plates that are moving slowly over the surface of the Earth. To the west, fresh crust produced by seafloor spreading at the East Pacific Rise is moving eastward, carrying with it the Nazca plate. This is being forced (subducted) beneath the South American plate. The subduction zone is marked by the deep offshore Peru–Chile trench. The movement has crumpled up the Andes and caused igneous activity along the chain, which has volcanoes and lava deposits along almost of entire length.

Eroded uplands and lowland sediments

On the eastern side of the continent are the stable shield areas of the Plateau of Brazil and the Guiana Highlands, made from Precambrian rocks over 600 million years old. On the edges of these areas are found the important metallic ores of South America: bauxites (used for aluminum) in Guiana, manganese in northeastern Brazil, iron in Minas Gerais in the southeast and in the Serra dos Carajás, south of the Amazon delta. Over hundreds of millions of years the shield areas were eroded almost flat, but they have been lifted up during more recent periods of mountain-building and now constitute the highland areas. The Guiana Highlands include Angel Falls, the highest waterfall in the world. South of the Plateau of Brazil the spectacular horseshoe of the Iguaçu Falls is higher and several times wider than that of North America's Niagara Falls.

Erosion of the old shield and the newer high mountains produced huge amounts of sediments. These have been deposited in the low plains and basins of South America, especially in the Amazon basin and the sub-Andean trough, which runs down the east side of the mountains and widens in the south to form the Gran Chaco in Bolivia and Paraguay. The rocks are mainly sandstones and conglomerates (the latter comprising rounded fragments set in a cement such as hardened clay). It is here that oil and gas are most likely to be found in the future.

The Pacific coast has only a few small rivers; not far away, on the eastern side of the Andes, flow the major rivers. The Amazon is only a few hundred kilometers from the Pacific as it starts its 6,570 km (4,080 mi) journey to the Atlantic. The Uruguay, Paraná, São Francisco and Orinoco rivers all rise on the shield areas.

Island mountains and deserts

Much of the shield area is relatively featureless, but where there is granite high, steep-sided hills called inselbergs may be found. In places these have been eroded into distinctive rounded forms; the most famous example is the Sugarloaf Mountain in Rio de Janeiro on the southeast coast of Brazil. Less well known are the *tepuís*, isolated island mountains on the border between Venezuela and Brazil. They include Neblina Peak (3,014 m/ 9,978 ft) and Roraima. Rising above the surrounding plateaus, these often inaccessible peaks have evolved their own distinctive plants and animals.

The Atacama is one of the driest deserts

The *altiplano* in Bolivia is a high plateau area between two chains of the Andes. At 3,700 m (12,000 ft) it is one of the highest permanently settled areas in the world, but the land is difficult to cultivate.

GREAT CATARACTS OF THE WORLD

Most waterfalls comprise several drops. The three highest single drops are those of the Angel (807 m/2,648 ft), Utigård (600 m/1,970 ft) and Mtarazi (479 m/1,572 m). South America also has the Iguaçu Falls on the Brazil–Argentina border, which total 4 km (2.5 mi) in width. Until submerged behind a dam in 1982 the Paraña river poured through the Sete Quedas falls on the Brazil–Paraguay frontier with a greater flow than any other on Earth.

	Total height	
	m	ft
Angel, Venezuela	979	3,212
Tugela, South Africa	948	3,110
Utigård, Norway	800	2,625
Mongefossen, Norway	774	2,540
Mtarazi, Zimbabwe	762	2,500
Yosemite, United States	739	2,425
Mardalsfossen, Norway	656	2,154
Tyssestrengane, Norway	646	2,120
Cuqueñan, Venezuela	610	2,000
Sutherland, New Zealand	580	1,904

The Angel Falls in the Guiana Highlands of southeast Venezuela is the world's highest waterfall. On one of the tributaries of the Orinoco, it drops from an isolated plateau block (*tepuís*) into the tropical forest below.

EL NIÑO – AN UNWELCOME CHRISTMAS GIFT

The Peru Current off South America's west coast is normally very cold. Every three to seven years, however, the sea surface warms up between October and March, and may stay warmer for up to two years. In 1982, for example, the Galápagos Islands west of Ecuador had temperatures 4.3°C (7.7°F) higher than usual and 10 times their normal rainfall.

The arrival of El Niño (the Child), as this event, which often takes place at Christmas, is called locally, is of concern to fishermen, as the warm surface water prevents plankton-rich colder water from reaching the surface, so the fish catch is drastically diminished.

The onset of El Niño may be related to a weakening of the trade winds that normally blow from the south and cool the ocean surface. El Niño itself causes the winds to slacken further, producing a kind of vicious circle. Volcanic activity on the Pacific seabed may also be responsible. The effects are not only local: temperature changes as far afield as Canada, western Europe and Australia have been traced to the El Niño phenomenon.

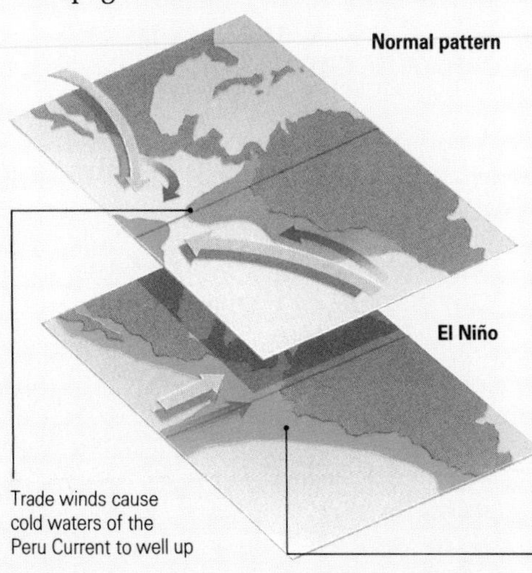

Trade winds cause cold waters of the Peru Current to well up

Periodically the trade winds weaken (they sometimes even reverse) and the cold upwelling is stopped as the warm countercurrent extends east

in the world. In some places, such as Arica in the north, no rainfall has ever been recorded. The landforms here and in the interior basins include great alluvial fans spreading out from the base of the mountains, and old lake beds called *playas* that are rich in salt. On the eastern side of the Andes, in Argentina, many of the rivers drain inland, and salt deposits are widespread.

The Amazon basin

Apart from deserts, all these landforms are to be found in the vast area drained by the Amazon. This great river basin drains an area nearly half that of the United States. The Amazon pours 180,000 cu m (6.4 million cu ft) of water into the Atlantic every second.

The basin is bordered to the west by the Andes, which rise steeply from the lowlands of the Marañón, Putumayo and Japurá. These and many lesser rivers flow east to from the upper reaches of the Amazon. The port of Manaus, where the

Negro joins the Amazon, is still 1,600 km (1,000 mi) from the sea. To the north are the shield lands of Roraima; to the south the Brazilian shield underlies the Plateau of Mato Grosso, whose great rivers Tocantins, Tapajós and Xingu all have rapids that provide sites for hydroelectric dams. In the center of the basin there are recent unconsolidated sands that make poor soils; in places they have been eroded into steep cliffs by the river. Here the waters of the Amazon and its tributaries dominate the landscape. The wide plain floods seasonally, and there is a complex of islands built up by the river Negro when it carried more sediment.

WATER, EARTH AND MUD

The water flowing in the great rivers of South America was once described as the "urine of the continents". Although perhaps rude, this is an apt description; a medical expert can learn a lot from a sample of urine, and river waters reveal much about the environments they drain.

Black waters and white waters

Seen from an airplane flying over the Amazon basin, there is a clear contrast between "black water" and "white water" rivers. Where great black and white water rivers meet there are dramatic contrasts in color and chemistry that may persist downstream for tens of kilometers before the waters finally merge. This mingling takes place, for example, where the Negro meets the upper waters of the Amazon (known in Brazil as the Solimões), and where the Caroní joins the Orinoco just above its delta.

The "black" waters are in fact coffee-colored, acquiring their color from acids

Mixing of the waters Two tributaries of the Amazon meet in southern Peru. The Ucayali, the larger of the two rivers shown here, is one of seven tributaries more than 1,600 km (1,000 mi) long. "Black water" and "white water" rivers drain geologically distinct areas.

THE BURIAL OF ARMERO

On 13 November 1985 a mixture of hot lava, ash, melted snow and ice flowed down a valley in western Colombia at incredible speed, wiping out much of Armero and half its population.

Armero, a town of about 22,000 inhabitants, lay in the valley of a tributary of the Magdalena river, which runs north parallel to the Cordillera Oriental in Colombia. About 50 km (35 mi) to the west rises the ice-capped cone of Nevada del Ruiz, one of the highest volcanoes in the Andean chain.

During 1984 residents learned that their slumbering giant, dormant for many years, had started to smoke. A few recalled folk stories of an eruption in 1845 producing a mixture of mud and hot volcanic ash that swept down the valley at great speed. Geologists call such phenomena lahars.

At first nothing was done, but as signs of an eruption continued to be reported hazard warnings were issued; they were generally ignored. By summer 1985 it was clear to the authorities that a major volcanic event was imminent. In September a map of hazards was commissioned. The mountain continued to erupt.

On 12 November the map was ready. It showed that Armero lay in the zone of probable damage. But it was too late. Within 24 hours disaster had struck, and 25,000 people had died under the mud flows.

The aftermath of catastrophe Almost nothing remains of the busy town of Armero two days after the devastating eruption of Nevada del Ruiz in 1985. The eruption inundated the town, the flood of mud and volcanic ash destroying virtually everything in its path and killing 25,000 people.

in the water derived from humus. They look black because they contain no sediment, which is a pale color, and therefore absorb rather than reflect light. These waters are chemically almost as pure as distilled water; they support hardly any life because they contain very few nutrients. Black waters such as those of the river Negro come chiefly from areas where prolonged weathering has resulted in soils poor in nutrients, and where the trees of the forest take out what few nutrients remain.

The white waters, by contrast, do carry sediment, are chemically rich and are full of animal life. When white waters flood the land they leave a thick layer of silt that provides nutrients for plants, whereas a black water flood washes out the few remaining minerals in the soil. There is a saying that people live on the black waters (which are largely free from insects) but farm on the white waters.

Large parts of the Plateau of Brazil, the Guiana Highlands and the hot wet lands of Amazonas (the middle and upper parts of the Amazon basin) have reddish laterite soils composed mainly of oxides of aluminum and iron, and quartz. Rainwater, which is chemically richer than the river water, loses its few minerals in the thick root mat of plants. Dying or burnt vegetation produces minerals, but these are quickly absorbed back again by the roots of neighboring plants. (Laterite usually lies just beneath the surface. If it becomes exposed it hardens and turns into a chemically inert crust that is difficult to remove. It has been suggested that removal of the forest will expose and harden the laterite and turn Amazonas into one huge, hard desert.) When water from the forest does have a high mineral content in the ancient shield areas, it is usually a sign that natural conditions have been disturbed by human activity.

White waters such as those of the upper Amazon river itself originate where the rocks and soils are rich in minerals, including feldspars and micas. They rise in the Andes, or in basins where the newer, soft rocks have not been exposed to so much weathering and where soil erosion and landsliding are common. The many nutrients in the water are washed into the rivers, which teem with life.

River flows on mountains and plains

The flow, or discharge, of the large rivers is controlled mainly by seasonal changes, often in the area of the headwaters. The Amazon has an enormous seasonal change in height; because of this, during the rubber boom of the last century a floating dock was built at Manaus to take ocean-going ships. The Negro flows into the Amazon; it is also connected to the Orinoco river to the north by the unique Casiquiare Canal, a 220 km (137 mi) river joining two great river basins.

The amount of rain that falls locally within the basin has little impact on such big rivers. By contrast, small streams in the forest rise almost immediately after a rainstorm because the soil is already saturated. As forest is cleared there are likely to be major changes to the seasonal and short-term responses of rivers to rainfall, because the forest will not be there to absorb the water, reduce runoff or hold the soil in place.

In the Andes the flow of the rivers is controlled by melting snow as well as by rainfall. Here the steep mountain slopes mean that avalanches, floods and mudflows are a constant danger. In 1877, for example, the eruption of Cotopaxi in Ecuador set off a mudflow that removed everything in its path for 240 km (150 mi). In 1970 near Huascarán in Peru, following an earthquake a great landslide swept into a valley, killing some 20,000 people.

The world's longest mountain chain

From the sunshine of the Caribbean to the icy waters off Cape Horn, the Andes extend over nearly 70 degrees of latitude and a distance of more than 7,250 km (4,500 mi). They form the world's longest unbroken mountain system.

The rocks of the Andes have been thrust upward and folded by the eastward movement of the Nazca plate. The process is still active, so the mountains are still being formed. The mountains occupy only a narrow zone, and there are rapid changes in altitude over short distances. From the offshore Peru–Chile trench to South America's highest peak, Aconcagua, the rise is 14,000 m (46,000 ft) – in just over 150 km (93 mi). These great changes of altitude cause processes such as erosion and landslides to act rapidly, simply because gravity is a relatively strong influence.

The Andes form part of the "Ring of Fire", a great arc of volcanoes that surrounds the Pacific. There are active volcanoes along most of the length of the Andes, especially in the northern section in Peru and Ecuador. They include the world's highest active volcano, Cotopaxi (5,896 m/19,344 ft) in Ecuador, Peru's highest peak, Huascarán, and Llullaillaco on the Chile–Argentina border.

Crossing the Andes

Changing altitude can have the same effect on vegetation as changing latitude if the length of the growing season is affected. Latitudinal changes are gradual, but in the Andes the changes in vegetation can be dramatic: over only 20 km (12 mi) there may be a transition from tropical rainforest in the Amazon basin, through drier oak forest at 1,000 m (3,300 ft), dwarf forest and scrub (2,500–3,500 m/8,200–11,500 ft) and the alpine bog vegetation of the *páramo*, to frozen wilderness above 4,500 m (14,750 ft).

Rainfall on the eastern slopes facing the Amazon basin may reach 5,000 mm (200 in) a year. The interior basins and gorges receive only between 500 and 1,000 mm (20–40 in) of rain a year, while the western coasts of Peru and northern Chile, though often blanketed in low cloud, receive virtually no rain at all. Farther south the western slopes have a Mediterranean climate, with hot, dry summers and warm, moist winters because the westerly wind belt moves north with the overhead sun. Farther south still, the westerlies bring rain throughout the year.

As rainfall decreases at lower altitudes, so temperatures rise. In the Peruvian Andes Vincocaya at 4,300 m (14,100 ft) has an annual average temperature of 1.9°C (35.4°F); in Cuzco at 3,500 m (11,500 ft) it is 10.7°C (51.3°F) and in Arequipa at 2,700 m (8,860 ft) it is 13.8°C (56.8°F).

In the far north the Cordillera Oriental forks and is divided by an inlet of the sea, Lake Maracaibo. Farther south the chain is simpler but also wider, at up to 600 km (370 mi). The Cordillera Oriental and Cordillera Occidental are separated by the high plateau, known as *altiplano* near Lake Titicaca and as *puña* to the south. Farther south still, the range becomes one, separated from the sea by much lower coastal ranges and the central valley of Chile.

Snow, glaciers and thin air

The Andes are narrowest in Tierra del Fuego in the far south, where they are penetrated by deep fjords. Here the permanent snowline is almost at sea level because of the extreme cold and the high precipitation brought by the Roaring Forties, the prevailing westerly winds. In the drier zone around 15–30°S, where the Pacific high pressure anticyclone is

dominant, the snowline rises to 6,700m (22,000 ft) on Llullaillaco. To the north the snowline falls again, to about 4,500 m (14,750 ft). Apart from the those in the far south, glaciers are scattered and small. During the last glaciation, which ended some 10,000 years ago, the Andean glaciers were much more extensive; they are now still smaller than they were 400 years ago.

The lower atmospheric pressure of the high Andes makes breathing difficult for visitors; local inhabitants – the guanaco, llama and alpaca as well as humans – have adapted to the shortage of oxygen.

GREAT MOUNTAIN RANGES

Asia has by far the largest share of great mountain ranges over 1,000 kilometers in length, an arbitrary limit that excludes, for example, the Alaska Range, the Alps and the Hindu Kush. Many ranges are parts of one major zone of mountain-building, such as the Andes and Rockies in the east Pacific's "Ring of Fire", or the system that extends from the Atlas and Alps to the Himalayas and into the islands of Southeast Asia.

	Approximate length km	mi
Andes, South America	7,250	4,500
Rockies, North America	4,800	3,000
Great Dividing Range, Australia	3,700	2,300
Trans-Antarctic Range, Antartica	3,200	2,000
Himalayas, China/India	2,500	1,500
Urals, Soviet Union	2,000	1,200
Atlas, North Africa	2,000	1,200
Altai, China/Soviet Union	1,600	1,000
Kunlun Shan, China	1,600	1,000
Carpathians, Europe	1,500	930
Tien Shan, China/Soviet Union	1,300	800
Caucasus, Soviet Union	1,200	750

The Cordillera Oriental or eastern range of the Andes in southern Peru forms the watershed between the eastward-flowing tributaries of the Amazon and rivers flowing south into the inland drainage basin that feeds Lake Titicaca.

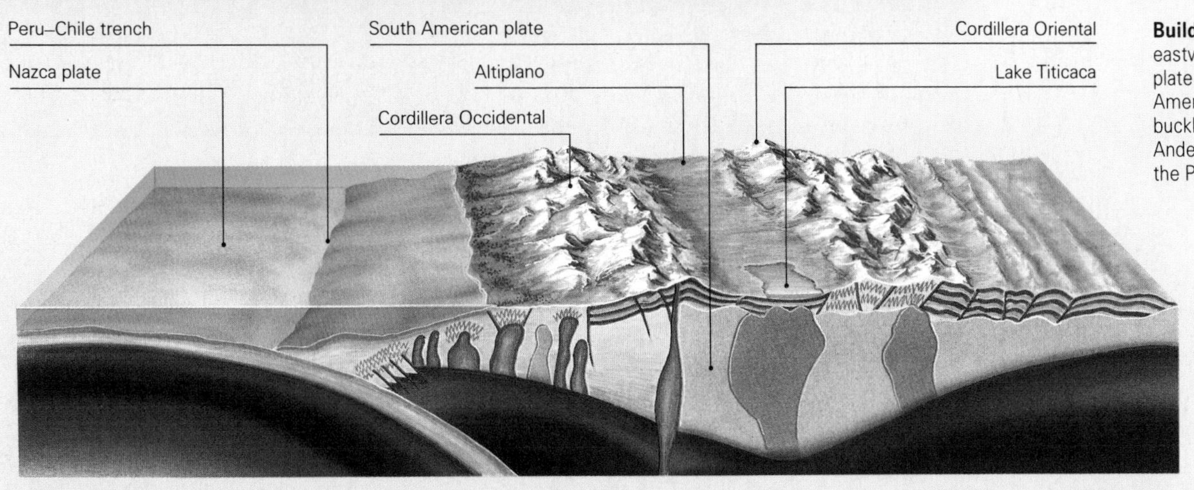

Building the mountains The eastward movement of the Nazca plate forces it beneath the South American plate, whose rocks are buckled, and pushed up to form the Andes. Where the two plates meet, the Peru–Chile trench is formed.

ICE-SCULPTED LANDSCAPES

LANDS OF THE MIDNIGHT SUN · ICE ON THE LAND · ROCKS AND LANDSCAPE

No region of the world has so many people living at such a high latitude as the Nordic countries. Warmed by the North Atlantic Drift current, the climate along the western coast of Norway is surprisingly mild. In contrast, the landscape is harsh, with snowcapped mountains, waterfalls, and many deep steep-sided fjords. Forested mountain slopes stretch up the spine of Norway and Sweden to the fells and plains of Lapland, where winters are long and dark, the subsoil is permanently frozen and the surface typically treeless tundra with peat bog. To the east the scenery is dominated by forests and lakes; only in the south of the region are there extensive areas of fertile lowlands. Some 900 km (570 mi) to the west, in the Atlantic Ocean, Iceland lies on the Mid-Atlantic ridge, where volcanic activity is still creating new land.

COUNTRIES IN THE REGION

Denmark, Finland, Iceland, Norway, Sweden

LAND

Area 1,255,017 sq km (484,437 sq mi)
Highest point Glittertind, 2,470 m (8,104 ft)
Lowest point sea level
Major features islands, fjords, mountains and high plateau in west, lakelands east and west of Gulf of Bothnia, lowlands in south

WATER

Longest river Göta–Klar, 720 km (477 mi)
Largest basin Kemi, 51,000 sq km (20,000 sq mi)
Highest average flow Kemi, 534 cu m/sec (19,000 cu ft/sec)
Largest lake Vänern, 5,390 sq km (2,080 sq mi)

CLIMATE

	Temperature °C (°F) January	July	Altitude m (ft)
Bergen	2 (36)	15 (59)	44 (144)
Oslo	−5 (23)	17 (63)	96 (315)
Stockholm	−3 (27)	18 (64)	11 (36)
Helsinki	−7 (19)	17 (63)	58 (190)
Reykjavik	0 (32)	11 (52)	126 (413)

	Precipitation mm (in) January	July	Year
Bergen	179 (7.1)	141 (5.6)	1,958 (77.1)
Oslo	49 (1.9)	84 (3.3)	740 (29.1)
Stockholm	43 (1.7)	61 (2.4)	555 (21.9)
Helsinki	49 (1.9)	68 (2.7)	641 (25.2)
Reykjavik	90 (3.5)	48 (1.9)	805 (31.7)

NATURAL HAZARDS

Cold, glacier surges, volcanic eruptions in Iceland

LANDS OF THE MIDNIGHT SUN

Almost all of the Nordic region lies north of latitude 55°N, and one third lies within the Arctic Circle (66° 32′N). More than 20 million people live in the region, where long winter nights are matched by long summer days. The Soviet Union is the only other country to have substantial numbers of people living at this latitude; in North America there are no major cities north of 55°N.

What makes the Nordic countries more populated than might be expected is their surprisingly mild climate. The warm waters of the North Atlantic Drift are blown toward Scandinavia by westerly winds, which absorb some of the heat and carry it inland. Their moderating influence is felt most strongly on the southwest coast of Norway. Moving to the north and east, continental influences become stronger. Winters grow longer and colder and summers shorter but warmer. Extremes are greatest in the east. In Finland (not part of Scandinavia, though a Nordic country) nowhere has an average January temperature greater than 0°C (32°F), but in July it can be as high as 30°C (86°F). The mountain climates of Norway and Sweden are also much harsher, with long, cold winters and permanent snow cover over 1,000 m (3,300 ft).

Over two-thirds of the region is susceptible to summer frosts, which are an ever present threat to agriculture. Growing seasons are correspondingly short, from 180 days in western Norway to less than 130 days in Lapland.

The westerly winds bring a lot of precipitation to the west coast of Norway, where the mountain barrier forces the warm air to rise and drop its moisture. Most areas by the sea receive at least 1,000 mm (40 in) a year, but higher up 4,000 mm (160 in) is common. To the northeast, in the rainshadow of the mountains, much less rain falls. In Lapland and at the head of the Gulf of Bothnia annual rainfall is less than 500 mm (20 in) on average. Much of the precipitation falls as snow. On the mainland, only in Denmark does snow lie for less than 40 days a year.

Where winter shapes the land

Scandinavia and Finland were covered by ice during the last glaciation, which ended only about 10,000 years ago. The land is slowly rising as it recovers after the

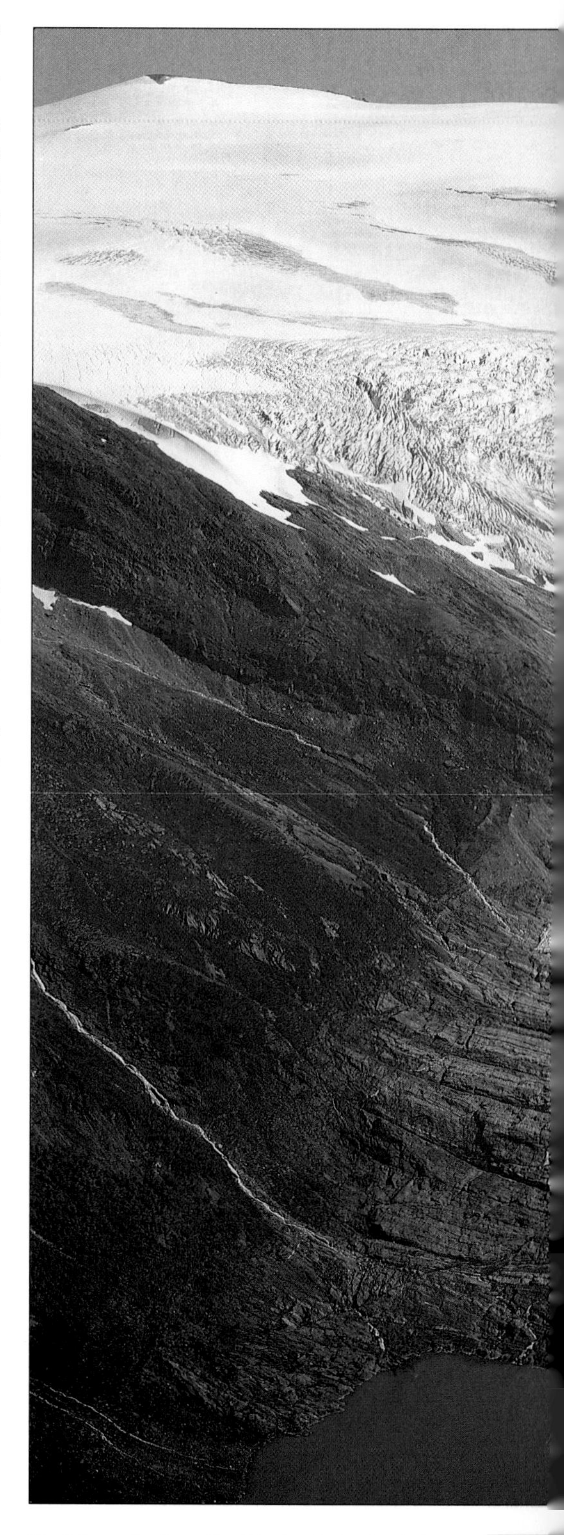

Glacial erosion in action The ice that extended over much of Scandinavia in the past has left its traces in many parts of the region. The valley of Svartisen just north of the Arctic Circle was once completely obscured by the ice, but is now occupied by only a small glacier.

Fertile plains in Denmark Southern Sweden and Denmark are different from the rest of the region because there are large areas of fertile lowland. These flat landscapes are the result of deposition rather than erosion, and with their milder, drier climate are intensively cultivated.

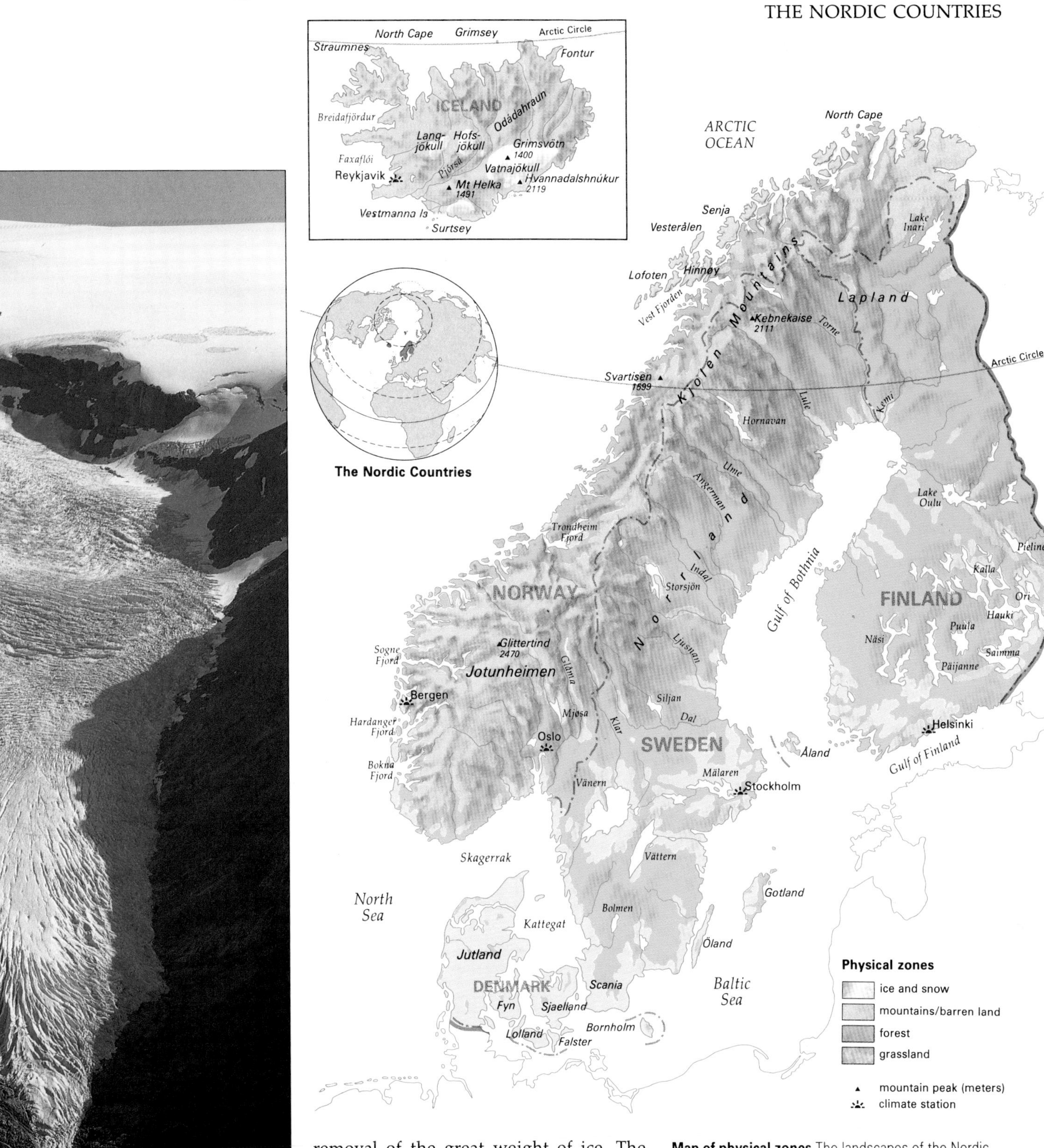

Map of physical zones The landscapes of the Nordic countries have been shaped by ice. Its influence can be seen everywhere – in the backbone of mountains that extends along the peninsula, in the tundra areas of Lapland, and in the fertile lowland plains of the south.

removal of the great weight of ice. The scenery of the region shows the influence of the ice and other landscape processes associated with cold climates. Rocks have been scraped bare and scored by the ice, the soils have had little time to develop so they are thin, and to the south there are moraines formed from debris deposited by the ice.

The western coast of Norway is very rugged, with bare uplands, permanently snowcapped mountains, long, deep fjords carved by glaciers, and over 150,000 rocky offshore islands. The coastline is so indented that if it were straightened out it would stretch halfway around the globe.

In the north of the region the undulating fells and plains of Lapland are largely tundra. Peat bogs form on the flat areas where the surface melts in summer, but the water cannot drain away because of the permanently frozen ground below. Forests of pine, spruce and birch cover much of the land farther to the south, and lakes are common. The lowlands of Denmark and southern Sweden have deeper soils and a milder climate.

ICE ON THE LAND

Winter still dominates the climate of the Nordic countries, but in the recent geological past its effects were much greater. During the ice ages of the past 2 million years the region has been in the freezing grip of a glacial period, or glaciation, on at least three major occasions.

At these times the whole region, including what is now sea, was covered by ice sheets and glaciers, possibly up to a depth of 3,000 m (10,000 ft). Glaciers in armchair-like rock basins (cirques) in the mountains fed into valley glaciers, which spread out at the foot of the mountains and joined with others to form slow-moving piedmont glaciers and eventually huge ice sheets. The advance and retreat of the ice cover has had a profound effect on the scenery. In some areas the ice has shaped the landscape by removing material from it (erosion), in others by adding to it (deposition).

Spectacular steep-sided valleys along the west coast of Norway extend from the interior plateau to the sea. These fjords were filled by thick glaciers whose bases eroded deep basins. When the ice retreated the valleys were flooded by the sea. The Aurland Fjord shown here is a tributary of the Sogne Fjord.

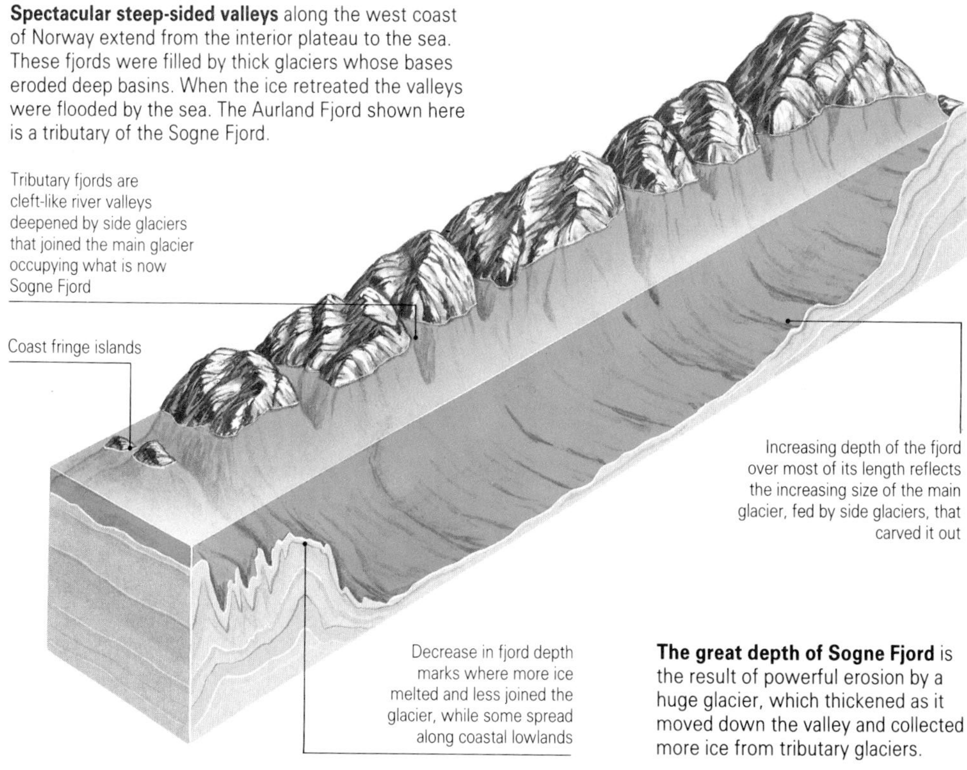

The great depth of Sogne Fjord is the result of powerful erosion by a huge glacier, which thickened as it moved down the valley and collected more ice from tributary glaciers.

Landforms eroded by the ice

As the ice moved across the land it scraped the surface bare of loose rocks and soil, so today soils in these areas are thin. The landforms produced range from small grooves and scratches (striations) where the bedrock has been scraped by glaciers with rocks embedded in them, to larger features such as the unevenly eroded asymmetrical mounds of rocks known as *roches moutonnées*.

The most spectacular features are where glaciers have eroded valleys in upland areas, making them deeper, broader and straighter. A broad U-shaped valley with tributary valleys hanging above it is classic glacial scenery, found widely in the mountain areas of Norway and Sweden. The fjords along the west coast of Norway are an extreme form of this erosion; they have been carved out over 150,000 years of glacial and interglacial periods. The deepest part of a fjord is well below sea level today; there is a shallow sill where the fjord meets the sea. When the ice retreated from the valleys the sea was able to invade them, creating a highly indented coastline. Sogne Fjord, the deepest, plummets to as much as 1,308 m (4,291 ft) below sea level, and extends inland for 204 km (127 mi).

Deposits on the land

Material eroded by the ice from the mountains and uplands of Norway and Sweden was deposited in lower-lying areas to the south and southeast. The movement of sediment took place on a huge scale, and in Denmark the deposits are typically between 30 and 50 m (100–160 ft) thick.

Glacial sediments are of two types. Those deposited directly by the ice contain angular fragments (called till), while material laid down by water from the melting ice is rolled along the streambed and becomes rounded, like gravel.

Moraines are composed of material deposited by ice. At the maximum limits of the ice and at points where it paused in its retreat, huge mounds or ridges of glacial rock debris were deposited. These terminal moraines cover a large part of southern Finland. In places, moraines that have been covered by the sea at some time since they were deposited have developed a rippled appearance (washboard moraines). Another common feature produced by glacial deposition are egg-shaped mounds. These can be some 400 m (1,300 ft) long and 20 m (66 ft) high, and are called drumlins.

In Sweden and Finland, water-worn sand and gravels form very distinctive ridges called eskers. They wind across the landscape for up to 300 km (180 mi) and can reach a height of 50 m (160 ft). They were laid down by streams running over, through or under glaciers. As long as the ice remained, the sand and gravel streambed was supported by its ice walls. When the ice eventually melted the deposits collapsed and formed ridges. The ridges become wider in places, marking where a delta formed at the mouth of a stream channel as it left the ice.

The Arctic fall Although it is well north of the Arctic Circle, northern Norway has a mild climate because of the influence of the North Atlantic Drift. The sea does not freeze in winter, and the growing season is long enough for deciduous silver birch trees to flourish.

A land of lakes

Sweden and Finland are well known for their lakes. Sweden has 90,000 and Finland 55,000 of them. In the upland areas of northern Sweden the lakes are long and narrow, occupying glacial valleys that have been blocked by moraines. In southern Sweden the lakes are remnants from a time when the area was much lower and completely flooded. The land is still rising as it readjusts following the removal of the great weight of ice that covered it. The many lakes of Finland fill shallow depressions that were made by the ice as it moved over the low shield of ancient rocks. At the end of each cold period of the recent ice age, the ice stopped moving and began to melt. As it did so it split into blocks, and the gaps between them filled with sediment. Once the block was completely melted the sediment around its edges formed the rim to a depression known as a kettle hole. Many of these have filled with water.

ECOLOGY AND CLIMATE IN BALANCE

The forest environments of northern Scandinavia are under threat. Some areas of birch forest on the Lapland plateau have been damaged to such an extent that new, treeless tundra areas may ultimately be formed. The culprits are the larvae of the geometrid moth *Epirrita autumnata*. In its voracious caterpillar stage it strips the leaves of the trees so thoroughly that in one area of 13,000 sq km (5,000 sq mi), 1,200 sq km (460 sq mi) of birch forest were affected.

The damage seems to follow years with a particular combination of climatic characteristics. The summer of 1964 was wet; that of 1965 both wet and unusually cold. The trees were defoliated by the caterpillars in the poor summer of 1965, and consequently lost all their reserves of energy. The following spring they had no buds, and no nourishment for new growth. The same poor summer also reduced the number of parasites of *Epirrita*, so the caterpillars flourished.

ROCKS AND LANDSCAPE

There are three major groups of rock in the Nordic countries. They are the ancient rocks of the shield areas, the old sedimentary rocks of the western mountain ranges (often metamorphosed and mixed with old shield rocks and igneous intrusions), and the younger sediments in the south.

Underlying much of both Sweden and Finland is the Baltic shield, with rocks generally more than 600 million years old. The shield includes metamorphosed rocks, that at 3,000 million years old are among the oldest in the world. There is a great variety of rock types, including granite, slate and other rocks that are very resistant. Evidence has been found of at least one ice age, which occurred some 600 million years ago.

The mountain range along the northwestern edge of the shield comprises old sedimentary rocks and igneous intrusions that were lifted up some 400–350 million years ago during the Devonian period. The folding was so intense that deep down the rocks were metamorphosed; with subsequent erosion, these are now often found at the surface. This area has been so modified by erosion and deposition that the shape of the surface today is completely unlike that of the original mountains, and is but a remnant from this mountain-building period, which affected Scandinavia and also northern parts of the British Isles.

The much younger sediments in Denmark and the south of Sweden (Scania) include sandstone, shale and mudstone and occasionally limestone. These most recent sedimentary rocks, themselves more than 60 million years old, are largely covered by deposits of sand, gravel and clay dating from the last glacial period. Iceland came into being in the Tertiary period (65–2 million years ago), the result of volcanic activity that continues today.

Scenic variety

The types of scenery in the region reflect the contrasts between the extremely old geological foundation and the most recent deposits of the past 2 million years, which often lie directly on top of them.

The western mountains occupy much of Norway and of northwestern Sweden. The *fjell* of Norway (*fjall* in Sweden) is a high plateau above the treeline with isolated mountain peaks and deep, steep-sided valleys. The highest peak, Glittertind, rises to 2,470 m (8,104 ft) and is fretted by the effects of glacial erosion. The deep valleys running from the plateau form fjords along the western coast. Parallel to it is a string of some 150,000 rocky islands or skerries, forming a natural breakwater known as the skerryguard. Some of the inner islands rise to over 1,000 m (3,300 ft).

To the southeast, a plateau extends from southern Norway through much of the spine of Sweden into northern Finland. It has a foundation of ancient shield rocks, overlain in part by glacial deposits. The Swedish part of the plateau, Norrland, lies between 200 and 500 m (660 and 1,600 ft) above sea level. It is composed in the north of plains with only isolated

ACID RAIN AND ACID ROCKS

In northern Europe precipitation (rain, sleet, snow and mist) has become between 10 and 80 times more acid over the past 30 years. (The popular term acid rain, excluding as it does some forms of precipitation, is not strictly accurate.) The culprits are the sulfur and nitrogen compounds emitted by fossil-fueled power stations and vehicle exhausts. These can be carried in the atmosphere for hundreds of kilometers before returning to the surface as weak forms of sulfuric and nitric acids.

Other parts of the world are affected by acid rain, among them northeastern North America, the Soviet Union and Japan. But the Nordic countries are particularly vulnerable to acid precipitation. Weather systems bring acid pollution from neighboring industrialized countries of Western Europe. The Precambrian rocks found near or at the surface contain no lime or other buffering chemicals to neutralize the acid. As a result water running off or percolating through the surface remains acid and makes the environment more acid. Of Sweden's 90,000 medium- to large-sized lakes, 18,000 have become between ten and a hundred times more acid, and 4,000 of them can no longer provide a habitat for the usual range of species. Large areas of forest have been damaged, and many trees are dying because they are weakened and less able to resist disease and attacks from insects.

The problems are made worse because acid water moving through the soil releases poisonous heavy metals such as cadmium, mercury and aluminum that are normally stable. They are taken up by plants, or find their way into watercourses where they damage plants and animals. Fewer nutrients are available to plants as they are also washed out of the soil.

The water moving through the environment is generally most acid in spring, when the snow melts. This delivers a deadly acid surge just at the time that young plants and animals are most vulnerable.

hills; farther south the scenery grades from open mountainside, through plateau country dissected by very active river valleys, to coastal plain.

The lowlands of central Sweden are separated from the northern uplands by a massive fault. The landscape of this lake district has been created by the splitting and faulting of the rocks, and the shapes of large lakes follow the pattern of these tectonic lines. There is a similar landscape in the southern part of Finland.

The uplands in southern Sweden are a gently tilted plateau of older rocks about 100 m (330 ft) above sea level and overlain by extensive glacial deposits. The lowland areas to the south occupy the whole of Denmark and the southwestern corner of Sweden.

Two types of landscape dominate Jutland (mainland Denmark). In the west much of the landscape is underlain by old glacial deposits, with intervening plains of gravel, sand and silt broken up by the old meltwater channels. Its sand beaches are the largest in Europe. The rest of Jutland has recent moraines forming a low hill country, and flat areas where the best farmland is found. Lakes fill depressions, while old tunnel valleys from the most recent period of glaciation cut through the countryside.

Scenery on the islands to the east of Jutland echoes these landforms, with sand dunes and marshes on the coasts and flats and other glacial features inland. An exception is the island of Bornholm off the south coast of Sweden, which is underlain by granite with a thin cover of glacial deposits.

The Finnish lake plateau averages about 100 m (330 ft) above sea level, and is scattered with innumerable lakes. In Finland one-tenth of the land surface is covered by lakes, and in the southeast lake water occupies more than a quarter of the surface. In the north and toward the coast more than three-fifths of the land surface consists of peat bogs. This landscape is very new. Since the ice receded there has not been enough time for a drainage system to become established, so there is much surface water, conditions in which lakes and bogs thrive. The Bothnian lowlands, inland from the shores of the Gulf of Bothnia, have also been affected by a rise in the level of the land, and old beachlines can be traced on the low-lying plain.

By contrast, the Faeroe Islands are underlain by basalt, while Iceland is mainly a bowl-shaped highland with coastal mountains and plateaus 500–800 m (1,650–2,600 ft) high in the interior.

The central spine of mountains that separate Sweden from Norway include those of the Sarek National Park, one of the largest wilderness areas in western Europe. The mountains were scraped bare of soil, and their contours rounded and smoothed, by the ice sheet during the last glacial period, at its height some 18,000 years ago. This area has been protected because of its outstanding natural beauty.

The tundra landscape of Stekenjokkfjall in southern Lapland at the height of summer. Lakes have been formed by the melting snow and ice, as the water cannot drain away through the permanently frozen ground beneath the surface. There is a flush of growth as the shrubs and grasses make the most of the short summer season.

The rising land

In 1621 Erik Sorolainen, bishop of the port of Turku in southwest Finland, predicted that the Day of Judgment was at hand. He cited as evidence the fact that the land was not stable after all, but rising. All round the shores of the Gulf of Bothnia harbor towns were rising and ports becoming unusable. At Pori, 120 km (75 mi) north of Turku, the harbor had to be moved seaward six times in as many centuries because it silted up as rivers cut through the rising land. At Vaasa, farther north, at the narrow "waist" of the gulf, where the land is still rising 87 cm (34 in) every hundred years, the port had to be moved to Vaskiluoto. Each year Finland gains 10 sq km (4 sq mi) of territory from the sea.

Like the Canadian Shield around Hudson Bay, the land around the Gulf of Bothnia is rising as a consequence of the melting of a great ice sheet that covered the land in the most recent glacial period. A depth of 3 m (10 ft) of ice has sufficient weight to depress the underlying land by 1 m (3.3 ft). It has been calculated that 2,000 m (6,500 ft) of ice depressed the land of Nordic countries by 600 m (2,000 ft).

After the ice sheet melted the land began to rise, at first by as much as 9 cm (3.5 in) a year in some areas. It can take between 15,000 and 20,000 years for the land to regain its equilibrium (isostasy). This readjustment began to take place 13,000 years ago. The rate at which the land is rising has declined ever since, but at the head of the Gulf of Bothnia it is still rising at 1 cm (0.4 in) a year.

It is not easy to distinguish the rise of the land from the effects of the rise of the sea (eustasy), but a series of stages has been identified using the evidence of former shorelines. Throughout Scandinavia there are shoreline features at levels often well above the present sea level. These old shorelines have beach ridges with remains of mollusk shells, driftwood and peat.

A history of the Baltic Sea

Because of the pattern of drainage, with the watershed in the mountains near the west coast of Norway, most of the water released as the Scandinavian ice sheet melted drained southeast into the Baltic. The Danish islands and the peninsula of Jutland created a dam across the Baltic Sea. Combined with rises in the land and in the sea level this produced a sequence of lakes and seas that were the ancestors of the present Baltic Sea.

Mollusk shells found in the high-level deposits have provided particularly important markers in reconstructing the sequence. *Yoldia arctica,* a saltwater species, *Ancylus,* a freshwater snail, and *Littorina,* a saltwater snail, have each given their name to one of the stages.

The freshwater Baltic Ice Lake formed between 10,500 and 10,200 years ago and was initially separated from the North Sea. At that time the ice sheet still covered a large area, and the rate at which the land rose was not great. When the ice sheet began to melt more quickly, the volume of meltwater in the Baltic Ice Lake basin increased and gradually overflowed into the North Sea through central Sweden.

This marked the beginning of the Yoldia stage, which lasted from 10,000 to about 9,500 years ago. The land was now rising at its fastest, and more rapidly than the rise in the level of the sea, so the Yoldia Sea was succeeded by the inland Ancylus Lake, which lasted until nearly 7,500 years ago. The Littorina Sea, the immediate ancestor of today's Baltic Sea, was formed when the sea flooded through the channels that had been created by meltwater in what is now Denmark.

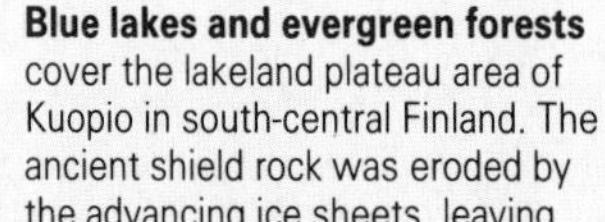

Blue lakes and evergreen forests cover the lakeland plateau area of Kuopio in south-central Finland. The ancient shield rock was eroded by the advancing ice sheets, leaving hollows and glacial deposits when they retreated. Since then the land has slowly risen; it now rises at a rate of about 1 cm (0.4 in) a year.

The Lofoten island group off the northwest coast of Norway was one of the first areas to be exposed when the ice sheets began to melt. The coastline has changed since then, as the sea level has risen and more of the once depressed land has gradually emerged.

The retreating ice The Scandinavian ice sheet was at its greatest extent about 18,000 years ago, when it covered northern Europe. By 12,000 BC the ice was retreating; 10,000 years ago the northwest coasts and the southern shores of the ice-blocked Baltic Ice Lake were exposed. Further melting led to the lake joining the North Sea. As the land continued to rise following the retreat of the ice the Ancylus Lake was formed. By this time the only remains of the ice sheet lay in northern Norway. As the land recovers from the weight of the ice that covered it for thousands of years during the recent ice age, its outlines gradually change.

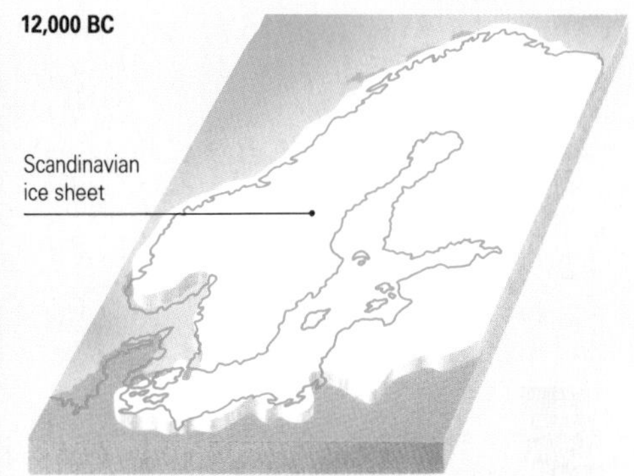

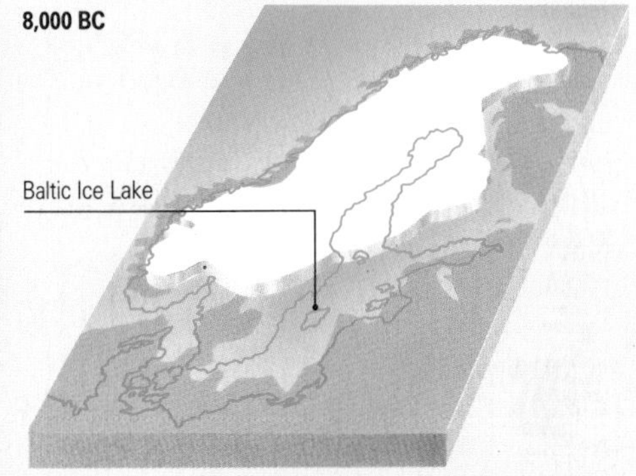

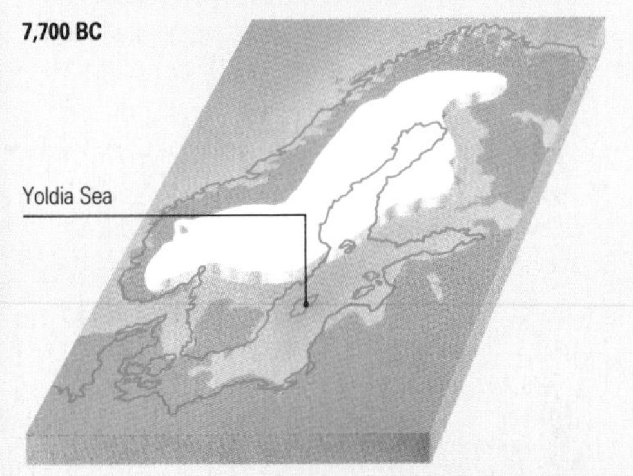

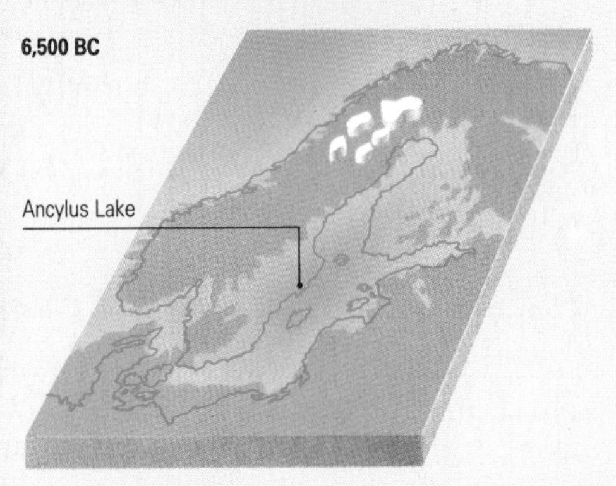

Land of ice and fire

The icy landscape of Iceland's Vatnajökull (*vatna*=water, *jökull*=glacier) belies the country's volcanic heart. This active volcanic island with its cold subarctic climate has a unique blend of landscapes.

Vatnajökull is Europe's second largest ice sheet. It stretches over 8,538 sq km (3,296 sq mi) in the southeast of Iceland. In places it rises to over 2,000 m (6,560 ft), and the ice can be as much as 1,000 m (3,300 ft) thick. Volcanic hotspots under the ice sheet cause the ice to melt, creating large lakes and rivers. Occasionally, the volcanoes erupt and send streams of water, ice and boulders cascading over the surrounding landscape. These floods are known as glacier runs or *jökulhlaupav*. Iceland is well known for its thermal springs and geysers, which are often found side by side with the glaciers themselves. They provide a source of heat for the Icelanders in a country that has no other natural sources of energy to combat the cold conditions.

Volcanic activity is associated with Iceland's location on the northern edge of the Mid-Atlantic ridge. Along the ridge new crust wells up as the North American and Eurasian plates of the Earth's surface layer – the lithosphere – move apart, a process known as sea-floor spreading. Occasionally this activity forms new land above sea level. Iceland was formed in this way, and new islands are still appearing. As recently as 1963 the island of Surtsey was thrust out of the sea off the south coast of Iceland in a fiery eruption.

The frozen landscape of Vatnajökull in Iceland covers a fiery, volcanic subterranean world.

NORTHWESTERN ISLES

A COMPLEX GEOLOGICAL HISTORY · WEATHER, WATER AND SOIL
ANCIENT UPLANDS AND SOUTHERN LOWLANDS

The long, indented coastlines of the British Isles encompass an unusually wide variety of rock types and landscapes. The high ground and rocky mountains in the north and west are formed from old rocks, and the landscapes show features of glacial erosion and deposition – reminders of the advance and retreat of the ice sheets that covered much of the region in the last ice age. The south and east have a younger, gentler terrain of sedimentary rocks folded and later eroded into the ridges and vales seen today. Always variable and difficult to forecast, the weather can be summed up as wet and windy. The west receives the worst of it from the prevailing southwesterly winds. The surrounding seas, and particularly the North Atlantic Drift, ensure that the British Isles have a relatively mild climate for this latitude.

COUNTRIES IN THE REGION

Ireland, United Kingdom

LAND

Area 314,385 sq km (121,353 sq mi)
Highest point Ben Nevis, 1,344 m (4,408 ft)
Lowest point Holme Fen, Great Ouse, –3 m (–9 ft)
Major features mountains chiefly in northern and western areas, with lower-lying areas in east and south

WATER

Longest river Shannon, 370 km (230 mi)
Largest basin Severn, 21,000 sq km (8,000 sq mi)
Highest average flow Shannon, 198 cu m/sec (7,600 cu ft/sec)
Largest lake Neagh, 400 sq km (150 sq mi)

CLIMATE

	Temperature °C (°F) January	July	Altitude m (ft)
Aberdeen	2 (36)	14 (57)	59 (194)
Dublin	5 (41)	15 (59)	81 (266)
Valentia	7 (45)	15 (59)	14 (46)
Kew	4 (39)	18 (64)	5 (16)
Plymouth	6 (43)	16 (61)	27 (89)

	Precipitation mm (in) January	July	Year
Aberdeen	77 (3.0)	92 (3.6)	837 (33.0)
Dublin	70 (2.8)	70 (2.8)	758 (29.8)
Valentia	164 (6.5)	107 (4.2)	1,398 (55.0)
Kew	53 (2.1)	56 (2.2)	594 (23.4)
Plymouth	105 (4.1)	71 (2.8)	990 (39.0)

NATURAL HAZARDS

Storms, floods

A COMPLEX GEOLOGICAL HISTORY

The British Isles – consisting of two large and numerous smaller islands off northwest Europe – cover a tiny proportion (less than a quarter of 1 percent) of the world's land surface. During the region's complex geological history continents have parted and come together and mountains have been formed and eroded, not once but several times, accounting for some of the most varied landscapes on Earth. It is most unusual for the remnants of so many geological events to be found so close together.

Some of the oldest rocks in the British Isles are found in the far northwest, southeast Ireland and the Channel Islands off the coast of France. They are over 2,500 million years old. The sediments that were later laid down in an early ocean, Iapetus, cover a larger area. Present-day Scandinavia, southern Britain, Ireland and the eastern maritime provinces of Canada seem to have lain close together on the southern margins of this ocean, which eventually contracted and closed. As it did so, some 400 million years ago, the sediments and also some volcanic rocks were deformed and pushed up to form the Caledonide mountain systems, running generally run southwest to northeast. As the mountains were eroded sandy sediment was deposited in nearby depressions, such as the central valley of Scotland in the north, or in shallow seas like the one that covered much of Britain's southwestern peninsula.

Gradually the sea spread and limestones formed during the Carboniferous period (360–286 million years ago). These are found in Ireland and make up the "mountain" limestones of the Pennines and Mendip Hills. Deltas of sandy sediments (now millstone grit) and organic swamp deposits (now the coal measures) were also formed. Another phase of land movements deformed rocks in southwest Britain and Ireland, producing a generally east–west alignment. Granite intruded into other rocks in the southwest, and minerals such as tin, lead, zinc and copper formed.

The emerging landscape

Since the Carboniferous period the British Isles have belonged geologically to Europe. At first desert-like conditions were present in the Triassic period (about 248–213 million years ago), and later extensive shallow seas with islands formed in the Jurassic period. About 180 million years ago a new ocean began to open up as the continents moved slowly apart. This is now the Atlantic Ocean. During this movement rocks were also folded and lifted up and there was some volcanic activity, evident in lava flows and dykes in northern Ireland and the North West Highlands.

A recognizable outline to the British landscape began to emerge with older rocks forming blocks and basins to the north and west, and younger sediments being laid down by rivers and in shallow seas to the south and east. However, to

A glaciated landscape The wild scenery of the Lake District was celebrated by 19th-century poets. It covers a small area of the Cumbrian Mountains of the northwest. Lakes such as Ullswater, shown here, fill steep-sided valleys formed by glaciation in the last 2 million years.

The low, rolling hills of the Lincolnshire Wolds look out toward the flat Fen lands on the edge of the North Sea. The Fens are alluvial areas laid down under the sea. When drained they provide fertile land, which is intensively farmed.

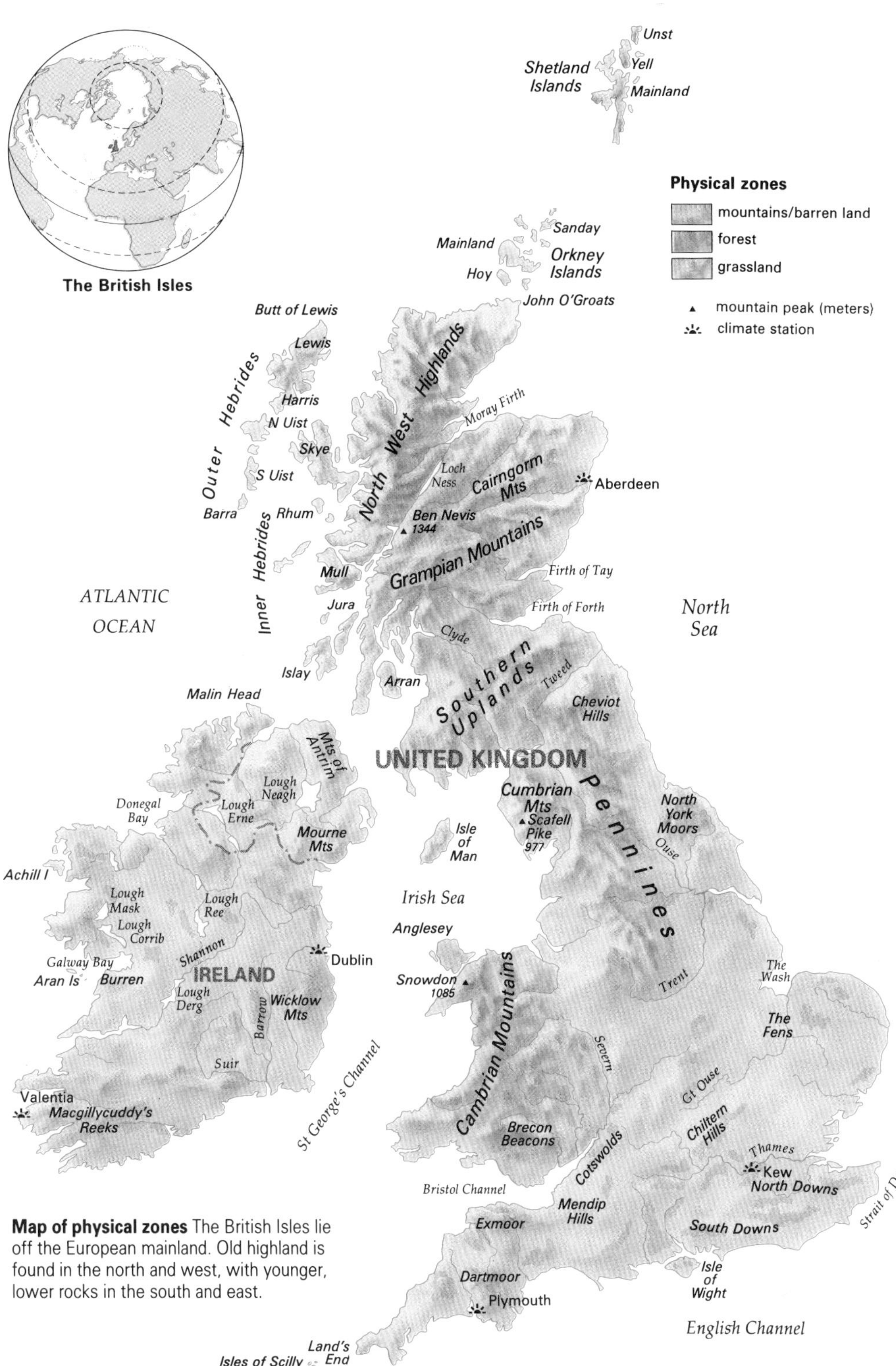

Map of physical zones The British Isles lie off the European mainland. Old highland is found in the north and west, with younger, lower rocks in the south and east.

the south of the region, between what is now Europe and Africa, another ocean, Tethys, was closing. Its sediments were being buckled in the Alpine mountain-building phase, causing faulting and gentle folding. Newer sediments in the southeast were pushed up to form the Weald, and down to form the London basin, now occupied by the river Thames.

Glacial Britain

In the last 2 million years climatic fluctuations have led to many advances and retreats of ice sheets across most of the British Isles. Glaciers eroded deep, U-shaped valleys in upland areas; the sediments carried by the ice were laid down either directly beneath the ice as it melted, or as glacial river deposits. The soils in the parts of the region affected by glaciation are derived from the skin of deposits laid down at this time.

Much of the landscape was shaped under cold climatic conditions rather than in the milder climates of the interglacial periods. The last cold phase was at its height about 18,000 years ago and ended about 10,000 years ago. It filled many upland valleys with ice. In the interglacial phase that has lasted since then the landscape has not changed dramatically, except as a result of human activity.

The British Isles' patchwork of rock types is the result of the sequence of landforming processes that have taken place: the building and destruction of mountains, the opening and closing of oceans, and changes in climate from arid conditions through to glacial. The older rocks are found mainly to the north and west. Here ocean sediments, volcanic rocks and desert sandstones have become set together in relatively rigid blocks. In the lowlands of the southeast newer sediments have been gently folded, and then eroded into an undulating landscape of ridges and vales in a succession of clay, limestones and sandstones.

WEATHER, WATER AND SOIL

The climate of the British Isles is more variable than that of other many regions because it is influenced by transient airflow patterns. Usually these dominate the weather for only a few days or weeks, though longer spells of consistent weather set in from time to time, such as very cold easterly airflow in winter.

In a normal year between 160 and 180 depressions, or low-pressure systems, tracking from west to east across the Atlantic Ocean, bring additional wind and rainfall. These are associated with well-known weather sequences involving frontal rainfall, sharp changes of temperature and shifting wind direction. Conditions change so fast that the exact weather pattern is extremely difficult to forecast accurately even with satellites and advanced computer technology.

The North Atlantic Drift, a continuation of the Gulf Stream current, moves up the east coast of North America and brings warm waters east across the Atlantic. Sea-surface temperatures are therefore relatively high.

A mild climate

The traveling weather systems and North Atlantic Drift establish a relatively mild climate for a region in these northerly latitudes. Weather systems cover only part of the region at any one time, and weather hazards are not very often a major problem. However, there are occasional exceptions.

A gale in October 1987 wreaked havoc in the southeast; at Lerwick in the Shetland Islands more than fifty gales a year have been recorded over a 30-year period. Coastal flooding resulting from a storm surge in the North Sea in 1953 led to a death toll of 300 and damage estimated at millions of pounds. Massive snowfalls can occasionally isolate communities, particularly in the west of the region. In eastern Britain thick fog may occur on 10 days or more in a year, and is the cause of some serious traffic accidents.

There are sharp regional contrasts in climates in the British Isles. The west is wetter and milder than the east. Most of Ireland has a yearly rainfall average of over 1,000 mm (40 in), parts of upland Britain over 2,000 mm (80 in), but parts of southeast England have less than 600 mm (24 in). Because of the latitudinal spread

Digging for peat Deposits of peat are found in the region's wet uplands and on poorly drained lowland in Ireland, the Fens and the Somerset Levels of the southwest peninsula (shown here). It was once widely used as a fuel in these areas. When the bogs are drained and cultivated the ground level may drop by several meters. The soils of the British Isles are locally varied: heavy soils on glacial clay deposits are waterlogged in winter and hardbaked in summer, sandy or gravelly rock produces stony, dry soils.

The Emerald Isle The greenness of Glanmore valley near Cork in southern Ireland is caused by high rainfall and mild temperatures.

(between 50° and 59°N), there is a marked difference between the north and the south in the number of daylight hours.

Climate is also affected by altitude. Temperatures become lower and wind speeds and rainfall higher as the land rises. This is partly because the rate at which air cools with altitude (the lapse rate) is steep in most of the air masses that affect the British Isles. A decrease of 6°C per 1,000 m (about 4.2°F per 1,000 ft) is regarded as normal. The cooler and wetter conditions of the uplands help to create their distinctive landscape of moorland soils, vegetation and blanket peat cover.

Water in the landscape

After rain falls, some of it evaporates. The rest percolates down as groundwater or joins rivers. In winter more rain falls than is lost through evaporation, and this is when river flows and groundwater replenishment are highest. When floods occur they may be on a broad regional scale, as in snowmelt floods after a severe winter, or happen very locally, as when a thunderstorm in summer affect, a small catchment area.

No river system in the British Isles is very large. The Shannon in Ireland is the longest river. On the mainland of Britain the Severn is the longest river in Britain and has the largest basin. In the north, the Tay, draining the wet highlands of Scotland, has the highest rate of flow. Many rivers in the uplands drain into lakes, the largest of which is Lough Neagh in Ireland. The Shannon has several major lakes on its course, which help to even out its flow.

Lakeland landscapes are most highly developed in the glaciated uplands of Scotland, north Wales and the Cumbrian Mountains (Lake District). Since the 19th century many of Britain's upland valleys have been dammed for public water supply or power generation, producing artificial lakes. The flow of rivers such as the Severn is much affected by regulation and the presence of reservoirs, reducing the severity of their floods.

Soils reflect the complex interaction between parent rocks and the influence of climate and vegetation. In general they developed under the mature deciduous woodlands that covered the islands after the last glacial period. Soils on valley floors can become waterlogged; on the hills they are thinner and more vulnerable to summer drought.

THE GREAT OCTOBER GALE

On the morning of 16 October 1987 southeast England awoke to a scene of devastation caused by the worst natural disaster to hit the area for over 250 years. During the night hurricane-force winds had battered the country for five hours, uprooting 15 million trees, killing 19 people and causing widespread damage. The cause was a depression that had tracked northeastward across the Atlantic accompanied by violent winds. These had been created by sharp differences of temperature and pressure between neighboring air masses moved by the rotation of the Earth.

The situation began to develop on 13 October as Hurricane Floyd reached the Florida coast from the Caribbean. It sent warm tropical air up into the atmosphere that met the cold high-altitude jet stream moving south. This created a 320 k/h (200 mph) "jet streak". On 14 and 15 October the jet streak moved east across the Atlantic Ocean and came into contact with a mid-latitude depression over northwestern France. This unusual meeting of weather systems released huge amounts of energy, which drove an exceptional storm toward England.

Winds of 151 k/h (94 mph) were recorded at ground level in London, and 177 k/h (110 mph) in the Channel Islands. Buildings collapsed, some 4,800 km (3,000 mi) of telephone cables were blown down and transport systems were disrupted. The cost to insurance companies was over £500 million, but the most lasting effect of the storm was the destruction of swathes of mature woodland.

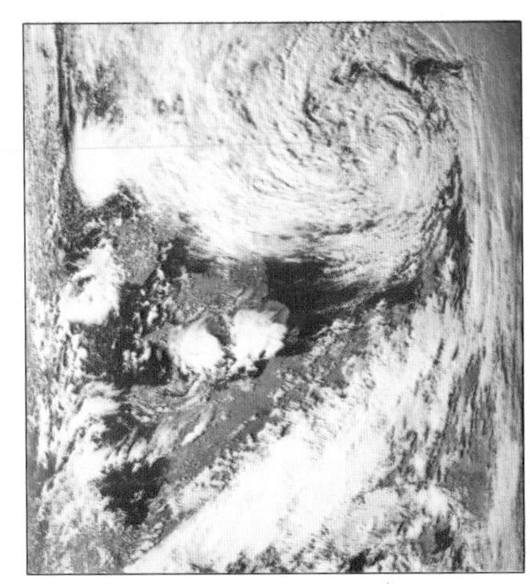

Swirling clouds, seen in this satellite picture, show the winds blowing into the center of the depression as it moved northeast across Britain, with devastating effect.

ANCIENT UPLANDS AND SOUTHERN LOWLANDS

There are two distinct types of scenery in the British Isles: the upland landscapes that have developed on rocks formed over 250 million years ago, and the lowland landscapes that have been formed on younger sedimentary rocks, found mostly in southeast England.

The uplands

The highest peak in the British Isles is Ben Nevis, in Scotland, at 1,344 m (4,409 ft); others in upland areas throughout the region reach heights not much less than this, such as Snowdon at 1,085 m (3,556 ft), and Macgillicuddy's Reeks in Ireland, 1,041 m (3,414 ft). Despite their small scale in world terms, they are all part of high-relief landscapes, which are truly mountainous in character with rocky summits bounded by deep, steep-sided valleys carved out by glaciers. Landscape features include armchair-shaped hollows, called cwms in Wales and corries in Scotland, often filled by lakes (tarns) dammed by deposits left by glaciers (moraines). Glacial deposits can also be found in the valleys, creating landforms such as the elongated hills called drumlins and the isolated boulders known as erratic blocks.

The underlying structure provides contrasts in the scenery. For example, within the Lake District are the craggy, fretted outlines of volcanic rocks as well as the smoother, rounded profiles of the shales. Moorland plateau landscapes, widely found in the peninsula of Wales to the west, the Pennines and the southern uplands of Scotland, are formed partly from the erosion of the lower bedrock.

In limestone country, special features have developed as acid rainwater has dissolved the rock over millions of years. Among the areas affected are the Peak District at the southern end of the Pennines, the Mendip Hills in the southwest, and much of central Ireland. The scenery includes underground cave systems, gorges formed by cave collapse, swallow holes and limestone pavements. These are found where soils and weathered rock on the surface have been removed, especially by glacial erosion, to expose the bedded limestone beneath. A spectacular example of this is the Burren, in County Clare in the west of Ireland.

The sedimentary landscapes

Southern and eastern England are very different from the main upland areas. Here the gently tilted sedimentary rocks form a series of cuestas, alternating steep scarp slopes, and gentler "dip" slopes. These cuestas are seen on Jurassic limestones such as the Cotswolds in the southwest and the North York Moors in the northeast, and on the Cretaceous chalk of the Chilterns and the Downs in the southeast. The smooth slopes of the chalk are thought to have developed during cold climatic periods in the past, when frost weathering and seasonal melting of snow and ice produced massive "wasting" of the land surface. Other evidence of climatic change is provided by networks of dry valleys, formed by running streams in a period when surface runoff was greater.

Between the cuestas and rock outcrops lie clay vales or broader basins and lowlands. The lowlands developed on Tertiary rocks up to 65 million years old, and have gravel terraces laid down by the major rivers. The center of London, for example, is built on a rising sequence of terraces laid down by the Thames, reflected in its many local "hills".

People in the landscape

Slopes and soil erosion are active on the upland slopes where the impact of human activity can be seen, for example in the sediments in lakes of the Lake District. Increases in the volume of river deposits are thought to be linked to deforestation and agricultural activity in the upland catchment areas. Rivers themselves have been altered by the impact of waste from mines (as in Cornwall in the southwest peninsula) reservoir construction, canalization for navigation and flood control, and by dense urbanization.

The surface soils and vegetation of the landscape are becoming increasingly affected by human activity, so it is more and more important that conservation principles are observed. These may involve protecting semi-natural landscapes for enjoyment or study, and ensuring that unnecessary and excessive erosion does not destroy vital agricultural soils or vulnerable landscapes.

A limestone pavement in the Malham area of Yorkshire in the Pennines. This striking landscape feature, known as karst scenery, is found in limestone country, where underground cave systems are also common. Pavement is formed where glaciers have scraped away weathered rock to expose bare bedrock, and it is therefore typical of mountainous areas at high latitudes. The bedrock is removed by the dissolving effect of acid rainwater fed through underground streams, so there is no deposition of sediment. This ash tree has managed to grow in soil that has accumulated in cracks in the rock.

RIVER EROSION

From its source in the Cambrian Mountains of Wales, the Wye follows a twisting course to the southeast. Just above its outflow into the Severn estuary the river has carved spectacular meanders some 250 m (800 ft) into the plateau, a striking example of the power of rivers to erode.

In their journey from uplands to the sea rivers carve out valleys and create landforms by depositing eroded rocks and soil. A river's erosive power is increased by the scouring effect of the load of debris it carries. The rate of erosion varies according to climate and the type of rock.

In upland areas, where the slope of river beds is steep, water is fast-flowing. The action of the rivers cutting down sharply into the bedrock produces V-shaped valleys and steep, ridged hills. Potholes and gorges may be carved out, and waterfalls are common. Downstream the river slows down as the gradient becomes less steep; tributaries may flow into it to increase its size. Much of its load is deposited at this stage, though it still carries smaller pebbles and mud. It erodes sideways rather than downward, producing a landscape of low hills and wide valleys.

Meanders form where the river undercuts the outer bank on a bend, while it deposits sediment on the opposite inside bank; over a period of years the river migrates sideways. Farther downstream still the river becomes even slower. Hills are eroded and valley floors are flattened out. Here deposition processes dominate, and the last of the river's load is dropped to form fertile alluvial plains.

A great number of rivers in the British Isles have been canalized to aid navigation and to minimize erosion by allowing floodwaters to flow away rapidly. Many watercourses are no longer entirely natural, particularly in urban areas, and there is conflict between environmental considerations and engineering needs.

The changing coastline

Great Britain has over 13,625 km (8,500 mi) of coastline. Rugged coasts in the far southwest, battered by Atlantic gales, contrast with serene marshes and estuaries in the east. The high, white chalk cliffs of Dover facing France to the southeast have a history and appearance very different from the inlets and sea lochs (lakes) of western Scotland at the northern end of the island.

The variety of Britain's coastal scenery, and its shape and form, closely reflect rock type and geological structure. Coastal landforms result from the processes of erosion, the movement of sand and pebbles along beaches and on and off the shore, and the deposition of sediment as beaches, deltas, estuaries and dunes.

The hard rock coasts of the north and west tend to have slowly eroding cliffs with beaches lying between rocky headlands, areas of coastal flats and dunes. Storm waves accelerate the drift of shingle and sand along the shore, and beach profiles change as winter and summer weather sets in. In the southeast, particularly on the soft clays, glacial sediments and other recent deposits, erosion at rates of several meters a year can alter coastlines. At Holderness on the North Sea coast whole villages have been lost to the sea in the past, but in other places, such as Dungeness in the extreme southeast, vast beaches of shingle deposits have accumulated.

Waves and tides are crucial to the movements of shingle and sand. Waves are generated by winds at sea. Storms can blow up steep waves near the coast, while swells are created by persistent ocean winds blowing in a constant direction far out to sea. On reaching the coast, waves "ground" on the sea floor, sometimes changing direction before they break on the shore. Depending on the type of wave and its direction, the sand and shingle may be swept onshore to form ridges, combed down into a long, smooth slope, or moved sideways by longshore drift as the waves strike the beach obliquely.

Tides are produced by the interacting gravitational forces of the Sun and the Moon; the highest and lowest tides take place when both are acting together. Tidal range in the open ocean is only about 50 cm (20 in), but it is magnified in the shallower waters of continental shelves, and especially in estuaries and coastal inlets. This explains Britain's high tidal ranges and exceptionally high tide waves, known as bores, such as those in the estuary of the river Severn.

The shingle on the shore

The sand and shingle on the shore reach it in several ways. Rivers bring sediment to the sea, which redistributes it along coastlines. Cliff erosion also produces material. Soft clay cliffs are liable to landslides and mudflows, and can erode by several meters a year; cliffs of hard rock erode more slowly through weathering and rock falls. The pattern of bays and headlands, and the detailed form of cliffs, may closely reflect the component rock types and the lie of rock strata.

Another source of beach sediment is the movement of shingle from offshore. This process often becomes apparent only when offshore shingle is removed by dredging and the local beaches are left without a source of new material.

Cliff erosion and the movement of beach material are closely linked. Beaches inhibit erosion, and at the foot of cliffs prevent them from become unstable. One

Spectacular columns of rock The Isle of Mull is one of the islands of the Inner Hebrides. The pillars of its soaring cliffs are made of columnar basalt, formed when magma, pushed between the rocks, cooled to make interlocking hexagonal columns.

way of controlling the rate of longshore drift of sand and shingle is to build groynes or jetties to retain it. However, this may deprive another stretch of coastline of materials, so its beach shrinks instead and erosion sets in.

The sea-level factor
Another vital factor in coastal processes is sea level change. In western Scotland raised beaches can be found well above the present shoreline because the land has risen since the ice sheet melted and its great weight was removed. By contrast, lines of former cliffs have been found beneath the surface of the Irish Sea. They were formed when worldwide sea levels were some 100 m (330 ft) lower than they are today because huge quantities of water were locked up in the ice sheets.

In the foreseeable future sea levels may rise still farther if the predicted global warming takes place. Low-lying coastal areas in many parts of the world would be inundated, and the shape of the British Isles would be drastically altered.

The scouring force of waves, driven across the Atlantic by westerly winds, has created this dramatic rocky landscape in south Wales. The cliffs recede as waves undercut them and remove debris, causing the upper layers of rock to collapse.

Erosion and deposition When waves reach the coast they can be destructive, eroding coastal features such as headlands, stacks and cliffs, or constructive, depositing material on beaches to build up sandbars, spits and dunes. These processes are interrupted when a river estuary enters the sea.

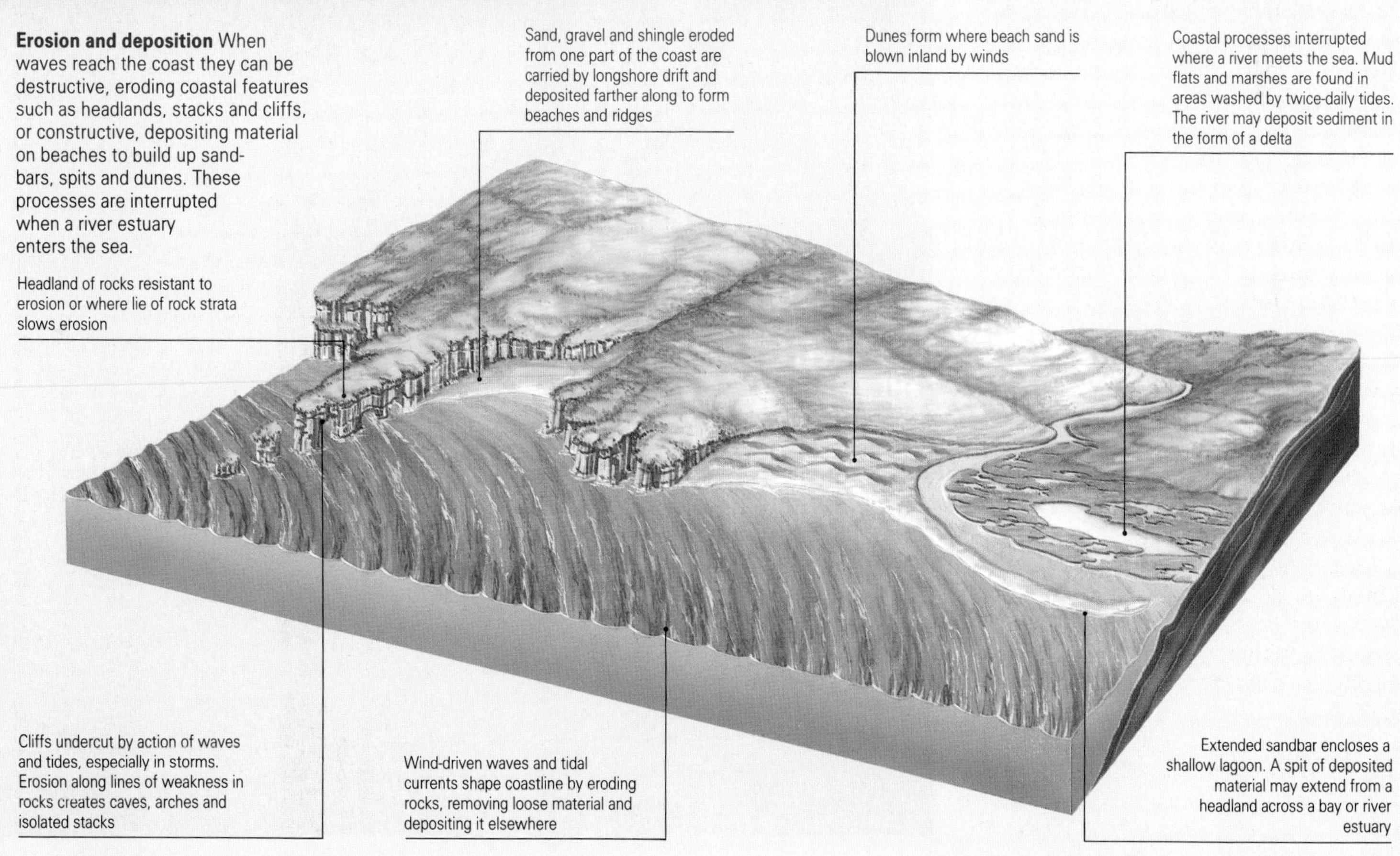

BETWEEN THE MOUNTAINS AND THE SEA

LANDSCAPES, COASTS AND CLIMATE · LANDS OF THE NORTH · SOUTHERN LANDSCAPES

Bounded on one side by the sea and on the other by mountains, France has a mixture of climates, rock types and soils. These combine to produce a surprising variety of landscapes that change from one small area to the next. The dramatic river valleys of the Dordogne and the Vosges contrast with the gentler *puy* country of the Auvergne, with its fossilized volcanic plugs. The long ridges of the Jura Mountains produce a softer landscape than the jagged peaks of the Alps and Pyrenees to the south and the west. The white cliffs of Calais and the rugged coast of Brittany, with its tiny coves, rocky headlands and rough seas, are very different from the wetlands of the Camargue, the warm, sandy beaches of Mediterranean France, and the extensive cultivated lowlands that spread across the fertile Paris basin.

COUNTRIES IN THE REGION

Andorra, France, Monaco

LAND

Area 547,492 sq km (211,272 sq mi)
Highest point Mont Blanc, 4,807 m (15,770 ft)
Lowest point sea level
Major features Pyrenees, Alps and Jura Mountains, Massif Central, Paris basin and basins of Garonne and Rhône

WATER

Longest river Loire, 1,020 km (630 mi)
Largest basin Loire, 115,000 sq km (44,377 sq mi)
Highest average flow Rhône, 1,500 cu m/sec (53,000 cu ft/sec)
Largest lake Geneva, 580 sq km (224 sq mi)

CLIMATE

	Temperature °C (°F) January	July	Altitude m (ft)
Brest	6 (43)	16 (61)	103 (338)
Paris	3 (37)	19 (66)	53 (174)
Strasbourg	0 (32)	19 (66)	154 (505)
Bordeaux	5 (41)	20 (68)	51 (167)
Marseille	6 (43)	23 (73)	8 (26)

	Precipitation mm (in) January	July	Year
Brest	133 (5.2)	62 (2.4)	1,126 (44.3)
Paris	54 (2.1)	55 (2.2)	585 (23.0)
Strasbourg	39 (1.5)	77 (3.0)	607 (23.9)
Bordeaux	90 (3.5)	56 (2.2)	900 (35.4)
Marseille	43 (1.7)	11 (0.4)	546 (21.5)

NATURAL HAZARDS

Storms and floods, landslides, avalanches

LANDSCAPES, COASTS AND CLIMATE

There are three types of physical landscape in France, reflecting three basic rock structures. The first is the ancient uplands of the northwest peninsula, the Massif Central, the Vosges mountains and Corsica, all of which have ancient, crystalline rocks that were folded in the Hercynian phase of mountain-building about 300 million years ago and subsequently much eroded. The second type is younger, and includes the highly folded rocks seen in young mountain landscapes of the Alps, Jura Mountains and Pyrenees, formed by folding in the past 65 million years.

The large sedimentary basins and lowlands of rocks laid down in this Cenozoic era form the third type of landscape. They can be seen in the Paris basin in the central northern area, which is drained by the river Seine, in the basins drained by the Charente and Garonne in the southwest, and by the Saône–Rhône system in the southeast; rocks of this period are also found on the plain of Alsace between the Vosges and the Rhine, and in the delta of the Rhône.

Coastal France

The long French coastline meets three seas: the English Channel, the Atlantic Ocean and the Mediterranean. Along the Channel coast in the northwest there are high white cliffs that correspond to those on the southern coast of the British Isles across the Channel. They are the continuation of the chalk country of the northeast. The indented coastline of the northwest peninsula was formed at the end of the last glacial period, when melting ice sheets caused the sea level to rise, flooding the river valleys.

Much of the Atlantic coast facing the Bay of Biscay is long and straight, with sandy beaches that are interrupted in the north by the Gironde, the deep estuary of the Garonne and Dordogne rivers. Farther south sandspits separate the sea from lagoons of which only one, the Arcachon

basin, remains open to the sea. Near the Pyrenees is the Côte d'Argent, or silver coast, named because of its beaches of fine white sand.

The Mediterranean coast from the eastern Pyrenees to the Rhône delta is also a sandy coast, with uninterrupted beaches and lagoons and many small harbors. It is an important area for tourism. East of the Rhône delta the coast becomes rocky where folded limestone sediments meet the sea. There are some outcrops of old crystalline rocks, and the beaches are often shingle or pebbles.

Climate patterns

There is a mosaic of climatic influences in France, which are accentuated by the detail of relief in the varied landscapes. There are great contrasts over quite small distances. For example, the western end of the Pyrenees a has maritime climate, the eastern end a Mediterranean climate.

The maritime climate of the northwest is particularly pronounced where low altitudes and smooth topography allow air masses to move inland from the Atlantic. The weather is cloudy, humid and wet, with prevailing westerly winds. There can be brisk and violent changes in the weather, and rain can fall for 200 days a year. Because it is warm as well as wet, vegetation growth is vigorous.

Eastward-moving fronts of low atmospheric pressure from the Atlantic bring the maritime influence to inland areas. In the Paris basin and on the western slopes of the Massif Central, the Vosges and the Pyrenees, the climate is influenced both by these fronts and by continental air masses. Here there is more seasonal variation. There is some rain in spring and the fall, but most of it comes in summer storms.

The continental climate of the interior has more extreme temperatures. There are cold, snowy winters, and hot summers with thunderstorms. The continental influence is strongest in low-lying areas, which are shielded from the maritime climate by the higher ground. In the eastern Vosges, Colmar receives less than 600 mm (23 in) of precipitation a year, though it is only 60 km (37 mi) from places in the western Vosges that receive 2,000 mm (80 in).

The influence of the Mediterranean Sea brings a subtropical climate characterized by mild, frost-free winters with moderate amounts of precipitation and warm, dry, sunny summers. This climate is found along the Mediterranean coast and also extends up into the Rhône valley, the southern Alps and the southern Massif Central. The climate supports the characteristic Mediterranean vegetation of evergreen oaks, olive and cork trees, herbs and other grasses.

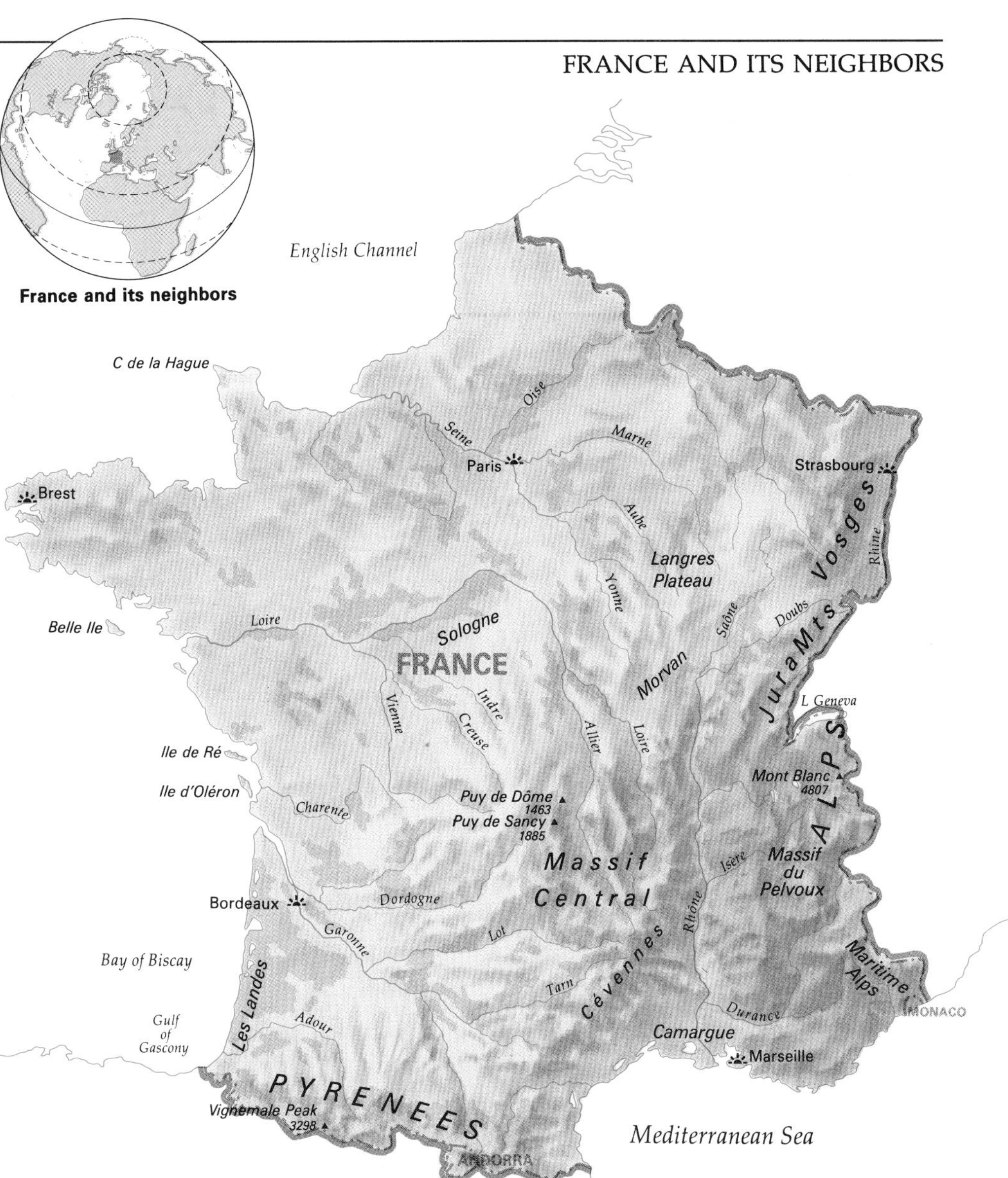

Map of physical zones France combines a long coastline, a mountainous fringe, extinct volcanoes and active glaciers with scenic upland massifs dissected by spectacular river gorges, and fertile lowland.

The morning sun strikes the summit of Mont Blanc, western Europe's highest mountain. It lies at the western end of the great mountain chain of the Alps, which is still being formed.

LANDS OF THE NORTH

France has a variety of distinctive landscapes. The greatest contrasts can be seen by dividing the country into two areas on either side of a line from the peninsula of Brittany in the northwest to the Maritime Alps in the southeast, separating it into northern and southern France.

The geological foundations of northern France are very old. Remnants of eroded Hercynian mountain chains can be found in the basement rocks that underlie its raised mountain blocks, or massifs. Here are the three major zones of the north: the massif of the northwest, the western Rhineland and the Paris basin.

The northwestern massif

The great northwestern peninsula of France is a massif, sometimes known as the Armorican massif. It consists of a platform of ancient igneous rocks such as schists, gneisses and granites. Its low hills and ridges, up to 200 m (650 ft) high, can be seen in the extensions of the massif to the east and beyond the Loire to the southeast. Many of the landforms here, especially toward the southeast, are very old indeed, and have changed little over tens of millions of years; other parts of the platform have been drowned by invasions of the sea. During the various changes in sea level during the Quaternary period (the past 2 million years), when levels were at times about 100 m (330 ft) lower than today, deep valleys were eroded that were later flooded. This created the highly indented coastline of Brittany, which now provides many safe harbors and fishing grounds.

Along the lower Loire river the basement rock subsided at the start of the Mesozoic era (248–65 million years ago), and was subsequently covered by thick sediments deposited in shallow seas. This produced the rock strata that now underlie the basins of north-central (and also southwestern) France.

The Vosges and the Rhine rift valley

The other principal massif in northern France, the Vosges, lies in the west, bounded by a section of the Rhine. The low northern Vosges are covered with sandstone, which yields poor soils, and are heavily forested with conifers. They are dissected by steep U–shaped valleys. In the high Vosges to the south, the peaks are rounded outcrops of eroded crystalline rock. During the cold periods of the Quaternary, glaciers here spread from an icecap, cutting the long, wide valleys that are now used for pasture. Between the Vosges and the English Channel lies the western tip of the Ardennes plateau which comes within France.

The Vosges separate the Paris basin from the Rhineland to the east. The French part of the Rhine river system includes the upper reaches of the Meuse, Moselle and Schelde rivers.

For more than 160 km (100 mi) the Rhine forms the eastern frontier of France. From the high Vosges the land drops away to the floor of the Rhine rift valley, a fall of some 1,200 m (3,900 feet) over a distance of 60 km (37 mi). The floor of the valley is as much as 32 km (20 mi) wide, with the Rhine flowing down the middle of it. Germany's Black Forest flanks the river to the east.

The Rhine rift valley was formed when the Alpine mountain-building period set up powerful tensions in the ancient basement rock, causing upward and downward movement of blocks along fault lines. The rift valley was lowered by subsidence along a series of fault lines, rather like the sagging keystone of an arch. The central lowered block (graben) in the Rhine rift valley has undergone two major periods of earth movement.

The first of these took place during the Oligocene epoch, 38–25 million years ago, when there was a semiarid climate. A narrow gulf, extending from the north, was closed off by the Jura Mountains, which were then in the process of formation. Very coarse alluvial fans were produced at the foot of the raised edge of the rift blocks, and in the central graben marls and rock salt (evaporite) were deposited. The second period of earth movement began in the Pliocene epoch, 5–2 million years ago, and is still continuing. The upper Rhine began to flow north through the graben, which was subsiding, and took its present course to the North Sea. Before this it had flowed west, then through the Saône plain to join the river Rhône as it flows south.

The indented coastline of Brittany at Crozon, on the tip of France's northwestern peninsula, which is part of the Armorican massif platform of ancient crystalline rocks. The vertical bedding planes of the rocks testify to major earth movements in the past.

THE PARIS BASIN

The Paris basin is a broad area where the layers of rock dip from the sides down toward the center. It covers some 120,000 sq km (46,000 sq mi), almost a quarter of France. Drained by the Seine, it is surrounded by the raised blocks known in France as massifs.

The basin is composed of sedimentary deposits laid on the basement rocks. There is a succession of rocks arranged rather like a set of saucers, with the smallest in the center surrounded by a series of larger ones. In the center are the youngest rocks, Tertiary limestones of the past 65 million years. Their horizontal beds give the characteristic flat landscapes. Toward the edges of the basin there are limestones of different ages, broken by outcrops of sandstone. In the valleys there are clays, marls and sands. To the east, steeper scarp slopes face outward from the basin. The range of rock types has produced a great variety of scenery. In the east are the ridge-and-vale landscapes of the chalk areas of Champagne and the limestone country of western Lorraine toward the Vosges. In the northwest is the chalk country of Normandy, cut by the river Seine, and in the north that of Picardy.

In the Beauce plateau near Orléans, fertile loess soils make good agricultural land.

Cirque de Gavarnie in the Pyrenees, on France's southwestern border. The cirque was hollowed out when a glacier eroded the rocks at the head of the valley.

The volcanic landscape of the Auvergne at Polignac in the Massif Central. The cones of these extinct volcanoes have been eroded away, but the more resistant plugs of solidifed lava still remain.

SOUTHERN LANDSCAPES

Southern France has many subtle changes of landscape, but there are three principal types: the young fold ranges of the Alps, the Pyrenees and the Jura Mountains; the older plateau landscape of the Massif Central; and the lowlands, valleys and basins, including the basin in the southwest and the Rhône valley.

Mountains of the south

The Alps in France extend in an arc from the Mediterranean coast to the Swiss border. They were forced up in several phases of mountain-building. The first Alpine phase, which formed the Alps and the Pyrenees, began about 100 million years ago in the Cretaceous period; the second took place during the Oligocene epoch (38–25 million years ago). The third phase, not yet completed, began 7–5 million years ago at the end of the Miocene epoch (25–5 million years ago), and continued in the Pliocene (5–2 million years ago) and the Quaternary. The Alps were lifted up to heights far above their present highest point of 4,807 m (15,771 ft) at Mont Blanc. Subsequent erosion was rapid, and the scenery of today shows many glacial landforms: glacial troughs, moraine-blocked lakes and ice-fretted mountain slopes. Glaciers are still active in shaping the scenery in some parts of the Alps.

The Pyrenees, a range of young mountains separating France from Spain to the southwest, are the oldest of the mountains formed in the Alpine mountain-building period. They are mainly crystalline rocks, covered in some areas by folded limestones. They rise to 2,000–3,000 m (6,500–10,000 ft), and were covered by icecaps during the glacial periods of the past 2 million years. Some spectacular cirques have been cut at the heads of valleys. The eastern end of the Pyrenees includes a series of subsided, faulted basins, which are filled with river sediments up to 2,000 m (6,500 ft) thick. The sediments were deposited when the basin of the western Mediterranean Sea was flooded after a time of aridity during the Miocene.

The Jura Mountains in eastern France consist predominantly of folded limestones above the rigid basement rock. The scenery is distinctive, with a regular arrangement of ridges and valleys that correspond to upfolds (anticlines) and downfolds (synclines) in the rocks. The valleys are connected by deep gorges; cut through the intervening ridges, they are known as cluses.

The Massif Central

There are four major areas in the Massif Central. To the south lies an extensive faulted plateau of limestones and some dolomites. This plateau, known as the Causses, was covered by thick sediments in the Mesozoic era, and has been lifted up by some 1,000 m (3,300 ft) over the last 10 million years. Rivers flowing from impervious crystalline and volcanic parts of the Massif have cut deep gorges such as the Tarn canyon through this limestone area. There are also some limestone karst features, but they are better developed underground in caves than on the surface.

In the Auvergne, in the north-central part of the Massif, there is a volcanic landscape. The rocks are covered with lake sediments and alluvium dating from the Miocene, which indicates that the underlying volcanic rocks must have resulted from the late Alpine mountain-building period. Older volcanic landforms have been worn down by erosion, but there are remains of the central plugs of volcanoes that became extinct only in the last 10,000 years, during the present

CORSICA

The Mediterranean island of Corsica lies about 160 km (100 mi) to the southeast of France across the Ligurian Sea. This mountainous island occupies an area of 8,277 sq km (3,195 sq mi). Its high point is Monte Cinto at 2,710 m (8,891 ft); there are several other peaks above 2,000 m (6,560 ft). In winter, snow is common in the mountains but almost unknown on the coast.

The western two-thirds of the island comprise old crystalline rocks that were strongly fractured during the Tertiary period, when the African and Eurasian plates collided. The intense dissection and fracturing of the basement rock caused by this tectonic upheaval has resulted in a very rugged terrain. Deep gorges have been cut into the rock. The upper parts of the island were shaped by glaciers during the recent ice ages, but cirques are the only glacial landforms that can be seen today.

The northeast peninsula is composed of folded limestones that form some spectacular coastal cliffs. On the eastern coast is a flat alluvial plain with a few low hills. There are sandy beaches, and lagoons dammed by sandspits. In the western uplands the shallow weathered soils support the dense scrub known as maquis, the fragrant natural vegetation for which the island is famous.

The Ardèche gorge on the eastern edge of the Massif Central. As it flows from its source in the Cévennes mountains to join the Rhône, the young, fast-flowing river has eroded the underlying rock, cutting gradually down through the limestone to form the vertical sides of the gorge.

Holocene epoch. They form a distinctive landscape, of which the famous peak of the Puy de Dôme (1,463 m/4,806 ft) is a typical example.

In the north and west are a series of broad, stepped crystalline plateaus. These flat, eroded uplands are drained by the rivers Garonne, Dordogne, Lot and Tarn. The rivers have cut gorges in some places. In the northeast is a fault-bounded block, the Morvan Massif, which has a radial pattern of drainage with some rivers flowing north to tributaries of the Seine and others south to the Loire basin.

The Aquitaine basin and the Rhône

The Aquitaine basin is drained by the Charente and Garonne in the southwest. Smaller than the Paris basin, it is flanked by the Pyrenees, the Massif Central and the Atlantic coast, and is a typical sedimentary basin in its north and northeastern areas. Here the rocks, mostly limestones with some marls, dip gently down toward the Garonne. Dramatic karst landforms have developed in the limestone areas, which are famous for their caves. In the south there is an area of deposits resulting from erosion of the Pyrenees. Outcrops of this material, including boulders, are found in the plateau between the upper Adour and Garonne rivers. The central part of the basin has subsided and has been invaded by the sea, forming the Gulf of Gascony. Sand covers the area of Les Landes, with its smooth, unbroken coastline south of the Gironde estuary.

The Rhône and its tributaries center on a major valley corridor that runs north-–south and separates the Alps and the Jura Mountains in the east from the Massif Central in the west. The Rhône valley gives way to a plain in the south. The river reaches the Mediterranean Sea through a large delta, the western part of which forms the marshy area known as the Camargue.

Local winds

Winds blow from one part of the lower atmosphere where the air pressure is high, to another where the pressure is low. On a global scale the movement of air masses gives rise to the general circulation that creates the belts of easterly trade winds, westerlies and others, and plays an important part in controlling the climates of the world. In the northern hemisphere winds near the land surface tend to blow clockwise round areas of high pressure, and anticlockwise round areas of low pressure. In the southern hemisphere the directions are reversed. Hurricanes, cyclones and typhoons are all winds that rotate around centers of low atmospheric pressure.

Local winds show the influence of features such as mountains. They are closely linked with local climates and weather conditions. The mistral is a cold, dry north or northwesterly wind. It affects the valley of the Rhône, particularly to the south of the point where the river is joined by the Isère. It is a strong, squally wind, most violent in the winter and spring. It blows over the delta for more than a hundred days a year, and can reach speeds of 100 km (62 mi) per hour.

The mistral arises when there is a high-pressure area of dense, cool air over central Europe and lower pressure over the Mediterranean, where the warmer air is less dense. Air is forced over the Alps and then literally flows under the influence of gravity to lower levels. When the mistral is blowing the temperature falls suddenly as cold air drains down the valley. Many buildings have small windows, or none at all, on their northward sides, which act as windbreaks; farms are often protected by a row of cypress trees or hedges of osier reeds to shelter them from the effects of the mistral.

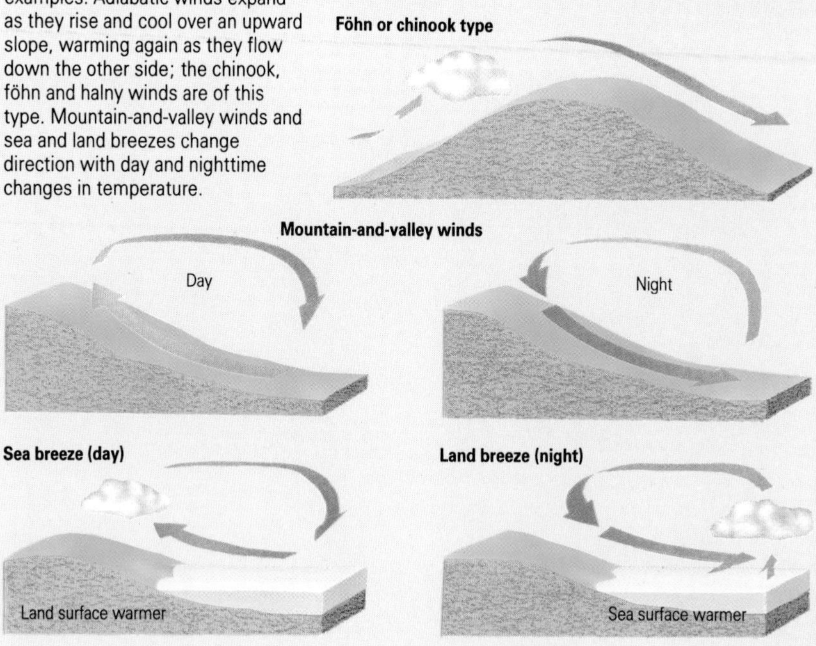

How winds form When cold upland air flows downslope, katabatic winds develop; the mistral and bora are examples. Adiabatic winds expand as they rise and cool over an upward slope, warming again as they flow down the other side; the chinook, föhn and halny winds are of this type. Mountain-and-valley winds and sea and land breezes change direction with day and nighttime changes in temperature.

THE BEAUFORT SCALE OF WINDS

The scale was devised by British naval officer Francis Beaufort in 1805. Wind speeds and effects on land were later added. Measurements are made by an anemometer 10 m (33 ft) above the ground.

Force	Name	Effects	Wind speed km/h	mph
0	Calm	Smoke rises vertically; water smooth	less than 2	1
1	Light air	Smoke shows wind direction; water ruffled	2–5	1–3
2	Light breeze	Weathervane moves, leaves rustle; wind felt on face	6–11	4–7
3	Gentle breeze	Loose paper blows around; flags extended	12–19	8–12
4	Moderate breeze	Small branches sway; small waves	20–29	13–18
5	Fresh breeze	Small trees sway, leaves blown off; white horses	30–39	19–24
6	Strong breeze	Whistling in telephone wires; sea spray, large waves	40–50	25–31
7	Moderate gale	Large trees sway; sea heaps up	51–61	32–38
8	Fresh gale	Twigs break from trees; foam streaks on seas	62–74	39–46
9	Strong gale	Branches break from trees; high waves	75–87	47–54
10	Whole gale	Trees uprooted, weak buildings collapse	88–101	55–63
11	Storm	Widespread damage; long foam patches	102–120	64–72
12–17	Hurricane	Widespread structural damage; air full of foam, spray	over 120	75

Winds of the world

There are many similar local winds, known as katabatic winds, in other parts of the world. The bora is a cold, dry, gusty northeast wind that blows down from the highlands of Yugoslavia to the Adriatic Sea. The oroshi wind in central Japan is of this type. Similarly, the more violent williwaw wind in Alaska falls from high snow-covered mountains to the coast.

A different kind of local wind is represented by the föhn, the warm, dry, blustery wind that blows down the lee side of mountain ranges in Switzerland and Austria in central Europe. When the föhn is blowing there can be rapid increases in temperature, clear visibility and blue skies. There are about 48 föhn days a year in valleys in northern Switzerland, when the snow can melt very rapidly. In the eastern Rocky Mountains of North America the same type of wind is called the chinook, which means snow eater. The chinook can raise the temperature by 21°C (38°F) in four minutes, and can melt away snow at a rate of 25 mm (1 in) an hour. On the other side of the Rockies, in southern California, blows the Santa Ana, another wind of this type. They are known as adiabatic winds.

Adiabatic temperature changes take place when wind passes over a mountain. When moist air rises on one side of the mountain it expands and cools at a rate of 1°C per 100 m (1.8°F per 330 ft) until it reaches condensation level. As the air continues to rise above this level, clouds form and rain or snow may be released. The air now cools more slowly. Once over the crest of the mountain, the air begins to warm as it falls. Having lost some of its

The Chamonix valley and peaks of the Alps, where the mistral develops. Cold air is forced up over the Alps, then blows violently southward under the force of gravity, and is funneled by the valley of the Rhône.

moisture, it warms more quickly as it descends than it cooled as it rose. These different rates of cooling and warming mean that the temperature of the air flowing down the slopes can be 10–20°C (18–36°F) warmer than previous air temperatures at the same site.

There are also local mountain-and-valley winds. These are best demonstrated in the summer, in valleys that are deep and straight and run north–south. During the day the air above the slopes and the valley floor is heated to temperatures well above those in the center of the valley. Because warm air rises, there is a warm flow of air up the slopes during the day, accompanied by sinking colder air in the valley center. At night the air above the slopes cools more rapidly, and flows down into the valley under the influence of gravity. Fog may form in the valley as a result.

There is a similar reversal of wind direction in coastal areas. Onshore winds or sea breezes during the day are caused by heat from the Sun warming the land surface more quickly than it does the surface of the sea. As the land also cools more swiftly, at night the wind direction changes and the wind blows offshore.

On a continental scale, a seasonal reversal of prevailing winds as air over a large landmass heats and cools is the mechanism driving the dry and wet monsoon winds of India and eastern China.

Sculptures of nature

The Causses area of the Massif Central in southern France has some spectacular interior decoration on offer for those prepared to go underground to view them. The Grotte de Clamouse is one of several caverns that have formed within the limestone rock, which was laid down over 300 million years ago to a depth of 790 m (2,600 ft) in a vast faulted depression. Reaching the caverns is often a challenge: some are near the surface and dry, but others are flooded and lie hundreds of meters below the surface. In the case of the Padirac cave, which is almost 100 m (330 ft) high, access is by lift down a pothole 100 m (330 ft) deep, a walk along a dark tunnel, and a boat ride across an underground lake. The largest of the caverns is Grand Dome.

The calcite sculptures that decorate the caverns have been deposited by water percolating down from the surface. The water is saturated with calcium carbonate dissolved from the limestone. The change of pressure and temperature in the cavern causes the calcium carbonate to be precipitated as the mineral calcite. Great curtains form where the water trickles down a sloping roof, there are columns where stalagmites and stalactites join up, fragile, needle-like stalactites from the source of a tiny drip and forests of stumpy stalagmites where drips fall on the floor.

These moist caves with their stable temperature also have practical uses. The caverns of Causses de Quercy were inhabited in prehistoric times, and the caves of Roquefort are used for the manufacture of the famous cheese made from sheep's milk.

Dramatic rock formations greet visitors to the Grotte de Clamouse in France's Massif Central. Calcite deposited by water percolating through the limestone has hardened into stalactites and stalagmites of all shapes and sizes.

WATERS AGAINST THE LAND

A PATTERN OF EROSION · CLIMATES, PLATEAUS AND PLAINS · HEATHLANDS, VALLEYS AND POLDERS

When an airplane touches down at Amsterdam's Schipol Airport, its wheels are several meters below the level of the North Sea. This is not unusual in the Netherlands, where about a third of the country lies below sea level. It is kept dry only by a complex network of dikes to hold back the water, and ditches from which water drained from the land is pumped out into the rivers and sea. The flat reclaimed land is characteristic of this northwestern edge of the North European Plain, which is crossed by the great rivers Meuse and Rhine. Tributaries of the Meuse rise to the south in the Ardennes, the only area of highland in the region. Rivers have cut deep valleys through the forested slopes of this ancient massif. The eroded material adds to the sediments being laid down in the delta of the Schelde and Rhine.

A PATTERN OF EROSION

In the Low Countries the principal contrast in landscapes and relief is between the Ardennes plateau in the south and the low-lying river valleys, delta and coastal plains in the north. The whole region provides an example of how erosion in one area must be accompanied by deposition in another.

Land uplift and erosion

The Ardennes is one of a number of mountain blocks in Europe that were raised during the Hercynian phase of mountain-building some 300 million years ago. Similar massifs are found in Brittany and the Vosges in France, and in the Black Forest and Harz Mountains in Germany. The Ardennes are formed of hard rocks more than 500 million years old that have been much folded and distorted through two phases of gigantic earth movements, the Caledonian and then the Hercynian. After the first phase they were covered by marine sediments, and then in the Hercynian movement great blocks were raised in the form of massifs bounded by faults in the rocks. The ridges and valleys of the Ardennes are generally aligned east–west or south-west–northeast, reflecting trends in the rocks produced by earth movements. However, the main rivers have cut down through the rocks, forming deep rocky gorges. The pattern of drainage on the land bears little relation to the underlying geological structure.

The whole massif, together with the

COUNTRIES IN THE REGION

Belgium, Luxembourg, Netherlands

LAND

Area 74,258 sq km (28,664 sq mi)
Highest point Botrange, 694 m (2,276 ft)
Lowest point in west Netherlands, –7 m (–22 ft)
Major features very low-lying areas in north, parts below sea level, Ardennes plateau in south

WATER

Longest river Rhine delta lies in west but most of Rhine's 1,320 km (820 mi) lie outside region, which includes the lower Meuse and Schelde
Highest average flow Rhine, 2,490 cu m/sec (88,000 cu ft/sec)
Largest lake IJsselmeer, 1,210 sq km (467 sq mi)

CLIMATE

	Temperature °C (°F) January	July	Altitude m (ft)
Eelde	1 (34)	16 (61)	4 (13)
Ostend	3 (37)	16 (61)	10 (33)
Brussels	2 (36)	18 (64)	104 (341)
Geulle	2 (36)	18 (64)	115 (377)
Luxembourg	0 (32)	17 (63)	330 (1,082)

	Precipitation mm (in) January	July	Year
Eelde	69 (2.7)	86 (3.4)	767 (30.2)
Ostend	41 (1.6)	62 (2.4)	598 (23.5)
Brussels	73 (2.9)	79 (3.1)	785 (30.9)
Geulle	63 (2.5)	85 (3.3)	763 (30.0)
Luxembourg	73 (2.9)	66 (2.6)	740 (29.1)

NATURAL HAZARDS

Tidal surges in coastal areas, flooding in flat river valleys

Map of physical zones The Low Countries are most appropriately named. Large areas of the land lie below sea level, and much of the Netherlands has actually been reclaimed from the sea. Behind the coastline are bands of sand dunes, flat fens and polders. The famous Meuse and Rhine rivers cross the lowlands on their way to the North Sea. To the south of the region rise the forested foothills of the Ardennes plateau.

younger sediments deposited over it, was tilted down toward the north and eroded to form a series of scarplands and plateaus to the north and south. Those to the north are covered by deposits of silt-sized loess soil dating from the ice ages of the past 2 million years. The entrenched valleys of the Ardennes, which cut deep into the gently sloping plateau surface, are a consequence of the erosion that was accelerated by tilting of the land. Much of the eroded material was carried north, where, over millions of years, it was deposited and formed most of what is now Belgium and the Netherlands.

Estimates of the rate of downcutting in the Ardennes have been made by dating layers of volcanic ash found in the terraces that mark the earlier courses of rivers. These suggest that rivers have cut down into the rocks of the plateau by as much as 11 m (36 ft) in the last 50,000 years, a rate of 2.2 cm (1 in) a century.

Sea levels and sediments

About 60 million years ago the North Sea basin was formed as a branch of the Atlantic Ocean rift that opened up as North America and Europe drifted apart. North of the Ardennes deposits were laid down under the sea, and these rocks too were raised and then tilted down toward the north by earth movements during the formation of the Alps in central Europe. These areas were also subsequently dissected by rivers.

The rivers formed a network of valleys that flowed northeast down the slope of the land exposed when the North Sea retreated. The valleys became partly filled by gravels, and gradually the drainage pattern began to adapt to the underlying geological structure, eventually evolving a southeast–northwest trend.

Where the rocks were more resistant, because the particles were coarser or better cemented together, the dissecting rivers left isolated patches of higher relief. In the Kempenland, south of the present course of the Meuse in northeast Belgium, thick gravel deposits on an ancient alluvial fan of the same river have protected the underlying rocks from erosion.

The covering of fine-grained loess deposited over most of the central part of the region was eroded and carried toward the north, filling in many valleys and producing a very subdued relief. About 40,000 years ago, in the last interglacial period between ice ages, when much of the world's ice sheets melted, the rising sea again flooded river valleys. Some 20,000 years ago, during the most recent ice age, the estuaries that formed when the sea level fell again were once more filled in with river deposits.

A polder landscape This typical Dutch scene shows a polder, an area of reclaimed land. Many polders are now intensively used for agriculture. In the background a village has grown up on a small hill, where the higher ground is less prone to flooding.

About 10,000 years ago, when most of the ice sheets melted once more, the sea rose again. Along the south coasts of the North Sea three zones of sediments were subsequently laid down: sandy beaches and coastal dunes in the west; behind these a zone of largely clay tidal flats and brackish lagoons; and, farther inland, a zone where freshwater peat was formed from plant remains decaying in swampy land. As the sea continued to invade, the zones slowly migrated eastward.

Today these recent sediments lie more than 20 m (66 ft) thick in the west of the Netherlands. Much of the area (over 30 percent of the Netherlands) is now below sea level at high tide, partly because the peat and clay lands, some of them reclaimed from the sea, have shrunk as they have dried and have become compacted. The present-day coastal zone consists of a series of barrier islands, tidal flats and sandbars, intersected by the estuaries of the various branches of the major rivers, the Meuse, the Rhine and the Schelde.

CLIMATES, PLATEAUS AND PLAINS

The Low Countries are part of the great western European climatic battleground, where continental air from the Eurasian and African landmasses is in constant conflict with tropical and polar sea air from the Atlantic and Arctic oceans.

This general climate pattern is modified by three factors more specific to the Low Countries. The North Atlantic Drift current brings warm tropical water from the Gulf of Mexico into the North Sea, appreciably raising temperatures along the coast in winter and lowering them in summer. Fogs also result when cold air passes over a warm sea.

Altitude causes very significant variations in rainfall, with areas near sea level receiving as little as 500 mm (20 in) a year, whereas the higher areas in the south receive over 1,250 mm (50 in). In the highest parts of Belgium and in Luxembourg much of the winter precipitation falls as snow, as temperatures decrease with altitude.

The influence of the continental interior increases toward the east, with seasonal variation and climatic extremes becoming more marked inland where the moderating effect of the ocean is reduced. A good illustration of this shows in the number of days on which frost occurs. There are as few as 30 in coastal regions, but 100 or more in the Ardennes. Although most of the major variations in the landscape of the Low Countries are due to geological and climatic history, the surface detail of these landscapes is the result of present-day climates and processes. For example, the bogs and streams of the Hohe Venn mountains, the highest part of Belgium, exist because of the high rainfall and low temperatures of this upland region. In contrast the active sand dunes of the Veluwe are a result, at least in part, of the fact that relatively little rain falls in the low-lying Netherlands.

The southern uplands

Landscapes in the Low Countries may be divided into types characteristic of seven different areas: scarp-and-vale in the far south, then the Ardennes, the lower plateaus farther north and the plains of Flanders, the heathlands, the great river valleys, and lastly the reclaimed polders. Their appearance, including their plant cover or natural vegetation, is governed partly by the climates of today, but more important influences are the climates of the past, combined with the history of erosion and deposition, rock types and geological structures.

The smallest and southernmost of these typical landscape areas is the well-formed scarp-and-vale district of Belgium and Luxembourg lying south of the Ardennes, between the upper courses of the Meuse and Moselle. It is crossed by three major escarpments of sandstone and limestone, which are about 200 million years old. Between these are clay and marl vales and an undulating plateau of sandstone that is punctuated by deep wooded gorges and high pinnacles, notably in the area known as Little Switzerland.

The Ardennes and the Flanders plain

The old rocks that underlie the gently rounded summits of the Ardennes were folded and then eroded, to produce a series of plateau-like surfaces between 400–600 m (1,300–2,000 ft), with ridges resulting from the underlying rock structure. Deep river gorges cut through the Ardennes, where many of the slopes are heavily forested.

Among the ancient slates, quartzites, limestones and schists of the northern Ardennes lies the Famenne. This flat-bottomed depression is below 200 m (600 ft), and was formed by the erosion of less resistant rocks. Two low plateaus to the north, the Condroz and the area between the Sambre and Meuse rivers, have also been isolated by erosion. They form a mosaic of woods and farmland, a zone of transition between the Ardennes and the low central plateaus farther north. Coal-bearing layers of rock are preserved in two adjacent low-lying areas bounded by faults; one is occupied by the Sambre and the Meuse, the other by the Meuse before its junction with the Sambre.

In the highest part of the Ardennes, the Hohe Venn mountains, periglacial landforms developed when tundra-like conditions prevailed during the ice ages of the past 2 million years. They include dry valleys, the remnants of mounds formed by the expansion of ground ice, and hollows formed by freeze–thaw action shattering the rock. The landscape is covered with heath, forest or upland bog.

To the north of the Ardennes, in the center of the region, lies an area of rich agricultural land on low, undulating plateaus that range between 50 and 350 m (165–1,150 ft). The underlying rocks are covered with deposits from the last ice age, particularly the fertile deposit of yellowish loess called as *limon*. This layer is up to 20 m (66 ft) thick in places, and is dissected by rivers except where the underlying rock is chalk, as in the Hesbaye and Herve plateaus and in the southern Netherlands, the location of Vaalserberg, which at 321 m (1,053 ft) is the Netherlands' highest point. The larger rivers have deposited eroded material in terraces that reflect the complex history of

HAZARDS OF THE MEUSE VALLEY

Climate and topography have often combined to wreak havoc on the inhabitants of the industrial cities in the valley of the Meuse where it flows northeast through Belgium.

The city of Liège stands where the Meuse is joined by the Ourthe. For 20 km (12 mi) upstream the valley is 1–2 km (0.6–1.2 mi) wide, and up to 80 m (260 ft) deep. In December 1930 a cold fog, bearing poisonous fluorine compounds emitted by local zinc works and factories, became locked in the steep-sided valley for three days by a combination of cold atmospheric conditions with sinking air, and a lack of wind to disperse the pollutants. Thousands of people then developed breathing and circulatory illnesses, and 60 of them even died as a result.

Five years earlier Liège had experienced the less lethal hazard of flooding. The major floods of 1925–26 were caused when meltwaters from the Ardennes joined the Meuse, already swollen by rainstorms. Much of the Meuse valley, from the Franco-Belgian border to the North Sea, was inundated. At the confluence of the Meuse with the Ourthe the floodwaters rose 8 cm (3 in) every hour. By the end of December the city of Liège itself resembled the islands within a meandering channel on which it was built. Some 4.5 sq km (1.7 sq mi) of the city were submerged under water that was in places 5 m (16 ft) deep; 2,500 houses were flooded.

A long history of such inundations is recorded on the walls of the city's cathedral and other churches, where notches have been made to mark the highwater levels of floods in 1571, 1643, 1726, 1740 and 1850.

upward movement of the land in this part of the region.

The monotonous plains of Flanders stretch from the central plateaus toward the low-lying polders of reclaimed land that border the coast. They are formed mainly of clays, and lie between 5 and 50 m (16–165 ft) above sea level, though there are a number of sandy hills that exceed 150 m (500 ft). River valleys are filled in with water-borne deposits, so the relief is markedly subdued.

Folded rocks Downfolded layers of sedimentary rock form a syncline above the banks of the Meuse in the Ardennes. The plateau is part of the mountain belt formed during the Hercynian; it has since been deeply eroded by rivers.

Forests and farmland Heavily wooded hills and rich pastureland make up the gently rolling landscape of the Ardennes plateau in southern Belgium. Here the river Semois has cut its way through the ancient rocks to form a winding valley.

HEATHLANDS, VALLEYS AND POLDERS

Much of the Low Countries forms the western edge of the great North European Plain that stretches through Germany and Poland into the Soviet Union as far east as the Ural Mountains. Sands and gravels deposited by rivers and glaciers cover most of the area.

The central heathlands

The heathlands have thin, sandy soils and characteristic sparse vegetation such as heather and gorse. They lie mostly between 50 and 100 m (165–330 ft) high. The Kempenland heaths stretch from the central plateaus across the Belgian–Dutch frontier as far north as the Meuse on its east–west progress across the Netherlands. The land is largely dry, because of the permeability of the rocks beneath the deposits that were transported and laid down by the Meuse and other rivers during periods when the flow of water was much greater. The Kempenland is crisscrossed by a dense network of valleys through which water once ran, formed when the ground was frozen and rain could not soak into the soil.

The Veluwe heathlands south of the IJsselmeer and the heathlands in the northeast of the region along the German border are covered with deposits left by the ice sheets that once extended from Scandinavia, and by their meltwater. The occasional, quite dramatic ridges are

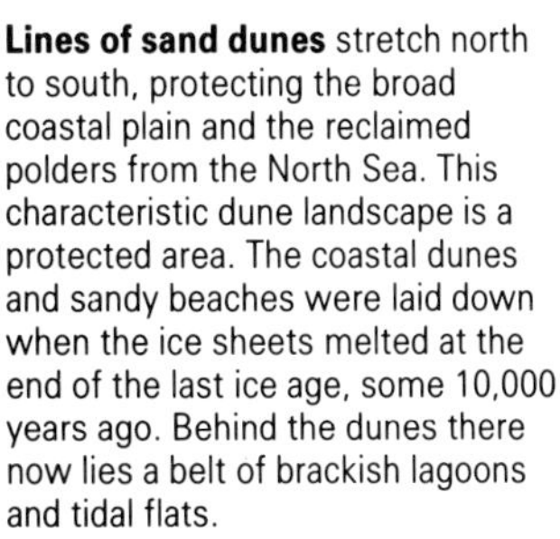

Lines of sand dunes stretch north to south, protecting the broad coastal plain and the reclaimed polders from the North Sea. This characteristic dune landscape is a protected area. The coastal dunes and sandy beaches were laid down when the ice sheets melted at the end of the last ice age, some 10,000 years ago. Behind the dunes there now lies a belt of brackish lagoons and tidal flats.

A polder in the early stages of reclamation Channels have been cut to drain the land, which is still covered with salt-marsh vegetation. It takes many years for the polder to dry out and become suitable for cultivation. The Dutch, experts at water management and drainage schemes, now live on and farm thousands of square kilometers that were originally flooded by the sea.

PEOPLE IN THE LANDSCAPE

The Low Countries are among the most densely populated regions in the world. Much of the landscape has been modified dramatically by human activity over the centuries.

Reclamation of marshes and peat bogs is perhaps the most dramatic change wrought by human activity on the region's appearance. "God made heaven and earth but the Dutch made the Netherlands," it is said. Some 3,000 sq km (1,160 sq mi) has been reclaimed from the sea and from freshwater marsh in the past 800 years, transforming the whole region.

However, even in the most heavily populated area of polderlands, lying between the Rhine and Meuse estuaries and the IJsselmeer, care has been taken to ensure that there are green spaces between the settlements, which house over 7 million people. In the central Netherlands to the east former peat, sand and gravel workings, now flooded, provide boating or other leisure amenities.

The southern parts of the region were one of the first industrialized zones in the world, a fact that has also left its mark. The coal measures in the Sambre and Meuse valleys have long been worked out, and most of the waste tips have been landscaped, but in the Kempenland coalfield to the north there are vast pyramidal spoil heaps, strikingly incongruous, that still emerge above the surrounding heathlands. Elsewhere, the worked-out quarries of limestone in the south of the Netherlands and iron ore in Luxembourg still scar the landscape.

moraines; the highest hills, which exceed 100 m (330 ft) in the Veluwe area, are thought to be a combination of several moraines that piled up along the front of advancing ice sheets.

Valleys of the great rivers

Crossing the plateaus and the heathlands are the valleys of the Meuse and of the Rhine and its branches. The Waal and Lek flow to the North Sea, and the IJssel enters the IJsselmeer. Huge amounts of material have been brought down by these major rivers and deposited across the lower parts of the region. Where the valleys are restricted in width, as in the Belgian plateaus and the southern extension of the Netherlands, old river courses can be traced in gravel terraces.

Barriers against the floods Lines of trees have been planted flanking the river Lek, the northern branch of the Rhine, which flows through the southern Netherlands. The high banks, or levees, have been artificially built up in order to contain the river. The banks are higher than the level of the surrounding land, and protect it from flooding.

Elsewhere the valleys are broad, with raised banks (levees) of material that is deposited when the rivers overflow their channels. Beyond them lie backswamps. Since the ice sheets melted some 10,000 years ago the rivers have changed their course many times. This is especially true of the Rhine. Despite the levees that line the lower Rhine and Meuse, the rivers pose a continual threat of flooding in the near-flat valleys.

The polderlands

The lowest-lying area of all in this low-lying region is the belt of remarkably flat reclaimed lands known as polders along the North Sea coast. Polders make up much of the coastal area, the islands in the estuaries of the Schelde, Meuse and Rhine, and virtually all the land to the north including that around the IJsselmeer and some areas in the northeast.

Along the coast the polders are often protected from the sea by sand dunes. Away from the sea, natural levees and artificial dikes are the major – indeed the only – features of the flat landscape. Some of the polderlands are covered with grass, particularly in the delta and along the Belgian coast, but most are intensively cultivated. Near or below sea level, the polderlands form a characteristic and largely manmade landscape.

A land reclaimed

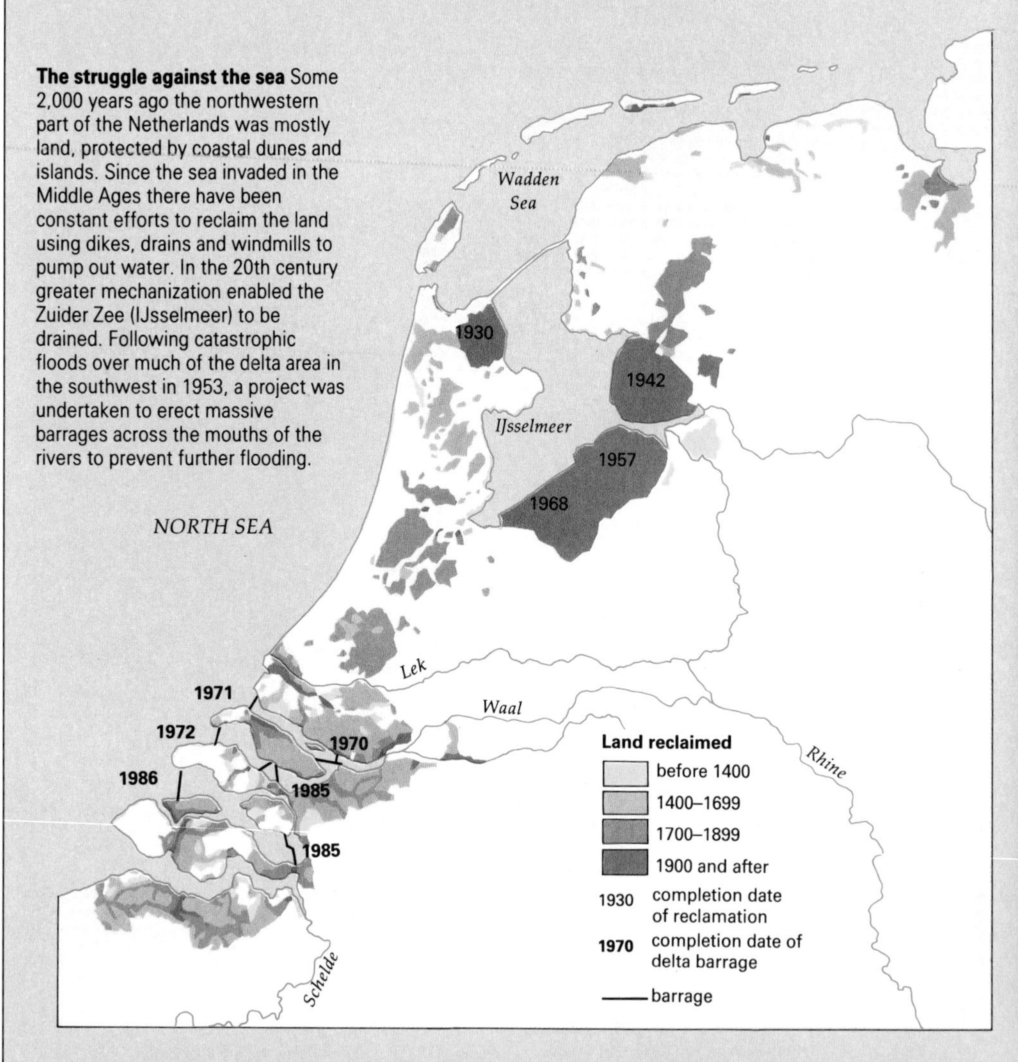

The struggle against the sea Some 2,000 years ago the northwestern part of the Netherlands was mostly land, protected by coastal dunes and islands. Since the sea invaded in the Middle Ages there have been constant efforts to reclaim the land using dikes, drains and windmills to pump out water. In the 20th century greater mechanization enabled the Zuider Zee (IJsselmeer) to be drained. Following catastrophic floods over much of the delta area in the southwest in 1953, a project was undertaken to erect massive barrages across the mouths of the rivers to prevent further flooding.

Were it not for resolute and sustained efforts to reclaim land, and to defend it against the ceaseless efforts of the sea to wrest it back, only about two-thirds of the Netherlands would exist. More than a quarter of the country lies well below sea level. Amsterdam's Schipol Airport is 5 m (16 ft) below sea level, built on an 18,000-hectare (45,000-acre) area of reclaimed land southeast of IJsselmeer known as the Haarlemmermeer polder. Only in the south of the country is there an appreciable area above 30 m (100 ft).

During the period of most recent glaciation some 20–15,000 years ago, a landmass linked Denmark to the British Isles where now there is the North Sea. As the climate warmed, the level of the sea rose. Sands were swept up to form a belt of dunes along the present southwestern coast and the West Frisian Islands. Behind these a large lagoon formed in the delta plain of the Meuse, Rhine and Schelde. In the 5th century AD gales breached the barrier of dunes, forming the deep inlet of the Wadden Sea and the islands of Zeeland, and enlarging the Zuider Zee (now called the IJsselmeer). Until the 7th century the people of the Frisian Islands built their settlements on clay mounds (*terpen*), which were less liable to flood. Then clay walls were built for the first time as a defense against the sea. Reclamation gradually continued by means of dikes and drainage ditches. The windmill, introduced in the 12th century, made it possible to pump water from low-lying land behind the dikes, and from marshes and lakes. The largest schemes, such as draining the Haarlemmermeer, had to await the 19th-century development of the steam engine. From 1600 onward, an average 800-1,200 hectares (some 2,000–3,000 acres) were added to the country each year by drainage and reclamation.

Draining the IJsselmeer

The greatest engineering feat was the closure and reclamation of much of the IJsselmeer. When it was approved in 1918, this scheme was intended to provide agricultural land and to reduce the amount of coastline exposed to the sea. However, since the 1950s reclamation has increasingly been used to relieve urban and industrial pressures on land elsewhere in this densely populated country. The scheme required the building of a 29 km (18 mi) dam across the mouth of the IJsselmeer, and the reclamation of five diked areas of polderland.

As the scheme progressed, the salty Zuider Zee, enclosed in 1927, gradually turned into the freshwater IJsselmeer (Lake IJssel), which is now used as a reservoir. Four polders totaling 168,000 hectares (420,000 acres) of land were drained between 1930 and 1968. The salt soils have been converted into fertile agricultural land; a whole infrastructure has been built on the transformed landscape, including roads, bridges, farms, villages and even a small city.

The fifth proposed polder, Markerwaard in the southwest, has not been reclaimed. Environmentalists objected to completion of the project because the distinctive freshwater plants and animals are a valuable ecological reserve and recreational asset. Recently the demand for agricultural land has also fallen.

Keeping the sea at bay Huge polder areas are created by shutting off the sea, sometimes with large barrier dams. The land is slowly pumped dry and gradually becomes less salty. Ninety percent of reclaimed land is intensively cultivated; houses, factories and roads are also built over the new land. The Netherlands is densely populated, and the polders are a valuable addition to the country's resources.

Linking the land A series of bridges links these old-established blocks of reclaimed land, which are separated by waterways. These rich wetland areas provide grazing for cattle and sheep, and have also been colonized by birds, insects and freshwater fish.

Coping with floods

So much of the Netherlands is below sea level that flooding is a constant threat. Major inundations took place in 1408, on a number of occasions in the early 16th century, in 1894, 1916 and 1953. All these are dwarfed by comparison with the 1421 Saint Elizabeth's Flood, when the Meuse dikes failed and 10,000 people lost their lives. This disaster also made permanent changes to the coastline, including the formation of the main estuary of the Meuse with a fenland at its head.

In January 1953 the high spring tides coincided with gale force westerly winds, which prevented the tide from ebbing. The incoming tide rose up to 3.6 m (12 ft) above the recognized danger level. Floodwaters forced 67 major breaches and some 500 minor breaches in the dikes, and inundated 16,000 hectares (40,000 acres), mainly in the islands of the southwest. The floods drowned 1,835 people and made 72,000 homeless. The flooded areas could be pumped dry only when breaches in the dikes had been repaired – yet every tide widened the breaches further. Sea sands were carried through the breaches, covering the agricultural soils and filling drainage ditches; it took a year for all the flooded areas to be reclaimed.

Later in 1953 investigations began into a scheme to close the sea arms of the Rhine–Meuse delta to prevent further flooding. Four broad, deep estuaries have been closed, and canals have been constructed to give continuing access to ports. Computer-controlled sluices determine the flow of water through the dikes. Another bold proposal is to build a dike linking the islands off the coast, and to reclaim the land between these and the mainland, which is at present covered by the Wadden Sea.

Future plans and current defenses are likely to be severely tested over the next hundred years by the consequences of the increase in the greenhouse effect. It has been calculated that sea levels will rise by about 1.5 m (5 ft) as the Earth warms and the polar ice caps melt. The Low Countries are one of a number of heavily populated low-lying areas of the world that would be affected by such a change.

PENINSULA OF THE MESETA

A PLATEAU LAND · THE PLAINS OF IBERIA · COASTS AND MOUNTAINS

The Iberian Peninsula lies between the Atlantic Ocean to the west and the Mediterranean Sea to the east. The snow-capped mountains of the Pyrenees isolate the peninsula from the rest of Europe, while to the south across the Strait of Gibraltar lies the continent of Africa. Much of the peninsula is an ancient high plateau, the Meseta, enduring searingly hot summers and freezing winters. Parts of the southeast are near desert. By contrast, along the highly indented coastline of the Atlantic in the southwest the fields are green and the climate wet and mild. Mountain ranges fringe the central plateau, from the Cantabrian Mountains in the northwest to the Sierra Nevada in the southeast. Lowlands are found in the coastal areas of the peninsula and along some river valleys, particularly that of the Guadalquivir.

COUNTRIES IN THE REGION

Portugal, Spain

LAND

Area 596,854 sq km (230,446 sq mi)
Highest point on mainland, Mulhacén, 3,482 m (11,408 ft); Pico de Teide on Tenerife, Canary Islands, 3,718 m (12,195 ft)
Lowest point sea level
Major features Meseta plateau in center, Cantabrian Mountains and Pyrenees in north, Sierra Nevada in south

WATER

Longest river Tagus 1,010 km (630 mi)
Largest basin Douro, 98,000 sq km (38,000 sq mi)
Highest average flow Douro, 312 cu m/sec (11,000 cu ft/sec)

CLIMATE

	Temperature °C (°F) January	July	Altitude m (ft)
Oporto	9 (48)	20 (68)	73 (239)
Lisbon	11 (52)	22 (72)	110 (361)
Santander	9 (48)	19 (66)	64 (210)
Seville	10 (50)	26 (79)	13 (43)
Ibiza	11 (52)	24 (75)	7 (23)

	Precipitation mm (in) January	July	Year
Oporto	159 (6.3)	20 (0.8)	1,150 (45.3)
Lisbon	111 (4.4)	3 (0.1)	708 (27.9)
Santander	124 (4.9)	64 (2.5)	1,208 (47.6)
Seville	73 (2.9)	1 (0.04)	559 (22.0)
Ibiza	42 (1.7)	5 (0.2)	444 (17.5)

NATURAL HAZARDS

Storms, flash floods, earthquakes in west

A PLATEAU LAND SURROUNDED BY THE SEA

The Iberian Peninsula is a huge block attached to the mainland of Europe by a wide neck of land along which stretch the Pyrenees. It originated as a fragment of the Hercynian continent, which formed across Europe in the Carboniferous and Permian periods – between 360 and 248 million years ago. During the Mesozoic era (248-65 million years ago) the continent broke up and large portions sank beneath the sea. Deposits laid down under the sea were crushed against this fragment by the northward push of the African plate in the Tertiary period (65–2 million years ago). They formed mountain ranges that include the Pyrenees and Cantabrian Mountains in the north and the Sierra Nevada in the south. Between them stretches the Meseta, a vast plateau which was protected from major folding by the tough underlying rocks of the ancient Hercynian continent. Parts of the Meseta were inundated by the sea some 8 million years ago at the end of the great period of Alpine mountain-building. Today limestone, marl and clay deposited by the sea form broad plains and the flat floors of basins amid the fringing mountains as, for example, to the west of the Sierra Nevada.

Maritime and continental climates

Although the peninsula lies mainly in the path of the westerly winds blowing from the Atlantic, there are great variations in climate. These are caused by the influences of the huge continental interior, the Mediterranean Sea, the mountains and the movement of the overhead Sun into the northern hemisphere in summer. Warm, dry winds periodically blow in from Africa's Sahara desert; during the winter all areas can experience cold conditions when cold air from the Meseta pushes down to the coasts.

In winter onshore winds bring cool, wet conditions to the Atlantic coasts. Before reaching the interior, the winds are forced to the north and south by a high pressure system that develops over the Meseta and keeps it dry and cold. "Nine months of winter and three months of hell," was how the poet Antonio Machado described his native town on the upper Douro river in the northern interior of the peninsula. The Mediterranean coast in the northeast receives most of its rain in the fall, especially in October, from winds blowing off the sea before the interior high pressure system becomes established. Farther south the coast lies in the rain shadow of mountain ranges and receives little rain.

The westerly wind belt moves northward in summer, following the movement of the Sun. The rain carried by these winds is restricted to the north and northwest of the peninsula. The Atlantic coast farther south is hotter and drier, because the stable air of the Azores high pressure system over the Atlantic to the west extends across the area. The interior warms up quickly in summer but remains largely dry. The rain in Spain falls mainly not on the plain but on the mountains, especially those in the northwest. However, though the mountains are the wettest areas, locally there can be great contrasts. The mountains in the southeast may have as much as 1,500 mm (60 in), but the neighboring coastal plains and the basins lying between ranges as little

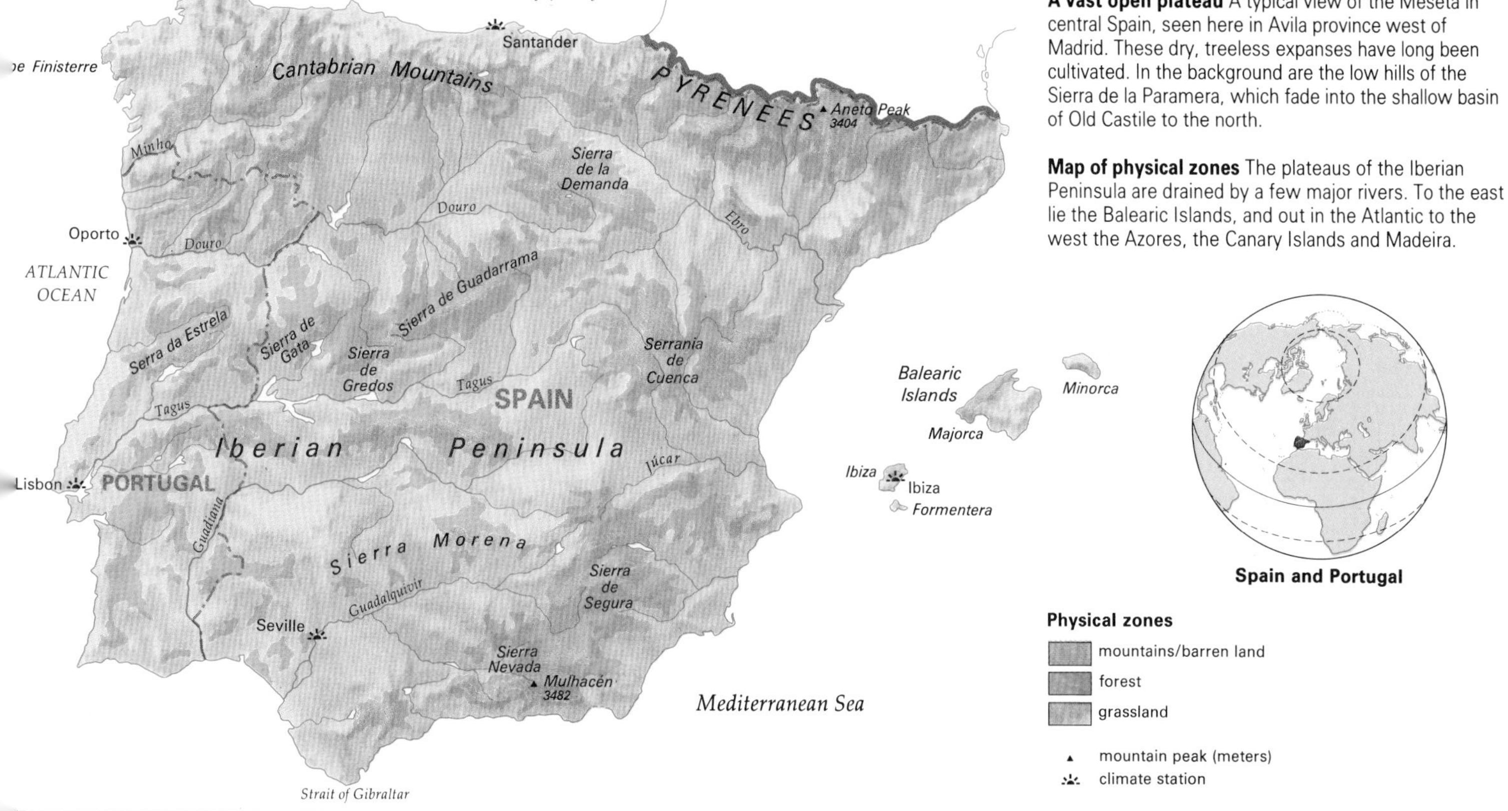

A vast open plateau A typical view of the Meseta in central Spain, seen here in Avila province west of Madrid. These dry, treeless expanses have long been cultivated. In the background are the low hills of the Sierra de la Paramera, which fade into the shallow basin of Old Castile to the north.

Map of physical zones The plateaus of the Iberian Peninsula are drained by a few major rivers. To the east lie the Balearic Islands, and out in the Atlantic to the west the Azores, the Canary Islands and Madeira.

as 300 mm (12 in). Such areas have a semiarid rather than a Mediterranean type of climate, with 370–520 mm (15–20 in) annual rainfall, mild, wet winters and hot, dry summers. On the high slopes of the Pyrenees snow lies all year; even in the Sierra Nevada in the southeast snow cover can last for seven months or more.

Changing vegetation patterns

During the cooler climates of the most recent ice age the uplands were probably grassland, much like the steppes of the Soviet Union today. When the climate became warmer and wetter woodland covered most of the area, but over the last 600 years this tree cover has been largely removed by farmers. The area seems to be becoming drier once more as warming caused by the greenhouse effect becomes more pronounced.

THE PLAINS OF IBERIA

Although tourists often favor the mountain and coastal areas, it is the vast expanses of the interior that dominate the Iberian Peninsula. The windmills that appeared as monsters to Don Quixote stood on a minor escarpment above the high plains of La Mancha in south-central Spain. They are still there today. As far as the eye can see there is a level plain dotted with windmills and small hills – some of them volcanic – that are crowned with huge castles.

The interior of Spain is a land mainly of plains and plateaus. The eroded surface of the Meseta, which flanks the Sierra de Guadarrama and the Sierra de Gredos, slopes very gradually away from the foot of the mountains across ancient granite rocks, to the centers of basins covered with gleaming white limestone laid down in the Tertiary period. Sometimes they are overlain by red soils derived from intensely weathered impurities in the limestone, reworked by the increased surface flow of water during warm spells in the ice age of the past 2 million years.

Hard, crystalline limestones form landscape features that are found in much of the peninsula. Some have become famous tourist attractions, such as the towers and caves of El Torcal some 80 km (50 mi) west of the Sierra Nevada, the caves of Majorca in the Balearic Islands, and the Rock of Gibraltar. In the spectacular Picos de Europa in the Cantabrian Mountains the

A lowland basin in southern Spain, formed when deposits were laid down by invasions of the sea. Towering above the plains are bare rock hills of crystalline limestone formed during the Tertiary period.

A natural amphitheater in the center of the Atlantic island of Madeira. A village clings tenuously to a block of rock, which has slipped from the slope above. The villagers have made use of the shallower slopes for their homes and terraced farmland.

limestones have been turned on end and deformed by earth movements associated with mountain-building. Intense mineralization took place in some areas, and almost invariably the limestones have been quarried for building materials. In the Sierra Morena, which marks the southern limit of the Meseta, there are deposits of copper, zinc, pyrite, lead and silver. In Estremadura in the southwestern interior the crystalline limestones have formed bleak, almost level plains.

The eroded plains of the peninsula are very ancient. Often, as in the Alentejo area of southeastern Portugal, the rivers have cut into the surface only slightly and the land slopes gently up to the mountains. Much of it is covered by red *raña* deposits, which are derived from weathered material transported from the distant mountains. Occasional isolated hills (inselbergs) stand above the plains. Two-fifths of the world's cork supply comes from trees in the Alentejo area, while its great plains are the breadbasket of the Iberian Peninsula.

Volcanic landscapes

The islands of Spain and Portugal have varied geologies. The Balearic Islands mirror the jagged limestones and plains of the mainland. By contrast the Azores, the Canary Islands and the Madeira Islands in the Atlantic are entirely volcanic.

In the west of the peninsula the platform of ancient, stable rocks has a high edge where the rocks are faulted and the major west-flowing rivers – the Minho, Douro and Tagus – fall steeply to the sea. The gorges of the Douro are especially wild. A large fault lies parallel to the Atlantic coast just to the north of the boundary between the African and European tectonic plates, which runs through the Strait of Gibraltar. Friction between these moving plates was responsible for the earthquake that devastated Lisbon in 1755. The town had to be completely rebuilt. Related to this tectonic activity are a number of volcanic structures, which form the attractive uplands of the southwestern Atlantic coast.

Rivers: a curse and a blessing

The rivers that drain the interior lands of Iberia include the Minho, Douro, Tagus, Guadiana and Guadalquivir flowing into the Atlantic, and the Ebro and Júcar flowing into the Mediterranean. They are both a curse and a blessing: a curse because floods are a serious problem in all parts of the country, a blessing because the rivers that rise in the mountains bring vital irrigation water to the plains.

The construction of dams along the major rivers has prevented much of the flooding, but the reservoirs in the drier areas soon fill with sediment carried from severely eroded lands. Even in the age of today's superdams a run of dry years can reduce reservoirs to dangerously low levels. In the drought of the early 1980s the reservoirs of southern Spain fell to one-eighth of their normal capacity.

In drier areas there is the problem of water becoming salty. As rainwater percolates through salt-rich rocks it removes the salts and carries them into the major rivers, such as the Ebro. Soils also become more saline when water used for irrigation evaporates from the surface. When too much irrigation water is applied the salts are flushed out into rivers and reservoirs. If underground water resources are overexploited near the coast, the sea may invade, making the water too brackish to drink or for irrigation.

RIVERS THAT COME AND GO

Early on a dark October morning in 1973, the villagers of Puerto Lumbreras in southeast Spain were at the market that had been set up in a dry riverbed. Suddenly, with a great roaring noise, a wave of water swept down the bed, drowning 143 people. At the same time, farther west along the coast at La Rábita an almost identical tragedy took place.

The major rivers of Spain and Portugal flow all year. However, many smaller ones flow only in the winter months or, in extreme cases, only every few years. A sudden storm in these areas can be disastrous.

This is what happened in 1973, when several catastrophic river spates were unleashed after a torrential rainstorm (*gota fría*). A long period without rain leaves the soils baked hard and the vegetation cover becomes thin. The heavy rain cannot soak into the soil. With little vegetation to slow it down, the water runs off the surface and a dry riverbed may suddenly fill. The water sometimes travels down the river bed as a bore, a steep wall of water that passes quickly but leaves behind enormous damage.

"Wadi" is the Arabic for the dry river courses that are subject to flash floods in North Africa and Arabia. In southern Spain they are called *ramblas*. They are characteristic of desert landscapes all over the world.

Though the river bed is dry, water often moves in the gravels beneath the surface. Centuries ago the Arabs in southern Spain developed complex systems of tunnels to drain this water for irrigation.

COASTS AND MOUNTAINS

A spectacular coastline Near Lagos on the southern coast of Portugal – an area of plains, low tablelands and few hills – sedimentary rocks outcrop on the coast. Rock arches have been shaped by the sea, and isolated stacks have formed where old arches have collapsed.

Jagged peaks and mountain lakes in the area of Montes Malditos in the central Pyrenees. These breathtaking mountains have been folded, then shaped by intense glaciation. The knife-edged ridge marks the rim of a cirque filled by Lake Gregueña.

In recent years the coastline of the Iberian Peninsula has been transformed by cement and asphalt, particularly along the Mediterranean shore, but there are still areas of great beauty and interest. Huge marshes at the mouth of the Guadalquivir provide important stopovers for birds migrating between Europe and Africa. Among the marshes are patches of permanently dry land that provide refuge for foraging animals. The river Ebro, which flows into the Mediterranean, has the largest delta on the peninsula. Until 1946 the delta was advancing at a rate of about 10 m (33 ft) a year, but the construction of reservoirs in the catchment area has dramatically reduced the sediment transported downstream. The delta is now eroding faster than the sediment arrives, and the coastline is receding as a result.

Raised beaches and drowned valleys

The hard rocks that form much of the peninsula's coastline preserve evidence of changes in sea level caused by the waxing and waning of the polar ice sheets during glacial and interglacial periods.

On the Mediterranean coast in the northeast, for example, there are wave-cut beaches 15, 30 and sometimes 60 m (50, 100 and 200 ft) above the present sea level. In the southeast, near the outlet of the river Almanzora to the east of the Sierra Nevada, the old beaches have been tilted by movements of the Earth's crust, so they increase in height toward the north.

Near the northwest coast the rivers cut down their valleys when the sea level was lower during glacial periods. As the melting polar ice sheets caused the sea to rise again the valleys were drowned, forming large estuaries called rias. These may penetrate 25–35 km (16–22 mi) inland. Between the Minho and Cape Finisterre the path of the valleys is determined largely by faultlines in the rocks; along the north coast, the rias follow the course of ordinary meandering valleys.

Shaping the mountains

During the ice ages of the last 2 million years snow and ice cover was widespread in the mountains and other high places. Glaciers were common, occupying valleys in the Pyrenees and the Cantabrian Mountains in the north and the Sierra Nevada in the south. The permanent snowline was around 2,000 m (6,500 ft) in the north, and rising to nearly 4,000 m (13,000 ft) in the warmer and drier south. It is now about 2,000 m higher.

Glaciation was most intense in the Pyrenees. The cooler, wetter northern slopes experienced the most ice action. Here jagged peaks and cirques abound in the headwater areas, and moraines block the valleys downstream.

Other mountainous areas, such as the Serra da Estrela in Portugal, and the Sierra de Guadarrama and Sierra de Gredos ranges in central Spain, were affected by a permanent cover of snow and ice as shown by the presence of periglacial features. These include strong contortions in the layering of soil, produced by ground ice, and the sorting of stone and finer material on the surface (referred to as patterned ground).

Pollen deposited in high peat bogs, for example those south of Grenada, provides a long record of climatic change. Windblown pollen is trapped in the bog each year, and a record of past vegetation, and consequently of climates, is preserved as the peat builds up in layers year by year. The deposits show that over the last 2 million years there has been a marked

GRANITE TORS

Huge boulders of granite, one stacked neatly on top of the other, rise out of hilltops or overlook valleys in the granite highlands of Portugal and the sierras of central Spain. From a distance they look like the towering ruins of bygone castles, but the size of the blocks suggests that only a giant could have handled them.

Such features are in fact completely natural, and are characteristic of granite landscapes in different parts of the world. In Britain they are known as tors, in southern Africa as castle kopje. On closer inspection the gaps between the boulders are found to be cracks or joints that separate a block from the surrounding rock. The vertical joints originally formed as the granite slowly cooled beneath the Earth's surface. As the overlying rocks were gradually removed by erosion, there was less weight on them and the consequent release of pressure caused horizontal cracks to develop. How these became enlarged and extended to form the separate boulders is a matter of debate.

The tors may have formed when the Iberian Peninsula was much colder. The cold conditions resulted in frost-shattering of the rocks and their transport away from the tor by muddy flows. The joints made it easier for freezing water to enlarge the cracks. Others believe they were produced by deep weathering in hot humid climates in the Tertiary period and then exposed by erosion, possibly during cold conditions.

fluctuation between a steppe vegetation of hardy grasses in cold, dry periods and tree cover in warmer, wetter periods. Tree cover would be the natural vegetation today over much of the region if most of the forests had not been cut down.

The highest mountains in the peninsula are in the Sierra Nevada in southern Spain. Here the snowcapped peaks that provide sun and snow for winter skiing reach nearly 3,500 m (around (11,500 ft) at Mulhacén before plunging steeply southward to the Mediterranean coast.

Distinctive mountain scenery is found in the gigantic limestone Picos de Europa in the middle of the Cantabrian Mountains. This area is famous for its caves – there are some of the largest and most complex underground cave systems to be found in Europe.

Spanish badlands

To travel beneath the northern slopes of the Sierra Nevada in southeast Spain is to pass through a landscape very much like desert. In places it rains on less than 20 days a year. Terraces on the dusty fields and check dams across dry riverbeds show the farmers' efforts to control the soil erosion that has been taking place in this dry region for thousands of years. Dry summers are followed by spasmodic winter rain, which causes brief but intense erosion. The problem is compounded by various factors.

The crumbly marl soils are easily eroded and broken down by flowing water; the mountainous slopes are very steep; and the heavy rain falls at the end of the dry season when the plant cover is very thin. In the past trees helped hold the soil in place, but they were cut down to supply fuelwood, mainly for metal smelting, and construction timber, especially for the Spanish navy. New growth was continually suppressed by grazing goats and sheep, so the area is now almost treeless. In some places virtually all the soil has been removed.

Catastrophic soil erosion

Erosion of the Earth is continuous and natural, but when soil is carried away much faster than it is formed by the weathering of the underlying rock the erosion is said to be catastrophic.

Erosion in the badlands of southeast Spain takes two forms. In sheet erosion water removes material more or less uniformly across the surface. Gully erosion creates deep channels as the intense rainfall is funneled down them, washing soil and loose rock into the beds of nearby streams. Gullies dramatically reduce farmland.

Gullies start where water has been naturally or accidentally concentrated. A leak from an irrigation canal or a deep tire impression may be enough to trigger the process. A gully may also form where

Lost lands A degraded environment near Baza in Granada province in southern Spain. Thousands of years of agriculture and deforestation have resulted in a sparse vegetation cover, which is unable to prevent the soil being washed away. Deep ravines result. Conditions in this naturally arid area are made worse by the lack of rainfall for long periods.

A dry, hilly landscape near Almería on the Mediterranean coast of southern Spain, where it rains on less than 20 days a year. Human activity contributes to the natural processes of desertification. Where deforestation has taken place water runs rapidly off the land because there are few plants to hold it back, causing V-shaped gullies to be formed.

rainwater has percolated into the ground and has collected below the surface to form pipes. The pipes become larger and may then collapse, providing a hollow for a new gully to develop. Unless action is quickly taken the gully will keep growing until there is no more space for it to grow. Eventually little farmland is left, and the land then has to be abandoned.

Creating deserts

In addition to erosion, good land may be changed into badland by the increasing incidence of droughts, as well as by the use of farming methods that are inappropriate for the prevailing conditions of soil and climate. Desertification is usually caused by a combination of both. Soil erosion in this part of Spain would naturally be very active, but the situation is aggravated by severe overgrazing, which reduces the vegetation cover. It is difficult to establish whether the climate is getting progressively drier as well, because good records of rainfall only go back about 50 years. There were very dry years in 1944–45, 1956–57, 1966–67, 1973–75 and 1979–83. Going farther back than this requires some detective work, such as examining newspaper reports and the price of crops at market, as well as searching for clues in soils and in the sediments of lakes and reservoirs.

Southeast Spain has been inhabited for thousands of years, and the combined effects of agriculture and deforestation have almost stripped it of vegetation. This has caused erosion during a period when the climate has probably changed very little. With summer temperatures expected to warm up over the next 30 years, the land will dry out even more, and severe erosion is likely to become more widespread unless action is taken.

The effects of erosion are not only felt locally. There is a risk of flooding farther down the valley. As the soil is lost the capacity of the land to hold water is diminished and plants cannot survive. Runoff is greater and more rapid, so rivers fill up quickly and overflow their banks. The eroded sediment also fills up reservoirs and shortens their life span.

Fortunately economic changes may now be working in the right direction. Dry cultivation of wheat and olives is becoming much less important and the number of goats and sheep is declining; mountain lands are increasingly being developed for tourism.

PENINSULAS AND ISLANDS

ROCKS AND HISTORY · CLIMATES OF PAST AND PRESENT · CHANGING LEVELS OF LAND AND SEA

An irregular coastline betrays the turbulent geological history of the central Mediterranean. From the mainland of Europe the peninsulas of Italy and Greece reach farther south than the northern limits of Africa. Italy extends from northwest to southeast, lying parallel to Greece with its more indented coastline and many islands. Most of the land is mountainous or hilly and, especially in the south, severely eroded. The steep, sun-bleached slopes of limestone rock have little soil cover. Italy's Po valley is the most extensive lowland. To the north the barrier of the Alps includes ice-covered peaks above 4,000 m (over 13,000 ft) in the west and, farther east, the jagged Dolomites. The natural woodlands have been deforested and aromatic, shrubby maquis or more open vegetation has replaced them.

COUNTRIES IN THE REGION

Cyprus, Greece, Italy, Malta, San Marino, Vatican City

LAND

Area 442,778 sq km (170,913 sq mi)
Highest point Monte Rosa, 4,634 m (15,203 ft)
Lowest point sea level
Major features Alps, Apennines, Pindus Mountains, Po valley, islands including Sardinia, Crete, Greek archipelago and Cyprus

WATER

Longest river Po, 620 km (380 mi)
Largest basin Po, 75,000 sq km (29,000 sq mi)
Highest average flow Po, 1,540 cu m/sec (54,000 cu ft/sec)
Largest lake Garda, 370 sq km (140 sq mi)

CLIMATE

	Temperature °C (°F) January	July	Altitude m (ft)
Genoa	8 (46)	24 (75)	21 (69)
Venice	2 (36)	23 (73)	2 (7)
Messina	11 (52)	26 (79)	54 (177)
Salonika	6 (42)	28 (82)	69 (226)
Athens	9 (48)	28 (82)	107 (351)

	Precipitation mm (in) January	July	Year
Genoa	79 (3.1)	40 (1.6)	1,270 (50)
Venice	50 (2.0)	67 (2.6)	854 (33.6)
Messina	146 (5.7)	19 (0.7)	974 (38.3)
Salonika	45 (1.7)	23 (0.9)	465 (18.3)
Athens	62 (2.4)	1 (0.04)	339 (13.3)

NATURAL HAZARDS

Earthquakes and volcanic eruptions, storms and floods, landslides, avalanches

ROCKS AND THE HISTORY OF THE CENTRAL MEDITERRANEAN

The nature and distribution of landscape types in the central Mediterranean region reflect the enormous power of forces within the Earth. Across the north of Italy, the Alps form a high arc that tops 4,400 m (over 14,500 ft) near Mont Blanc on the border with France and at Monte Rosa, Italy's highest peak, on the border with Switzerland. The commonest rocks are resistant granite, gneiss and schist. Limestone is more usual to the east and south, and forms the Dolomites.

Below the towering Alps spreads the plain of the river Po. The flat, low-lying expanses are naturally prone to flooding. Huge quantities of sediment have been washed into what was once a shallow arm of the Adriatic Sea. The basin has subsided; the accumuated sediments of clay, sand and gravel are now up to 10 km (6 mi) deep. Similar, but smaller, lowlands are found elsewhere in Italy and Greece.

The rest of the region is dominated by the Apennines, whose characteristic limestones form the backbone of the peninsula, and by the similar terrain of the Pindus Mountains and the Peloponnese in Greece. Clays in many areas produce gentler contours, as in Sicily. By contrast, in the Promontory of Gargano jutting east from Italy into the Adriatic Sea, in other areas of southeast Italy and in the islands of Malta and Gozo to the south, the beds of limestone are almost horizontal and form plateaus.

Two other landscape types are the volcanic terrains of the island of Thira (Santorini) in Greece, also found around Vesuvius in central Italy and Etna in Sicily, and the rugged mountain scenery of the generally older igneous and metamorphic rocks of Sardinia.

The mountainous land that occupies two-fifths of the region has a general trend from northwest to southeast. This can be seen in the 1,000 km (600 mi) long Italian peninsula, but is clearest in the shorter, more massive peninsula of Greece. Here individual ranges of the Pindus Mountains extend like fingers of land into the Aegean Sea, often continuing as chains of small islands that arc east, then northeast. Crete, the largest of the Greek islands, stretches across the southwestern entrance to the Aegean. Its mountains are remnants of a block that was thrown up and tilted in Tertiary times, 65–2 million years ago. The relatively straight lines of the Italian coast indicate that the land has been raised, while the more indented coastline of Greece points to invasion by the sea as the land has sunk.

A complex collision course

The central Mediterranean region consists of zones of great disturbance in the Earth's crust, separated by areas of relative stability. The basin of the Mediterranean Sea lies astride the boundary of two massive tectonic plates: the Eurasian plate to the north and the African plate to the south. About 100 million years ago immense forces within the Earth began to push the African plate gradually north, forcing its leading edge beneath the Eurasian plate and bringing the harder rocks of the African continent toward those of Europe. The igneous rocks of the high mountains in southwest Cyprus were formed as molten lava was forced to the surface by these movements. Finally the Tethys Sea between Africa and Europe was closed when the continental

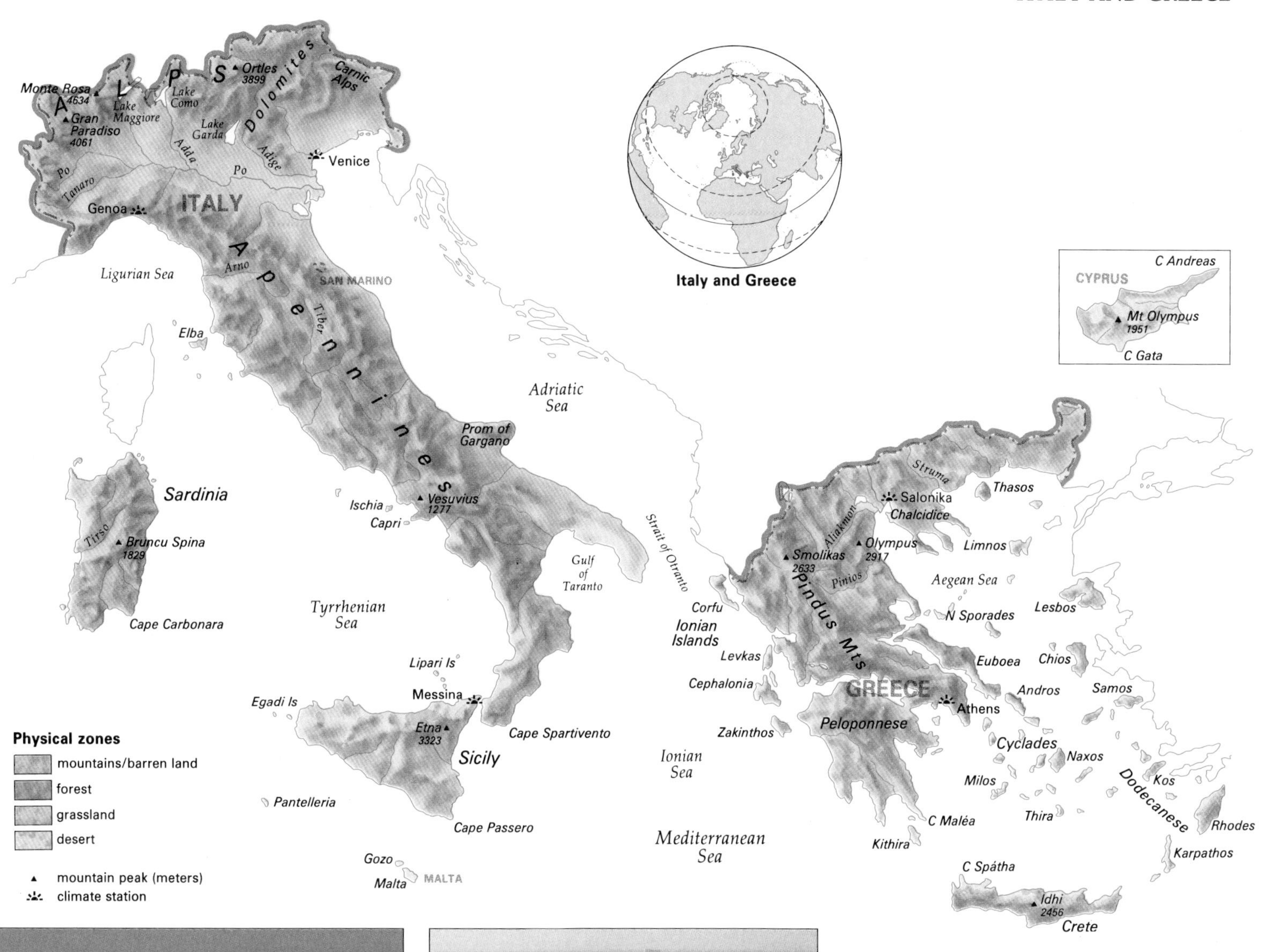

Map of physical zones The two peninsulas of Italy and Greece stretch out into the Mediterranean Sea. Their complex coastlines and numerous islands demonstrate the long history of volcanic activity in the region.

The Blue Grotto of Malta is a feature of limestone coasts such as those found on the islands of Malta and Capri. The grottoes acquire their rich blue color from phosphorescent organisms on the sea bed and sunlight shining through clear water.

Spring in Tuscany, in the northern part of the Italian peninsula. Gently folded sandstones form the higher land; clays and alluvial soils are found in the valleys, where vines and olives have long been cultivated.

blocks met, compacting its limy sediments into rocks. These rocks were then crumpled, about 30 million years ago, to form the high Alpine ranges that stretch from the Atlas Mountains of northern Africa, through the Apennines, Alps and Pindus Mountains of Europe to the mountains of Turkey and southern Iran. The Mediterranean basin, successor of the Tethys Sea, dried and filled up several times before the Atlantic burst in for the last time some 5 million years ago.

The contorted shape of the Alpine mountain belt shows that it was no simple collision. The African plate rotated against the southern edge of the Eurasian plate, so the boundary zone fractured and fragmented into a number of small plates that ground against each other. The forces of compression caused rocks to melt at

The Vouraikos valley in the Peloponnese runs north into the Gulf of Corinth through the bare limestone hills of the Arcadian plateau. The valley is typical of the many small lowland areas found between the mountains that dominate the Greek landscape.

depth; magma has erupted at the surface to form volcanoes. This tectonic activity still continues, as the widespread occurrence of earthquakes shows.

Patterns of change

The Italian peninsula once lay adjacent to France and Spain, but it has rotated anticlockwise through nearly 90°, and continues to move northeast at about 2.5 cm (1 in) a year. Meanwhile, the shallow Adriatic Sea to the east, with a depth generally less than 200 m (660 ft), is becoming smaller, and the Tyrrhenian Sea to the west wider and deeper, with a maximum depth of 3,371 m (10,060 ft), even though it is filling with sediment.

The Ionian Sea, lying between southern Italy and western Greece, is the deepest part of the Mediterranean, with a maximum depth of 4,901 m (16,080 ft). It covers a basin that is subsiding in front of the Aegean plate as the plate moves gradually southwest. The rocks of southeast Sicily include a small portion of Africa that has become detached, providing further evidence of rifting and rupturing. Sardinia, like its northern neighbor Corsica, is an uplifted splinter of the most ancient rock foundations.

CLIMATES OF PAST AND PRESENT

It is a popular myth that the central Mediterranean enjoys an enviable climate with summers hot enough to grow oranges and lemons, and winters warm enough to attract tourists in search of the winter sun. Although certain regions do have a pleasant, mild, sunny climate – the Riviera coast of northwest Italy, Sicily and Malta, for example – the reality is very different. These are large areas of mountains where conditions are harsh, capricious and extreme.

Climatic conditions in the region are influenced by latitude, by longitude and by altitude. The latitudinal position determines the overall character of the Mediterranean climate. The hot, dry summers are associated with the subtropical high-pressure belt that moves north with the overhead sun and brings anticyclonic conditions. When the anticyclone moves south in winter it is replaced by generally lower pressure, bringing rain-bearing winds from the Atlantic in the west.

However, there are strong north–south variations in climate. Northern Italy adjoins the European mainland, but Sicily and Malta – more than 10 degrees to the south – are almost part of Africa, and a mere 200–300 km (120–190 mi) from the searing heat of the Sahara. The contrast is more noticeable in winter, when Sicily and Malta can be bathed in warmth while temperatures in the Po valley drop below freezing.

The longitudinal position affects the balance between maritime and continental influences. In general winters become colder and drier from west to east as the maritime influence of the Atlantic Ocean decreases. The continental influence is most evident when bursts of intensely cold air, known as the maestrale in Sardinia and the bora on the Adriatic coast, escape southward from central and eastern Europe – winds in stark contrast to the burning sirocco, which blows north from the Sahara in spring and summer.

High land complicates the pattern still further. The winding 1,200 km (750 mi) long spine of the Apennines has many ridges and peaks above 1,500 m (4,900 ft). The range continues into the high Alps in the north. These mountains form a barrier to moisture-bearing air moving in from the Atlantic. Precipitation increases with altitude on the western slopes, and because of the low temperatures the snow often lies as late as May; lower ground in the rain shadow to the east has less rain. The pattern is similar over the lower mountains of Greece.

The central Mediterranean climate, therefore, is far from benign. Heatwaves and cold spells are common, and heavy downpours frequently ravage mountains whose rocks have been weakened by the baking of the sun and the splitting action of intense frost.

Legacies of climatic change

The present climate contains extremes, but it has been harsher in the past. During the cool or cold phases of the last 2 million years, snowfields and small ice caps developed in the mountains. The legacy of the ice age can still be found. Some 114 sq km (44 sq mi) of ice remains in the Italian Alps, and glacial features such as armchair-shaped rock basins (cirques), U-shaped valleys, scooped-out basins such as those occupied by Lakes Como and Maggiore, and ramparts of glacier debris (moraines) such as those that dam Lake Garda all testify to widespread glacial activity. Glaciers also developed farther south, as can be seen from cirques in the northern Apennines above 1,400 m (4,600 ft) and in the central Apennines above 1,600 m (5,250 ft).

Meltwaters from the glaciers shaped

The jagged peaks of the Dolomites in northeast Italy are part of the Alpine mountain system that forms the northern frontier. Some 65 million years ago the steep limestone cliffs lay on the sea bed.

landscapes all along the Alpine fringe, while farther south the cold phases resulted in widespread freeze–thaw (periglacial) activity, including rock shattering and the downslope movement of soil and rock debris (solifluction).

Two further factors stimulated erosion. The repeated accumulation of huge ice sheets in polar latitudes caused Atlantic depressions to track farther south, which had the effect of increasing precipitation in the region. At the same time sea level was 100–140 m (330–460 ft) below where it is now, stimulating the rivers to cut deeper through the rocks.

A LAND STRIPPED TO THE BONE

"Like a dead animal with bones showing" was one description of the bare limestone hills of Greece. It is not the only place, however, where the damage from the combined impact of human settlement over more than 5,000 years and of the forces of nature can be seen. Overgrazing by sheep and goats and clearance of the forests have degraded the natural vegetation that once covered the uplands and mountains. With little to protect the soil, the vagaries of climate and the generally steep terrain have led to widespread and severe erosion as well as frequent landslides. What caused the "bare bones" of Greece has also created the *calanchi* (badlands) extending over 2,500 sq km (970 sq mi) of clay deposits in Italy.

The debris produced by this considerably accelerated erosion has been deposited either as valley infills or along the coast, where it forms coastal lowlands. Investigation of these deposits has revealed how sheets of gravels, sands and clays were often laid down in phases, some of which reflect historic changes in land use.

Over longer periods of time, tectonic movements have raised the level of older deposits, which have later been eroded by rivers so that two or three separate infills are exposed. These deposits often contain distinctive human artefacts, and they can therefore be dated by archeologists.

CHANGING LEVELS OF LAND AND SEA

Climate, relief and sea level are dominant factors in the shaping of land surfaces. In the central Mediterranean basin the level of the land and of the sea has changed significantly over the past million years. Evidence of their former relative positions can still be clearly seen.

Land surfaces are changed by wind, water, ice and gravity working on rocks and exploiting lines of weakness. These weaknesses become the ridges, hills and valleys of the landscape. Such features are created because glaciers, rivers and landslides all tend to move material downhill. This tendency is increased if the land surface is raised and slopes become steeper, so the vertical movements of the Earth's crust in the region have accentuated the activities that sculpt the surface of the land.

Earth movements

The Alps have been lifted by some 4,000 m (13,000 ft) over the last 30 million years, while the valley of the Po adjacent to them lies in a subsiding basin. The Apennines were created when sediments were compressed by parts of the Earth's crust moving toward one another 20–3 million years ago, and subsequently lifted by huge vertical movements of up to 3,000 m (10,000 ft). In the far south massive blocks bounded by faults tower over nearby lowlands where the land sank and filled with sediment.

The influence of these movements on the coastline has been profound. The square shape of the Gulf of Taranto, which separates the heel of Italy from the toe, is defined by faults, as is the Promontory of Gargano. Most of Italy's west coast has an irregular pattern of faults created when the western half of the original Apennines collapsed beneath the Tyrrhenian Sea. The frequent earthquakes show that the Apennines are still actively evolving. Much of the southern Apennine landscape is formed of jagged summits, crags and steep-sided, narrow valleys drained by torrents, because erosion has not kept pace with the rise of the land.

A third of Italy's population lives in areas classified as being at high seismic risk. Damaging tremors take place on average once every four years, and huge shocks periodically cause many deaths and extensive devastation. More than 85,000 people were killed at Messina in northeast Sicily in 1908; on the mainland, 927 died in Friuli in the northeastern in 1976, and 2,909 at Laviano in the Apennines west of Vesuvius in 1980.

The effect of tectonic movements can clearly be seen in the indented coastlines of Greece and its archipelago, which show the impact of local subsidence. Elsewhere the land has been much higher than it is today. Dramatic evidence of this comes from southern Italy, where staircases of raised beaches are found. One excellently preserved staircase on the

The island of Thira in the Aegean has undergone many eruptions. It is the remnant of an old caldera that collapsed in a huge eruption about 3,500 years ago, creating clouds of ash and devastating waves.

The highly prized marble of Carrara in northern Italy has been quarried for hundred of years. It is found in bands in a bed of limestone about 900 m (3,000 ft) thick. Marble is a coarse-grained crystalline rock, composed principally of calcite and dolomite, that is formed when limestone is metamorphosed.

THE THREAT TO VENICE

In November 1966 St Mark's Square in Venice was submerged under more than a meter of water. A combination of torrential rain, a high tide and the northerly sirocco wind had caused the flooding. High-tide levels have risen some 25 cm (10 in) since 1900, and parts of the city are now flooded up to forty times a year.

Venice is built on more than a hundred islets in a lagoon at the head of the Adriatic Sea. Its fate has always been bound up with the sea; the floods of 1966 pointed to a dramatic change in relationship. There are four possible reasons. Land subsidence due to tectonic activity in the Earth's crust is probably negligible, as is compaction of the soft sediments of the Po delta under the weight of buildings. However, the pumping of water from rocks beneath the lagoon caused about 14 cm (5.5 in) of subsidence; this water-source was exhausted in 1975. Perhaps most important is the global rise in sea level caused by melting ice sheets. Currently estimated at just over 1 mm per year, it accounts for another 10 cm (4 in) rise during the past century. This eustatic change is continuing, and the rate may accelerate if the predictions about future global warming are proved correct.

Venice must be protected, but the question is how. The most likely solution involves the construction of four raisable barriers to link the line of islands forming the outer limit of the lagoon. This would allow shipping through to the port of Marghera, but maintain the lagoon at a safe level and keep out the high-tide surges.

Ionian Sea coast has at least six broad platforms, which rise to 60 m (200 ft) above sea level.

Such features could not have been caused by a worldwide drop in sea level. If all the ice on the globe were to melt, the level of the oceans would rise by some 50 m (160 ft). This has almost certainly never happened during the ice ages of the last 2 million years, and it is unlikely that sea levels have been much more than 10 m (33 ft) higher than those at present along the coastlines that have been stable. In parts of southern Italy deposits that were laid down by the sea early in the Quaternary period can be found at heights of up to 945 m (3,100 ft), and evidence of 200 m (660 ft) of land uplift is widespread. These figures average out to give a rise of just 0.03–1.5 mm per year, but major uplift probably took place in spurts.

The town of Pozzuoli near Vesuvius is built in the Phlegraean Fields, an active volcanic area of low craters, hot springs, and vents in the ground known as fumaroles from which gases, acids and steam escape. It was a thriving port in Roman times, with a monumental market place whose ruins are today known as the temple of Serapis. Surviving columns bear marks made by shellfish in a band that is between 4.5 and 7 m (15–23 ft) above the present high-water level. This led to the suggestion that the sea level first rose 8.5 m (28 ft) and then fell by 7 m (23 ft) in the last 2,000 years.

However, it is now known that Pozzuoli provides a unique example of abrupt changes in the level of the land rather than the sea. These changes are the result of molten rock (magma) rising through the crust of the Earth. The magma bubbles cause the ground surface to bulge. Between 1969 and 1983 the port rose by 1.7 m (5.6 ft), and between June 1983 and January 1985 it rose another 1.8 m (5.9 ft), leaving the harbor quay 3.3 m (10.8 ft) higher than in 1970. Fears of a volcanic explosion led to the temporary evacuation of 40,000 people, though the bulge now appears to be subsiding. The shellfish boreholes in the temple of Serapis show that such changes in the level of the land have happened for at least 2,000 years. They may have been taking place ever since volcanic activity began in the area some 35,000 years ago.

Volcanoes of the Mediterranean

In terms of the hazards of nature, Italy is probably the most dangerous place to live anywhere in Europe. Over a third of the population is at risk from earthquakes, volcanoes, landslides, seismic sea waves (tsunami), floods or, in the northern Alpine regions, avalanches. Some 135,000 people have died in natural catastrophes during this century alone.

Volcanoes look awesome, but compared with earthquakes they cause few deaths, except when they erupt without warning. Despite more or less continuous activity the first deaths caused by Etna since 1843 were in 1979, when nine tourists were killed by an unexpected explosion on the flank of the volcano.

Volcanic activity in the Mediterranean is associated with the later stages of mountain-building, when molten rock (magma) migrates upward along fractures in the rocks. During the last 2 million years volcanoes have erupted in many parts of Italy: between Lake Garda and the Adriatic Sea; in the northwest on the other side of the peninsula; in the vicinity of Vesuvius; and in the south in the "instep" of the boot-shaped peninsula. The fire god Vulcan was once worshipped on Vulcano, southernmost of the Lipari Islands; farther south looms the huge mass of Etna in eastern Sicily; two small volcanic islands lie to the southwest.

Volcanoes mark a point where molten rock from the Earth's interior has forced its way to the surface. When a major eruption is imminent more smoke comes from fumaroles; then pressure blows out the old plug that blocked the vent.

In the past Greece has also had its fair share of volcanic activity on or near the island arc that extends southeast from the mainland. The most famous instance was the catastrophic explosion, in about 1500 BC, of Thira (Santorini), whose sudden destruction may have given rise to the myth of the lost civilization of Atlantis, said to have been engulfed by the sea.

Virtually all the volcanoes in the Mediterranean are strato-volcanoes. They have a central vent surrounded by a steep-sided cone formed from layers of lava and ejected debris, such as ash. Most of the molten rock beneath the surface of the Earth in this region is dark in color, rich in iron, relatively poor in silica and relatively fluid. This magma normally erupts fairly quietly in the form of lava flows, but as it rises it can be contaminated by the melting of local rocks, changing its chemical composition so the eruptions become more violent and eject debris ranging from ash to huge volcanic bombs that weigh several tonnes.

Mainland Europe's only active volcano, Vesuvius, is probably one of the best-known volcanoes in the world. Volcanic activity began some 10,000 years ago, since when there have been four different volcanoes on the site. The most devastating eruption on Vesuvius came in AD 79, when huge volumes of debris were ejected, and the falling ash (tephra) buried the towns of Pompeii and Stabiae and their inhabitants. Nearby Herculaneum was overwhelmed by a glowing cloud of gas, dust and mud that rolled down the slopes of the volcano. The average interval between eruptions on Vesuvius is 24 years; the last eruption was in 1944. The volcano shows few signs of life, but it poses a major threat to the growing city of Naples just to the west.

Eastern Sicily is dominated by Etna, a huge pile of lava and ash that rises to 3,323 m (10,902 ft) and has a circumference at its base of about 150 km (90 mi). Etna has been continuously active for at least 2,500 years; a catastrophic eruption took place in 1669. The economic wealth of this part of Sicily owes much to the volcano, and its lower flanks support more than thirty-six towns and villages, some of which are built on lava flows only 70 years old. The volcano invites settlement because of the highly fertile soils derived from the volcanic rock, its wetter climate, its groundwater for irrigation and its negligible soil erosion. What makes living on Etna possible is the volcano's gentle style of eruption, with lava flows that slow to walking pace within little more than a kilometer (about a mile) of their source.

Attempts are being made to reduce the likelihood of damage from future eruptions. There have been investigations into diverting lava flows using dams, barriers, channels, dynamite and cold water. Scientists also monitor the volcano. They test the gases it emits, record vibrations caused by rising magma and, by using radio signals from satellites, measure swellings even of just a few millimeters in the land around Etna. Dangers can also be forecast using records of previous eruptions. A hazard map has been drawn up that alerts the inhabitants to possible risks, and offers guidance to those planning development nearby.

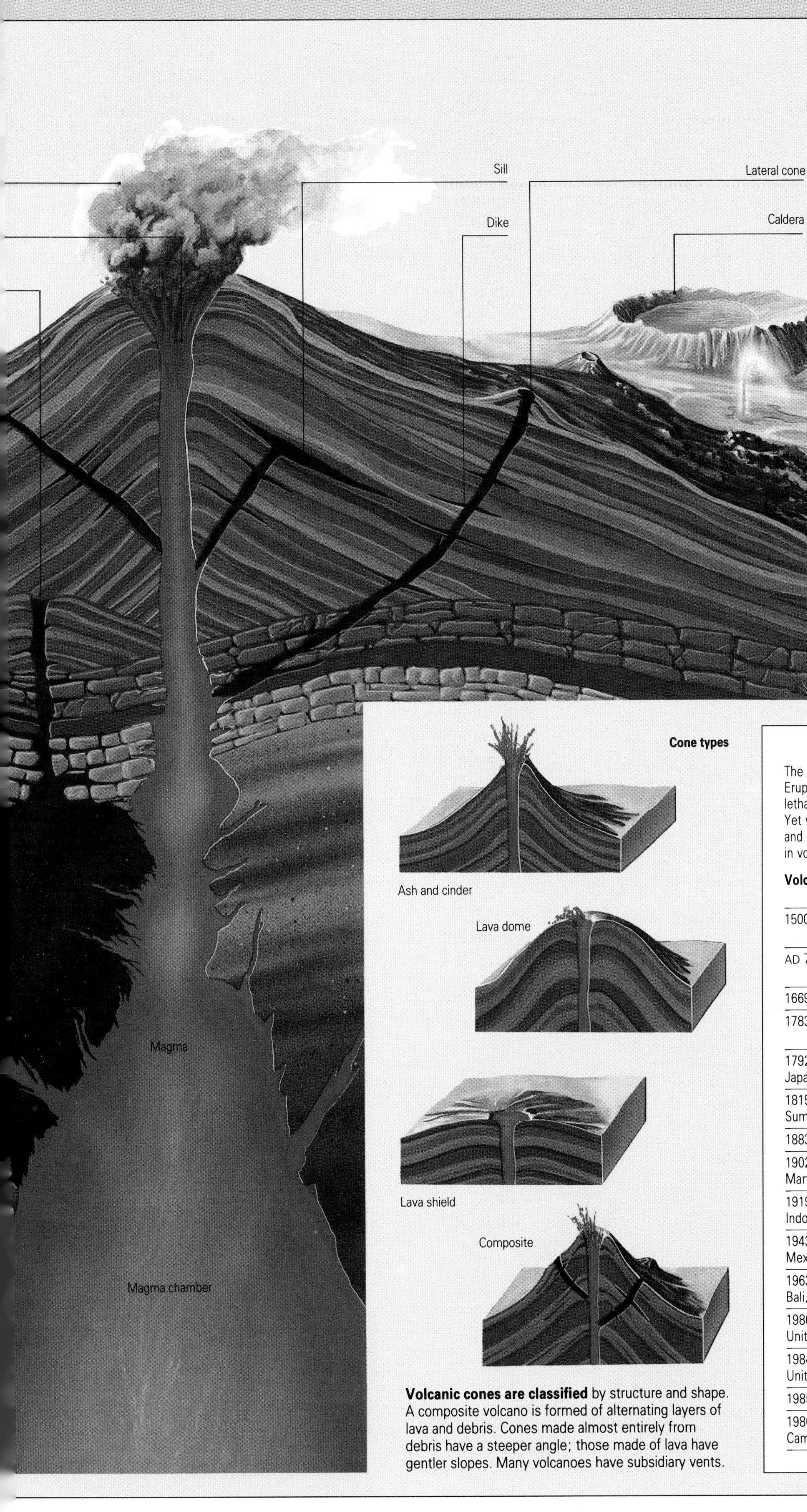

A glowing eruption of Etna, Sicily's famous volcano. It is one of many active volcanoes in the region; unusually, it never lies dormant but is continuously active. Its eruptions are usually predictable, and pose little threat to the people who live on its slopes.

Volcanic cones are classified by structure and shape. A composite volcano is formed of alternating layers of lava and debris. Cones made almost entirely from debris have a steeper angle; those made of lava have gentler slopes. Many volcanoes have subsidiary vents.

GREAT ERUPTIONS

The destructive power of volcanoes is awesome. Eruptions are often accompanied by earthquakes, by lethal mudflows or by seismic sea waves (tsunami). Yet volcanoes provide fertile soils, mineral deposits and geothermal energy, so many people choose to live in volcanic areas.

Volcano	Deaths	Contributory factors
1500 BC Thira, Greece	Life extinguished	
AD 79 Vesuvius, Italy	2,000	hot gas, dust, mudflow
1669 Etna, Sicily, Italy	20,000	earthquake
1783 Laki, Iceland	10,000	gas poisoning, famine
1792 Unzen, Kyushu, Japan	10,000	
1815 Tambora, Sumbawa, Indonesia	56,000	tsunami, famine
1883 Krakatau, Indonesia	36,000	tsunami
1902 Mt Pelée, Martinique	30,000	hot dust, gas
1919 Kelud, Java, Indonesia	5,500	mudflow
1943–52 Paricutín, Mexico	1,000	new volcano
1963 Gunnung Agung, Bali, Indonesia	1,600	
1980 Mount St Helens, United States	60	
1984 Mauna Loa, Hawaii, United States	none	lava flows
1985 Ruiz, Colombia	22,000	mudflow
1986 Lake Nyos, Cameroon	1,700	gas bubble from crater lake

Pillars of history

Atop one of the 24 giant pinnacles of Meteora in northern Greece is perched the monastery of Roussanou. There are several similar monasteries in the area that have provided havens for monks over the centuries, though few are permanently inhabited today, and access is more often by a new path than by the block and tackle of former times.

Meteora (Greek: "in the heavens above") is situated on the eastern flank of the Pindus Mountains, on the western edge of the fertile plain of Thessaly. Millions of years ago thick deposits of sandstone and conglomerate (pebbles bound together by a natural cement) were laid down under a sea. About 60 million years ago the area was pushed up by movements of the Earth to form a high plateau. The movements fractured the rocks, exposing new surfaces to the elements. The rocks weathered along these lines of weakness, and the more resistant blocks have remained as towers. They average about 300 m (1,000 ft) in height.

Weathering has helped to reveal the history of the rocks. On exposed surfaces the pebbles of the conglomerate stand proud as the fine-grained "cement" binding them together is eroded away. Weathering along the bedding planes of the sandstone towers has revealed the individual layers of rock, sometimes horizontal just as they were laid down under the sea and sometimes tilted as a result of Earth movements.

The famous peaks of Meteora in Greece tower above the plain of Thessaly. These ancient rocks reflect a long history of Earth movement and erosion.

MOUNTAINS AND RIVERS OF EUROPE

LANDSCAPE HISTORY AND CLIMATE · RIVERS, UPLANDS AND LOWLANDS · THE ALPS – THE BACKBONE OF EUROPE

The landscapes of central Europe give a snapshot of processes involved in the shaping of the entire continent. South of the shores and estuaries of the North and Baltic seas, the flat plain is interspersed by occasional heathlands. Sand, gravels and fertile loess soils were laid down by ice sheets in the recent geological past. The volcanic landscapes of Germany's central uplands, which were affected by very cold tundra conditions, have since been dissected by the wine-producing valleys of the Lahn, Moselle, Neckar and Rhine. In the far south of Germany and curving in an arc east–west across Switzerland and Austria, the Alps are a geologically younger land with spectacular mountain, meadow and lake scenery shaped by glaciers. The Rhine, one of Europe's great waterways, flows from the south into the North Sea.

COUNTRIES IN THE REGION

Austria, Liechtenstein, Switzerland, West Germany

LAND

Area 373,756 sq km (144,270 sq mi)
Highest point Monte Rosa, 4,634 m (15,203 ft)
Lowest point sea level
Major features High Alps, Bohemian Forest, Black Forest and other uplands in center, lowlands in north, Rhine rift valley

WATER

Longest river most of the Rhine's 1,320 km (820 mi) length and 252,000 sq km (97,000 sq mi) basin are in the region, also the upper part of the Danube.
Highest average flow Rhine, 2,490 cu m/sec (88,000 cu ft/sec)
Largest lake Constance, 540 sq km (210 sq mi)

CLIMATE

	Temperature °C (°F) January	July	Altitude m (ft)
Hamburg	0 (32)	17 (63)	14 (46)
Zurich	−1 (30)	18 (64)	569 (1,886)
Lugano	2 (36)	21 (70)	276 (905)
Munich	−2 (28)	18 (64)	528 (1,732)
Vienna	−1 (30)	20 (68)	212 (695)

	Precipitation mm (in) January	July	Year
Hamburg	57 (2.2)	84 (3.3)	720 (28.3)
Zurich	75 (2.9)	143 (5.6)	1,137 (44.8)
Lugano	63 (2.5)	185 (7.3)	1,744 (68.7)
Munich	59 (2.3)	140 (5.5)	964 (38.0)
Vienna	40 (1.6)	83 (3.3)	660 (26.0)

NATURAL HAZARDS

Avalanches and landslides in mountains

Reflection of a mountain peak in the still waters of Backalpsee near Grindelwald, Switzerland. Glaciated scenery is a feature of these Alpine fold mountains; a short glacier can be seen creeping into the lake. Erosion by ice and by frost-shattering has exposed rock faces and formed knife-edge ridges.

LANDSCAPE HISTORY AND CLIMATE

The rocks, relief, and details of scenery in central Europe evolved in three main phases, the first two of which involved major disturbances of the Earth's crust. In the first phase, which took place about 300 million years ago, sedimentary rocks such as coal, sandstone and limestone were lifted, fractured and folded as the southern supercontinent of Gondwanaland gradually drifted north into Europe. The great pressures and heat generated caused the formation of metamorphic rocks such as slate. The phase is named Hercynian after the Harz Mountains in northeast Germany, which were formed at the time. The new mountains were gradually eroded, and eventually large areas were inundated by the sea. Here new rocks were laid down, including those of the Jura Mountains on the Swiss–French border.

The second major phase of mountain-building, the Alpine, reached its greatest intensity about 30 million years ago as Africa continued to push into Europe. The Alps were at the center of these disturbances; distant areas were also affected. Layers of sedimentary rock were folded in the Jura Mountains. The rift valley of the Rhine, bounded by fault lines, developed in what is now southwest Germany.

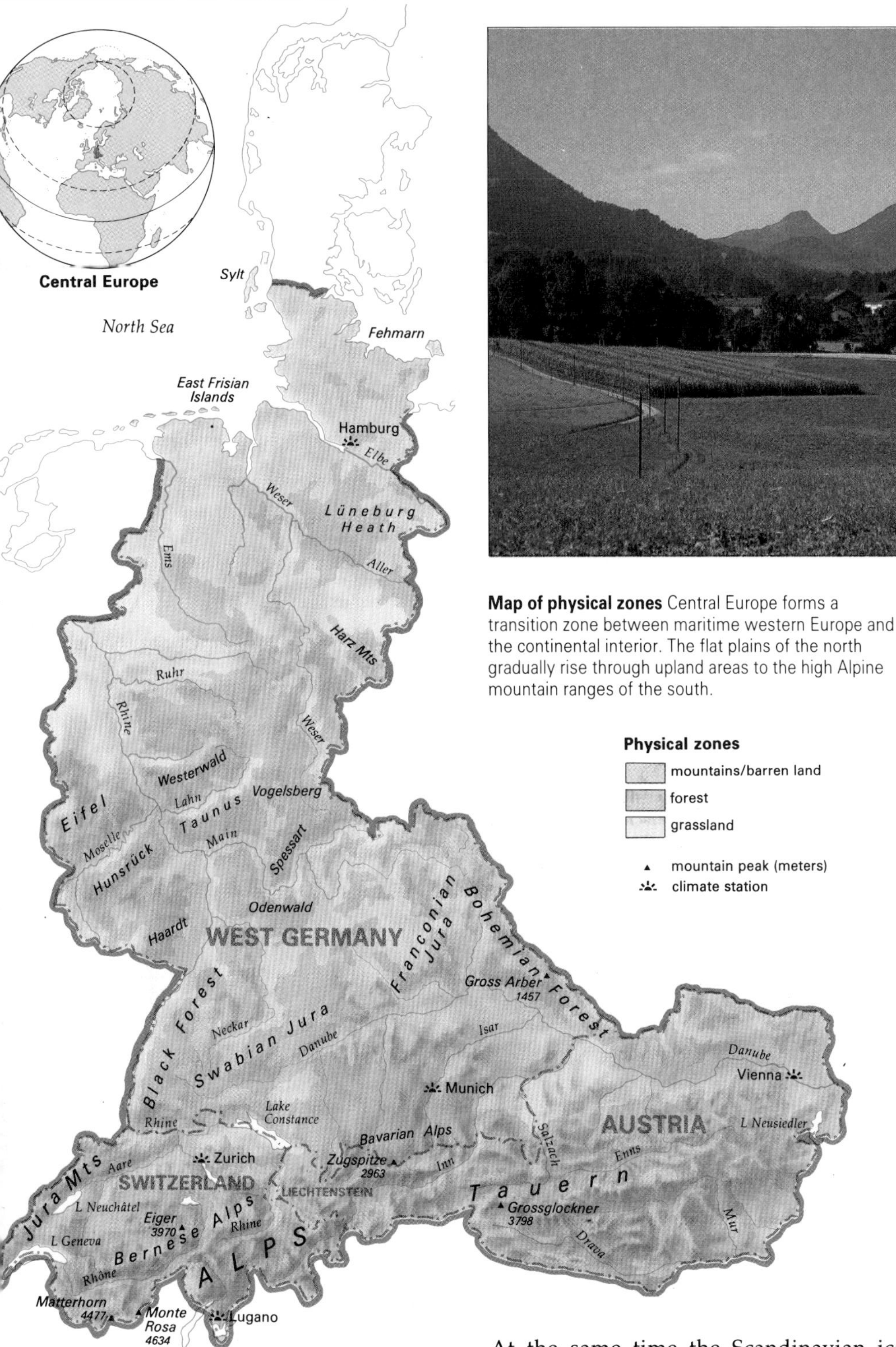

Map of physical zones Central Europe forms a transition zone between maritime western Europe and the continental interior. The flat plains of the north gradually rise through upland areas to the high Alpine mountain ranges of the south.

The Samerberg area of Bavaria Between the Rhine and the Danube rivers the land rises gently to the foothills of the Alps. Fertile glacial till has been deposited on the land; this and subsequent erosion by water have softened the landscape.

Quaternary climate changes

The third and most recent phase of landscape evolution has taken place during the Quaternary period of the past 2 million years. The surface features have been formed during a succession of alternating cold and mild climates. During the colder periods ice sheets and glaciers extended well beyond their present limits. Ice repeatedly overran many parts of Switzerland and Austria as glaciers in the high Alps spread onto the lower ground. Eroding ice carved U-shaped valleys and smoothed rock surfaces. Much of the eroded material was deposited on lower ground, as in central Switzerland. At the same time the Scandinavian ice sheet extended south over large areas of northern Germany.

Even when the Scandinavian and Alpine ice sheets of the last glaciation were at their maximum extent about 18,000 years ago, there was a large area between them that, while virtually free of ice, experienced very cold (periglacial) conditions. Frost was the main agent molding the landscape, assisted by snow and wind. The growth of the ice sheets meant that there was less water in the oceans, and at times much of the present North Sea became dry land.

In recent centuries human activity has become an important contributor to the evolution of landscapes, especially at medium and low altitudes. In northern Germany, for example, peat bogs have been drained for agriculture, and substantial areas of sandy heathland have been planted with forests.

Climates of today

There are two main climatic trends within central Europe. One is the change from maritime climate in the northwest to an increasingly continental climate farther east. The second is, to the fact that there is a steady rise from north to south in the height of the land surface from the North Sea to the Alps. The north German plain is low and fairly flat, and there are no barriers to air movements. Its main climatic variations relate to distance from the sea rather than any change in altitude. The effect is strongest in winter – in January Cologne is on average 3°C (5°F) warmer than Berlin. Precipitation also decreases from west to east.

In the central uplands of Germany and in the Jura Mountains climate shows a stronger correlation to changes in altitude and aspect. There is a general decrease of temperature and an increase of rainfall and snow cover with height. The climate of the Alpine foreland to the north of the Alps is influenced by the mountains. In the Swiss Mittelland cold air draining off the higher ground becomes trapped and often causes fog. In the Alps themselves, climate, vegetation and other aspects of the environment are zoned according to altitude, and range from almost subtropical to glacial. The area can experience marked climatic differences within short distances. The Alps separate the temperate regions to the north from areas with a warm Mediterranean climate to the south.

A view from the Black Forest, looking south over Wiesenthal and the Rhine valley to Switzerland. The forest has grown on a deeply dissected plateau that is about 300 million years old, where the Danube and the Neckar rivers have their source.

RIVERS, UPLANDS AND LOWLANDS

The Danube and the Rhine are two of Europe's longest rivers, the Rhine its busiest inland waterway. All but the lowest reaches of the Rhine lie within the region. The twists and turns of its course and the landscapes through which it passes are a record of its complex evolution; it includes parts of what were once the basins and courses of other rivers.

The story of the Rhine

Rising in Switzerland, where it occupies a straight trench in the Alps, the Rhine follows an arc through Lake Constance before dropping over the 20 m (65 ft) Schaffhausen Falls. Near the head of Lake Constance a tributary captures water from the Danube basin as the river emerges from its underground journey through limestone rocks.

The Rhine turns north from the Swiss border into a wide, flat, steep-sided rift valley that was let down between two fault lines at the time the Alps were being formed. It is flanked to the east by the uplands of the Black Forest, a deeply dissected plateau 300 million years old that has dense forest cover. Farther north the uplands, including the Odenwald to the east, are also extensively forested. The flat, fertile valley floor is punctuated by occasional hills, some of them volcanic.

The river turns through two right angles before pursuing its northerly course through a picturesque gorge, past castles, the legendary Lorelei rock and terraced vineyards. Two tributaries of the Rhine, the Lahn and the Moselle, have also eroded deep trenches in the highlands, cutting down as fast as the land was rising. The Rhine and its tributaries have maintained the same general course during both the rise of the uplands and the sinking of the rift valley.

The middle Rhine highlands stretch away from the gorge. Formed at the same time as the Alps, they are compact and generally level. To the east, contours are broken by outcrops of resistant rocks, as in the upper reaches of the Weser river, and by volcanic features including the Vogelsberg, Europe's largest basalt upland. There are smaller lava flows in the Westerwald. To the west are the small volcanic cones and crater lakes (*Maare*) of the Eifel. The Harz Mountains to the northeast are as old as the uplands of the Black Forest.

Farther downstream the Rhine enters a plain (part of the North European Plain, which stretches from the Low Countries to the Ural Mountains in the Soviet Union) and then passes through the industrial region of the Ruhr coalfield before crossing into the Netherlands. Here a line of coal and brown coal (lignite) deposits marks the northern edge of the central uplands. The Rhine's course has shifted over the past 2 million years in response to climate changes that have altered the amount of sediment it carries and the level of the sea. The river's rate of flow and its channel are now regulated, near the coast to prevent flooding, and upstream to assist navigation.

The south and east

From the southern end of the Black Forest a continuation of the French–Swiss Jura, known as the Swabian and Franconian Jura, swings away in a great arc, first northeast and then north. This is a bleak open country of underlying limestone rocks, with little surface water. In the center of this area is the Ries depression, which may have been formed by the impact of a giant meteorite. The steep scarp slopes of the Jura face northwest toward the basins of the Neckar and Main rivers, both of which are tributaries of the Rhine. The gentler dip slopes drop southeast to the Danube.

Farther east, along the West German border and extending into Austria, the deeply dissected plateau of the Bohemian Forest rises to 1,457 m (4,780 ft) at Gross Arber. The ancient rocks beneath the forest cover seem to have escaped any significant glaciation, but frost, snow and wind have sculpted the landscape over the past 2 million years.

From its source in the Black Forest the Danube flows east through a wide valley that separates the Jura and the Bohemian uplands to the north from the foothills of the Alps to the south. It passes through the low-lying areas of northeastern Austria, including the Vienna basin. Some parts of these areas are as little as 120 m (390 ft) above sea level. The flat landscapes of the extreme southeast, which extend farther into the plains of Hungary, the puszta, are the part of Austria with the most continental climate and vegetation. There are extensive deposits of fine-grained, wind-blown loess.

The north German lowlands

The North Sea coast is shallow, and silted with large tracts of salt marsh. It is cut by the Ems, Weser and Elbe rivers and fringed by the Frisian Islands. The sandy Baltic coast has many bays and inlets. To the south lies the third of Germany that is part of the North European Plain. Although it is very low and predominantly flat, the plain has some local diversity, such as the undulating, sandy Lüneburg Heath and the low-lying peat bogs.

The surface of the land has been formed largely during and since the recent glaciations, through the action of ice sheets, rivers, frost, snow and wind on land near the ice (periglacial effects). The landscape shows the effects of several advances of the Scandinavian ice sheet, which at its maximum extended as far south as the northern slopes of the central uplands. The features are mainly depositional, and include ground moraines, drumlins, lake basins and outwash plains that are up to 200 m (660 ft) deep in places. Prominent features include broad, shallow valleys (*Urstromtäler*) running from east to west; their courses, some used by today's rivers, mark the fronts of successive advances of the ice sheet. In the smaller area overrun by the last ice advance, north and east of the Elbe, the landscape is one of new-looking hills and lakes and is sometimes known as Holstein Switzerland. The most important legacy of the periglacial past is probably the fertile belt of loess soil that was deposited between the lowlands and the central uplands.

The meandering Saar The river once flowed sluggishly over a wide, flat floodplain. As the land gradually rose, the river continued to cut down through the rocks, maintaining its original course.

SURFACE PATTERNS FROM A FROZEN PAST

Periglacial features of the landscape are the product of severe frost, aided by wind and snow. Many of these features date from the colder phases of the geologically recent past. During the glaciations of the past 2 million years landscape-forming processes in areas adjoining ice sheets and glaciers were seasonal, with the land thawing out for only 4–5 months of the year.

In north-central Europe the spring meltwater flowed in numerous small valleys, many of which are now dry. Ice wedge polygons, typical of sand and gravel pits, are a large surface pattern formed by the joining of vertical wedges of material that differ in color and texture from the surrounding deposits. This material filled spaces that were left when wedge-shaped masses of ice in the ground melted. In cross-section, sand and gravel pits may also reveal deformities in the layers of deposits caused by freeze–thaw processes; their contorted appearance has suggested the name "flame structure".

In northern parts of the region deposits of weathered material that have moved over a layer of permanently frozen ground (permafrost) are also common. This soil-flow (solifluction) still takes place in the high Alps; it can be recognized by terracing on slopes and by the movement of individual blocks. When frost weathering of rocks becomes intense and freezing causes the land to heave, the loose material on gentler terrain is often rearranged into smaller patterns of polygons or circles. On slopes these may elongate into stripes.

Periglacial processes still take place today – in polar and subpolar regions, and at high altitudes in middle and low latitudes.

THE ALPS – THE BACKBONE OF EUROPE

From the southwestern part of central Europe the Alps extend for more than 800 km (500 mi) in a crescent shape through Switzerland and northern Italy, Austria and southern Germany, as far as northern Yugoslavia. The range was forced up 35–25 million years ago as the continent of Africa gradually drifted into Europe. The scenery is dazzling, with snowy peaks, deep valleys and lakes. Among the most awesome challenges for the climber are the mighty Matterhorn (4,477 m/14,688 ft) and the Eiger (3,970 m/13,025 ft).

The Alps occupy the southern three-fifths of Switzerland. They are the country's main physical region and have a complicated geology. Rock types range from the sedimentary limestones of the Bernese Alps in the north to the glittering schists of the central southern chain, rocks whose original characteristics have been greatly altered by heat and pressure (metamorphosed). Earth movements have also led, to many rocks being contorted and folded in an intricate fashion.

The landscape resulting from this tectonic activity has dramatic changes in altitude over quite short distances. At 372 m (1,221 ft) Lake Geneva in the southwest is only 100 km (62 mi) away from the soaring peak of Monte Rosa at 4,643 m (15,203 ft) – a span of more than 4,000 m (13,000 ft). There is a correspondingly wide range of climates. It is almost subtropical around the lakes on the Italian border, and glacial in the mountains. The mountain environment is prone to hazards such as flooding, landslides and avalanches. These result from steep slopes, loose material on the surface, and high precipitation, including snowfall.

One-fifth of the Swiss Alps lies under permanent snow and ice. The greatest areas of ice are in the Pennine (southern) and Bernese Alps. The latter contain the longest glacier in the Alps, the Aletsch, which covers 24 km (15 mi). In terms of earth history the glaciers have so recently retreated from the land that it is still recovering. The upper valleys of the Rhône and Rhine form impressive east–west trenches that separate the northern and southern chains of the Alps.

In Austria the Alps comprise the rugged northern and southern limestone ranges with the Tauern, including Grossglockner (3,798 m/12,461 ft), in between. The three zones are separated by trenches such as those of the upper parts of the Inn and Salzach rivers: they are the Austrian equivalents of the Rhône and Rhine valleys of Switzerland.

Although in Austria the Alps are not so high and have a less complicated geology, they are nevertheless very similar to the Swiss Alps, with the same wide range of environments and subject to the same hazards. On the northern borders of Austria the Bavarian Alps include the Zugspitze, at 2,963 m (9,721 ft) Germany's highest peak.

The Jura Mountains

Along the northwest border of Switzerland lies another range, the Jura Mountains. Their marine limestones date from the Jurassic period (213–144 million years ago), which was named for the region. Later, when the Alps were formed, the area was folded into a striking series of parallel ridges and troughs. The upfolds of the rocks, which form the ridges, are the areas of highest rain and snowfall. The porous limestone allows water to disappear underground, so there is little surface water. What rivers there are have in places eroded through the ridges to produce spectacular gorges. The Jura lose height from the southwest to the northeast, where plateaus and deep valleys form the so-called tabular Jura.

The middle land

The central region of Switzerland, the Mittelland, lies between the Jura Mountains and the Alps. The surface rock is formed from material deposited by Alpine rivers flowing into the area from north and south. There are two types. The larger rocks and boulders were laid down nearest the Alps, and these formed conglomerate rocks. The finer material was carried farther before being deposited; it is now the area's sandstone and marl.

During the recent Quaternary period the action of rivers and glaciers increased the diversity of landscapes. Rivers have eroded their valleys and formed terraces; glaciers have hollowed out lake basins, molded hills and laid down deposits. The scenery is varied, despite the relatively low altitude, which rises from 400 m (1,300 ft) near the Jura to some 800 m (2,600 ft) at the foot of the Alps. It is sometimes called the Swiss Plateau or Foreland. The equivalent area in Germany is the Bavarian Foreland, which extends north from the Bavarian Alps to the Danube. Here debris from the Alps was deposited to a depth of hundreds of meters thick, and in the south was subsequently covered by glacial deposits.

Austria's counterpart to the Mittelland lies immediately north of the eastern Alps. It too is relatively low in altitude (250–800 m/800–2,600 ft), and similarly consists of Tertiary rocks, predominantly sandstone, which have been shaped by glaciers and rivers.

GLACIAL COMINGS AND GOINGS

The world's glaciers went through major expansions and contractions during the ice ages of the last 2 million years. Less well known are the much smaller fluctuations of historic times, yet these have had a significant effect both on the environment and on the people living at the time.

From the later Middle Ages to about 1860, Alpine glaciers were relatively large, often 1–3 km (0.6–2 mi) longer than today. This colder period in Europe has been called the Little Ice Age. Evidence not only of the growth in glaciers during that time, but also generally more severe winters and colder conditions, comes from old documents and maps, from paintings and drawings, field measurements, and studies of vegetation and glacial deposits.

Today the margins of most Alpine glaciers are retreating. However, in the past expanding glaciers have sometimes overrun fields, roads and buildings. In Switzerland they interfered with the irrigation system, while in Austria they contributed to the decline of mining in the high mountains. Where a glacier ended on steep ground, ice could occasionally break away and avalanche into the valley below, causing untold destruction. Sometimes, as a glacier waxed and waned, a lake formed and subsequently emptied rapidly, causing extensive floods. A number of villages in Switzerland have been devastated several times in this way.

Nowadays the glaciers have a more beneficial function. Meltwater from glaciers is harnessed to drive hydroelectric generators that provide two-thirds of the electricity needs of Austria and Switzerland. Much less water flows in the winter months, but summer meltwater can be stored behind large dams, to be released in winter when demand is greatest.

Millstätter Lake occupies a glaciated valley in the mountains of southern Austria. It is 12 km (7.5 mi) long and 140 m (462 ft) deep. Glacial lakes are a temporary feature of the landscape. Some gradually fill with sediment deposited by streams entering the lake; others are drained as they breach the barriers behind which they were formed.

The Gorner glacier winds through the Monte Rosa mountains of southern Switzerland. The mountains form part of a horseshoe chain around the town of Zermatt, opening out to the north. The glacier has retreated during the last hundred years, but its snout is still well below the snowline.

Avalanche!

A huge avalanche of snow may be an awesome sight, but avalanches can be deadly and cause tremendous damage. Not only the snow itself is dangerous. An avalanche can generate winds of up to 300 km/h (185 mph), which can fling people through the air and make buildings in their path collapse and disintegrate under the impact of their shock waves. A Swiss forestry worker once regained consciousness after being caught by such an avalanche to find that deep snow had broken his fall. He had escaped with only a few fractures, but the avalanche had carried him for a kilometer through a vertical fall of nearly 700 m (2,300 ft).

A cascade of snow

There are tens of thousands of snow avalanches in the Alps every year. Most of them take place in the high mountains during bad weather, and go largely unnoticed. The few avalanches that reach the lower areas, where people live, cause considerable concern, and a great deal of research has been undertaken to try to predict and control them.

In Switzerland many avalanches begin at high altitudes, around 2,000–2,500 m (about 6,500–8,000 ft). Here the snow is usually quite thick and slopes fairly steep. Shaded slopes, depressions in the ground and gullies all act as collecting places for deep snow. Any steep slope can hold a certain amount of snow without it slipping; the amount depends on the surface conditions. The biggest avalanches tend to form on slopes between 30° and 40°. The snow does not build up so much on steeper slopes, or travel with such devastating effect on shallower ones. Slopes with long grass are very slippery and the snow can start to slide before it has built up to any great depth, whereas shrubs and trees on slopes help to anchor the snow and allow it to accumulate.

There are several factors that can trigger a sudden cascade of snow down a mountain. The depth of snow may increase during a snowfall or blizzard, which may also redistribute snow that has already fallen to such an extent that the snow can no longer maintain its hold on the slope. Water from melting snow or rain can act as a lubricant. The warm föhn wind blowing down the northern slopes of the Alps can cause very sudden rises in temperature and rapid melting, putting large areas at risk with little warning. Once conditions are suitable an avalanche

An avalanche of snow can be set off by the smallest disturbance. Detonation is often used to cause an avalanche under controlled conditions; it minimizes the damage to power lines and buildings, and greatly reduces the risk to life.

Snow fences are built across slopes to stabilize the snow at particularly susceptible sites.

The effects of an avalanche on Davos in eastern Switzerland. The devastation demonstrates the strength of moving snow, which can cause houses to collapse. The risk of avalanches has increased as forests are felled to make ski runs.

can be started by nothing more than a passing skier or the noise of a branch breaking under its load of snow. The authorities often use detonators to start an avalanche under controlled conditions in order to prevent a sudden, unexpected and more damaging avalanche.

A major avalanche may be up to 1 km (0.6 mi) across and move at tremendous speed for several kilometers. Avalanches of wet snow move more slowly than ones of cold, powdery snow but they can be more devastating, felling trees and even cutting swathes through forests. Avalanches frequently cause death or damage to property. A death toll of 50–100 has been recorded on a number of occasions in the Alps; the greatest disaster took place in World War I, when some 10,000 soldiers on the Austrian–Italian front lost their lives on one day in a series of snow slides.

Parrying the blow

Physical obstacles may be used either to prevent avalanches from starting or to impede their progress. Dense woodland, appropriately located, can prevent avalanches, but in many parts of the Alps forests have been removed for timber or, more recently, to clear the ground for ski slopes. Artificial means such as fences now have to be used to stabilize snow at sites where avalanches might begin. Electricity pylons and buildings downslope of a known avalanche zone can be protected by structures that deflect the oncoming snow. Covered sections divert avalanches safely over roads or railways, and triangular structures split the flow so that it passes harmlessly each side of the object being protected.

In Switzerland landscapes are zoned according to avalanche risk. This information is taken into account when building developments are planned. Radio warnings are broadcast during times of avalanche danger. A daily task for Swiss railroad officials is to inspect sites where avalanches may occur and take steps to ensure that services are not disrupted.

FROM THE BALTIC TO THE BLACK SEA

HISTORY OF THE LAND · ICE-FASHIONED FEATURES · DIVERSE LANDSCAPES

From the sands and cliffs of the Baltic Sea in the north Eastern Europe extends some 1,700 km (1,050 mi) to the rocky shores of the Adriatic Sea. A similar distance separates the Harz Mountains in the west of the region from the Black Sea in the east. The continental climate of the north, with its lakes, grasslands and forests, gives way to subtropical southern lands. The line of mountains that winds through the region includes the Ore Mountains, the wooded and snowcapped Carpathian Mountains and the Dinaric Alps. Petroleum and gas deposits lie in the foothills of the Transylvanian Alps, while the plateau of Kras in Yugoslavia is famous for its limestone scenery. The Elbe and Vistula rivers cross the northern lowlands; most of the south is drained by the Danube, the second longest river in Europe.

COUNTRIES IN THE REGION

Albania, Bulgaria, Czechoslovakia, East Germany, Hungary, Poland, Romania, Yugoslavia

LAND

Area 1,275,191 sq km (492,224 sq mi)
Highest point Musala, 2,925 m (9,594 ft)
Lowest point near Gulf of Gdansk, –10 m (–33 ft)
Major features northern lowlands, Carpathian Mountains, Balkan Mountains, Dinaric Alps, Great Alföld, Danube valley

WATER

Longest river Danube, 2,850 km (1,770 mi)
Largest basin Danube, 773,000 sq km (298,000 sq mi)
Highest average flow Danube, 6,430 cu m/sec (227,000 cu ft/sec)
Largest lake Balaton, 590 sq km (230 sq mi)

CLIMATE

	Temperature °C (°F) January	July	Altitude m (ft)
Gdansk	–1 (30)	18 (64)	12 (39)
Dresden	0 (32)	19 (66)	129 (423)
Prague	–3 (27)	18 (64)	374 (1,227)
Tiranë	7 (45)	25 (77)	89 (292)
Bucharest	–3 (27)	23 (73)	82 (269)

	Precipitation mm (in) January	July	Year
Gdansk	31 (1.2)	73 (2.9)	499 (19.6)
Dresden	42 (1.7)	120 (4.7)	680 (26.8)
Prague	23 (0.9)	82 (3.2)	508 (20.0)
Tiranë	132 (5.2)	28 (1.1)	1,189 (46.8)
Bucharest	43 (1.7)	55 (2.2)	578 (22.8)

NATURAL HAZARDS

Earthquakes, landslides and floods

HISTORY OF THE LAND

Eastern Europe straddles two major geological zones that meet along a line from the delta of the Vistula on the Baltic Sea to the delta of the Danube on the Black Sea. To the northeast lies a stable zone of Precambrian rocks more than 600 million years old, covered by younger deposits. To the southwest is a belt of fold mountains. In the far south these are subject to earthquakes, as at Skopje, Yugoslavia, in 1963, and at Bucharest, Romania, in 1977.

Against this background there are three main types of scenery – mountains, uplands or plateaus, and lowlands. The most dramatic landscape feature of Eastern Europe, the arc of the Carpathian Mountains, was lifted up at the same time as the Alps, some 30 million years ago. The Sudetic Mountains, were formed in the Caledonian and Hercynian periods of mountain-building (respectively some 400 and 300 million years ago), and continue this arc to the northwest.

Next to the mountain ranges lie uplands and plateaus that were themselves mountains some 250 million years ago. They were eroded, then lifted up at the same time as the Carpathian Mountains. Now dissected by major river valleys, they include the uplands of western Czechoslovakia and southeastern Poland, and the Transylvanian Alps.

The lowlands include part of the great North European Plain, which stretches from the Low Countries to the Soviet Union. They bear the traces of invasions by glacial ice on several occasions in the

past 2 million years. More enclosed by upland areas are the lowlands of western Hungary and Yugoslavia, the central plain or Great Alföld of Hungary, which is crossed by the Danube and Tisza rivers, and the lowlands between the Transylvanian Alps and the Danube.

The river Danube links all these landscapes. This great waterway has maintained its course through periods when the land has been uplifted; the many deposits along the river and its tributaries record the river's history.

Local variations in scenery reflect changes in the underlying rocks. Southwest of the Danube the Dinaric Alps extend along the eastern side of the Adriatic Sea. Here in Dalmatia parallel bands of limestone and flysch (shale and slate) have eroded at different rates, forming a ridge-and-valley pattern. Because of invasion by the sea, the mountain summits now rise from the water as lines of islands parallel to the coast. The name Dalmatian is given to such coasts wherever they occur.

Patterns of climate and vegetation
The northern lowlands are susceptible to the influence of air from the northeast (cold and rather dry) and the northwest (warmer and more humid). The Carpathian and Sudetic Mountains form a barrier to these influences and to warmer air masses from the south. By contrast, the south is subtropical, as it is warmed by air flowing north across Greece from the Aegean Sea.

In the mountains there are considerable temperature changes both daily and over

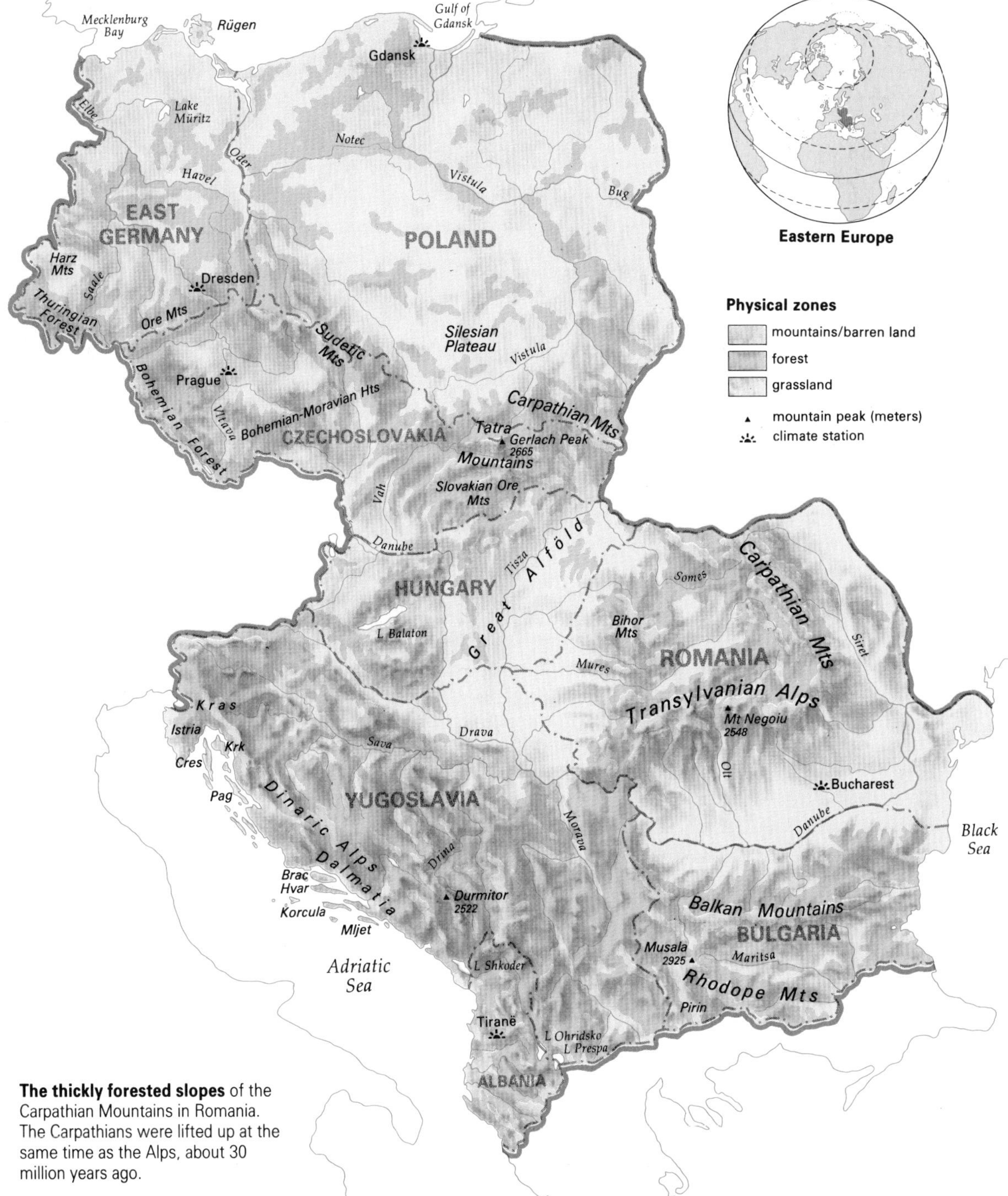

Map of physical zones Eastern Europe has large tracts of deciduous and coniferous forest, the open steppes of the Hungarian puszta, extensive mountain ranges and the dramatic karst country of Yugoslavia.

The thickly forested slopes of the Carpathian Mountains in Romania. The Carpathians were lifted up at the same time as the Alps, about 30 million years ago.

the year. Mountain basins in the south have a dry climate, with an annual temperature range of up to 30°C (52°F). Annual variations in temperature on the northern plain increase from west to east: in the Oder valley in Poland the range is 17°C (31°F), in the Bug valley 400 km (250 mi) to the east, 23°C (41°F).

Pressure differences between air over the Dinaric Alps and over the Adriatic give rise to strong winds, such as the cold bora or yougo winds that flow south from the Balkans to the sea, usually in winter. (The southern part of the region includes much of the Balkan Peninsula.) The warm halny wind sweeps down the northern slopes of the Tatra Mountains. Most rain and snow falls on the windward, northern side of the Carpathians. Many rivers flood twice a year – in spring when the snows melt and in summer after intense rainstorms. In much of the region the rain is heaviest in June and July, but more falls in August on the Baltic coast, in October on the coast of Yugoslavia, and in November in the southeast.

ICE-FASHIONED FEATURES

The outlines of Eastern European landscapes are dictated principally by the structural zones and the major rock types, but some dramatic changes to the scenery have taken place during the last 2 million years as ice sheets and glaciers expanded and contracted.

Ice has advanced over the land in Eastern Europe on four occasions (glaciations) during the past 2 million years. At its greatest extent the ice sheet extended from Scandinavia as far south as the Carpathian and Sudetic Mountains. During glacial periods the valley glaciers in mountain regions also expanded. The northern part of Poland and East Germany, the mountains and areas adjacent to them all show signs of glacial erosion and deposition. In northern Poland it was the most recent phase of the last glaciation, the Pomeranian, that made the greatest impact, producing the diverse and picturesque scenery of the lakelands. Areas not directly affected by the most recent incursion of these ice sheets and glaciers experienced cold conditions and produced distinctive periglacial ("near ice") landforms and deposits.

In the mountains the glaciers carved out amphitheaters (cirques) and U-shaped valleys. They transported debris and deposited it, forming moraines, notably in the Tatra Mountains but also farther south, for example in parts of the Transylvanian Alps and even in the Pirin range south of the Balkan Mountains. The Scarisoara ice cave in the Bihor Mountains in northwest Romania, with its ice stalagmites and stalactites, is probably a relic of the ice age.

Valleys and eskers

Glacial effects on the landscape are not produced only by erosion and deposition. Ice can obstruct drainage and rivers to impound lakes, and where the ice has melted great torrents of meltwater can change entire landscapes. During successive phases in the retreat of the ice as it

A RIVER IN THE PAST

It is often thought that the landscape-forming processes of today have been going on for thousands of years more or less unchanged. However, a closer look at landforms and the layers of successive deposits in a river valley reveals that the present is just one of many phases. During each one of these a river will give a distinctive imprint to the landscape it flows through.

Since the end of the last glaciation the river Vistula and its tributaries have modified their courses many times as they have drained the major part of Poland. A study of former beds and terraced deposits has helped in understanding the history of the land in these plains of northern Europe.

Three major phases can be identified in the recent history of the Vistula and other rivers of northern Poland. During and immediately after the last glaciation much more water flowed in the rivers than today. The river channels were broader, and carried more sediment from glacial meltwaters. The Vistula was braided, with many channels and intervening coarse gravel bars.

Once the ice sheet had all melted and the climate became more like it is now (about 7,000 years ago), the river flow was reduced. The channel of the Vistula became narrower and its course more meandering, with finer deposits.

Originally much of the Vistula basin was forested, but in the last 2,000 years a third phase in recent geological history has taken place as the forest has been cleared by humans. Without tree roots to impede surface water, the river flow has increased again, and larger amounts of sediment have been transported as a result. The rivers of northern Poland have become wider and more unstable in their courses, sometimes braiding once again.

Alpine meadows at Zakopane on the north-facing slope of Beskidy Zachodnie in Poland's Tatra National Park, which covers the highest area of the Tatra Mountains, at the northern end of the Carpathians.

A view upstream along the Elbe as it winds through the valley near Dresden in East Germany. Rising in Czechoslovakia, the river flows northwest across Germany on its journey to the North Sea.

melted, rivers flowed into one another, swelled, spilled over low watersheds and formed vast valleys parallel to the edge of the ice sheet. After the next period of warming, water could flow farther north, only to meet the obstruction of the retreating ice sheet once again and so form another great valley.

These enormous ice-margin valleys run east–west across the north of the region. In width they are from 2 km (1.2 mi) up to as much as 50 km (31 mi). The valley bottom is generally flat, but the sides are relatively well defined, steep and high, and often marked by terraces formed in different periods. The valleys are much too wide for the river flows of today, but the low watersheds between them have been used to establish navigable canal links between rivers.

Signs of past glaciation are also present, on a much smaller scale, in the glacial drainage valleys and long sand and gravel ridges that are often found together in the north of the region. The valleys were probably eroded by streams running beneath the ice sheet, or possibly by ice below the main sheet. The valley floors are uneven, and now often contain deep, narrow lakes, such as Lake Hancza in the Masurian Lakes of northeast Poland, which covers only 3 sq km (1.2 sq mi) but is over 100 m (330 ft) deep.

Long winding ridges known as eskers often run parallel to the valleys. These casts of sediment-laden streams were formed either when debris carried by a stream in, under or on the ice in a valley was deposited as the ice melted, or where streams flowing beneath the ice deposited at the retreating ice margin some of the material they were carrying.

Periglacial features

Areas not directly affected by glaciation but situated near ice sheets experienced periglacial conditions, prolonged periods of freeze–thaw. The uplands of central Poland show typical periglacial landscape features. Soil sections show "flame structure" distortions that reflect the pressure when an unfrozen layer is trapped between frozen layers of soil. Ice wedges have developed where wedges of ice that occupied and enlarged cracks in the soil were subsequently replaced by other deposits, often of gravel. Other landforms seen today are the many small, now dry, valleys (*Dellen*) that were formed when the land had to cope with much greater flows of water during the spring and summer snow melt than it does today.

Cold winds blowing off the ice sheets picked up fine glacial and river sediments. As in many other parts of the world, they were later deposited as loess.

DIVERSE LANDSCAPES

The northern part of eastern Europe includes part of the North European Plain. Here three belts extend east–west: lakelands, lowlands, and plains and uplands.

South of the Baltic Sea there are many lakes, giving exceptionally picturesque landscapes. Finger lakes in long, narrow glacial drainage valleys are typical. The slopes of these valleys are steep and the lakes usually deep (up to 100m/330 ft). Ice-marginal lakes now fill depressions between the moraines left by glaciers. These are usually bigger but shallower – the largest, Lake Müritz, covers 117 sq km (45 sq mi) and is just 33 m (108 ft) deep.

The maritime influence in the northwest of the region gives a relatively mild climate. Farther from the moderating influence of open-sea waters it is more continental, with higher summer and lower winter temperatures in the Masurian lakelands of the northeast.

To the south of the lakelands are the vast plains and lowlands. Apart from the ice-marginal valleys and sandy plains, the flat landscape is relieved only by river valleys and by the small, dry valleys dating from times of periglacial climate.

There is a tremendous diversity in the landscapes of the upland belt. In areas covered with fine-grained loess, fertile, dark chernozem soils have developed, for example in the uplands of southeastern Poland. The edges of this area are cut by steep-sided, flat-bottomed ravines.

Mountain altitude zones

The inner or western side of the Carpathian Mountains is composed of crystalline rocks, and the outer part of alternating shales and sandstones, where petroleum deposits are found.

A special feature of the Carpathians is that climate and vegetation are determined by altitude. The land is farmed up to 900–1,000 m (2,950–3,280 ft). At this level and up to 1,200 m (3,940 ft) beech woods predominate, or mixed beech and fir woods in the Transylvanian Alps. Between about 1,200 and 1,500 m (4,920 ft) are spruce and fir woods, while the mountain dwarf pine is typical between 1,500 and 1,800 m (5,910 ft), with rhododendrons in the Transylvanian Alps. There are mountain meadows between 1,800 and 2,200 m (7,220 ft), though in the east Carpathians these meadows, called *polonina*, start as low as about 1,200 m (3,940 m). The upper range of forests has been lowered artificially through cattle grazing. Above in meadows, vegetation is sparse. Gerlach Peak (2,665 m/8,745 ft) in the Tatra Mountains is the highest point; some other peaks to the south are nearly as high.

West of the Carpathians the upland of western Czechoslovakia is bordered by the Sudetic Mountains and the Ore Mountains and by the Bohemian Forest. It has a continental climate and relatively few forests compared with eastern Czechoslovakia. In its northwestern part

The Dalmatian coast in Yugoslavia, where the Dinaric Alps run parallel to the Adriatic coast. The sea now fills valleys where rocks have been eroded away.

Perched boulders at Rozen, Bulgaria. The pointed pinnacles have been shaped by water gradually eroding the soft rocks.

THE HALNY WIND OF THE TATRA

The halny ("alpine meadow") wind of the Tatra Mountains is a warm, dry wind that descends the leeward slopes of the range, facing toward the north. The halny is similar to the föhn of the European Alps and the chinook of the North American Rockies. It arises when winds blow north from a high-pressure zone over Czechoslovakia toward a low-pressure zone over Poland. As the humid air climbs the southern slopes of the Tatra it expands, losing 0.6°C (1.1°F) of temperature for every 100 m (330 ft) it ascends, and so deposits much of its moisture as rain or snow. Accelerating down over the northern slopes at speeds up to 150 km/h (93 mph) as a result of the increasing pressure above it, the air contracts, warming 1°C (1.8°F) for every 100 m (330 ft) it descends. These fluctuations in temperature are known as adiabatic changes. (By contrast, the cold, northerly wind known as the bora, which descends from Yugoslavia to the Adriatic Sea, is a katabatic wind, falling from cold highlands to an adjacent warmer zone.)

At 1,000 m (3,300 ft) on the southern slopes of the Tatra the temperature may be 7°C (13°F) and the relative humidity 100 percent; as the wind breasts the peaks at 2,500 m (8,200 ft) air temperature may drop to −1°C (30.2°F) and humidity may still be high; by the time the air reaches 1,000 m (3,300 ft) on the northern slopes its temperature will have risen to 21°C (69.8°F), its humidity falling to 25 percent.

The halny may blow for hours or days, and may raise temperatures by more than 10°C (18°F). Its high speed can cause serious damage to forests; its warmth can cause snowmelt and trigger dangerous avalanches.

are well-known health spas associated with mineral water springs.

Farther south a basin with an unusually warm climate lies betweeen the mountains. It was formed as a result of an inland sea filling up with sediments. Lake Balaton, all that remains, covers 614 sq km (237 sq mi) but is only 11 m (35 ft) deep.

The Transylvanian upland lies between the east Carpathians, the Transylvanian Alps and the Bihor Mountains. Its hilly surface is cut by the deep valleys of the Olt, Mures and Somes rivers. In the warm climate of this area vines and fruit trees are cultivated, and there are numerous mineral springs.

South of the Carpathians

The Olt river flows south in a gorge through the Transylvanian Alps to a lowland area extending to the Danube. Soils are poor in its western part but fertile in the loess-covered land to the east, an important farming center. The Danube, which is navigable for much of its length, has been an important trade route for hundreds of years, especially for countries with no access to the sea – Hungary, Czechoslovakia and Austria. The delta of the Danube, on the Black Sea, is fast-growing, swampy and dotted with numerous lakes.

South of the Danube lie the Balkan Mountains, which were formed at the same time as the Alps and have since been much eroded. They are covered with beech and fir forests up to about 1,800 m (5,900 ft); above that level there are extensive dry pasturelands. Between the Balkan Mountains and the Rhodope Mountains to the southwest lies the Maritsa river valley.

The Dinaric Alps run south parallel to the Adriatic Sea from the Alps proper through Yugoslavia and Albania. The rugged fold mountains and plateaus include, in the northwest, the Kras plateau. Vegetation is minimal, except on slopes facing the Adriatic.

Karst scenery in Yugoslavia

The Kras plateau (German: *Karst*), in the Dinaric Alps just inland from the northeastern tip of the Adriatic Sea, has given its name to a type of dry, barren, deeply eroded limestone scenery. On the surface bare "pavements" are dissected by gullies (grikes) between limestone blocks or ridges (clints). Some gullies become enlarged and funnel-shaped, turning into sinkholes. Others develop into flat-bottomed depressions. Almost all the surface water forms streams that disappear into the ground and emerge at the surface farther on.

Karst scenery is widespread in Europe and elsewhere. In parts of the world where a hot, wet climate speeds up the chemical weathering process, the formations may be on a larger scale, as in the spectacular limestone towers of southwest China and the hummocky "cockpit" country of Jamaica. Karst landscapes may also contain labyrinthine underground passages like those of the Mammoth Caves in the Appalachian Mountains of North America.

On the Kras plateau itself the flat-bottomed, steep-sided basins range from a few meters to kilometers across, with an area of betweeen 10 and 600 sq km (4–230 sq mi) and sides up to 800 m (2,600 ft) high. At the village of Cerknica the basin contains a lake; the limestone floor is covered by river mud. In the spring thaw or after heavy rains, water flows from caves and surface gutters of the surrounding limestone, filling the lake in a few days. The waters take as many weeks to drain, through a mud-clogged sinkhole. They travel 3 km (1.9 mi) underground before emerging into a river, then disappear into another sinkhole before reaching the Adriatic.

Limestone is a very common type of rock. It is formed chiefly from the skeletons and shells of tiny sea creatures deposited in seas many millions of years ago. The main mineral is calcite (calcium carbonate). Another type of limestone, dolomite (containing magnesium carbonate) also outcrops over one-fifth of the surface of Yugoslavia, in thicknesses of up to 9,000 m (29,500 ft).

Rainwater can absorb carbon dioxide from the atmosphere or from the soil. The carbon dioxide changes the water into weak carbonic acid, which dissolves the limestone as it percolates along the joints and bedding planes of the rock. In parts of Yugoslavia, mountain folding after the limestone was laid down has shifted these bedding planes into diagonal and vertical alignments, allowing the water to penetrate deep into the rock.

Corridors and caves

Limestone scenery underground is less well known but even more extraordinary – kilometer after kilometer of channels and galleries that open out into caves with rising stalagmites and hanging stalactites in columns, curtains and other forms. These beautiful calcite formations are deposited when the water loses carbon dioxide through evaporation, which reduces the amount of dissolved calcium carbonate it can carry .

There are about 7,000 known caves in Yugoslavia. The longest system of all, at Postojna, has more than 16 km (10 mi) of corridors. The largest cave, known as the Concert Hall, is 38 m (125 ft) high, 65 m (210 ft) long, and 40 m (130 ft) wide.

Limestone cave systems are still being discovered all over the world. Humans have used them for shelter for thousands of years. The cave paintings of Lascaux in central France have been dated to between 15,000 and 20,000 years ago. In Yugoslavia shepherds have long sheltered and watered their flocks in caves. Folk stories relate how sheep disappear in one cave and later emerge from another some way off. Caves also support their own special animal life. Bats and cockroaches are common; there are also blind white fish and, in Yugoslavia, a blind cave amphibian called an olm.

Stepped gorges and waterfalls at Plitvice, Yugoslavia. Surface stream channels run over the limestone rock, though most of the water has already found its way underground. The area is covered with beech, fir and pine forest.

GREAT CAVES

The great cave systems have been fashioned by the action of carbon dioxide-bearing water dissolving the calcium carbonate of limestone rocks. Limestone is the most widespread soluble rock on Earth, and caves are found in many places throughout the world.

Deepest caves	m	ft
Jean Bernard, France	1,535	5,036
V. Pantjukhina, Soviet Union	1,508	4,948
Laminako Ateak, Spain	1,408	4,619
Sima del Trave, Spain	1,381	4,531
Sniejnaya-Mejennogo, Soviet Union	1,370	4,495

Longest cave systems	km	mi
Mammoth Cave system, United States	530	329
Optimisticheskaya, Soviet Union	153	95
Hölloch, Switzerland	133	83
Jewel, United States	114	71
Osyorhaya, Soviet Union	108	67

Largest cave chambers
The Sarawak Chamber in Malaysia is 700 m (2,297 ft) long, 430 m (1,410 ft) wide, and 120 m (394 ft) high. Other giant chambers are found in Belize, France, Oman and Spain.

The intricate shapes and patterns of the Postojna caves have taken many thousands of years to form. Between 3,000 and 10,000 years may pass while a minute crack in the rock enlarges to a pencil-sized opening. As more water passes through the crack the weathering process speeds up; nevertheless, a small cave only 1–3 m (3–10 ft) in diameter will take another 10,000–100,000 to develop. At this stage calcite deposits will start to form stalagmites, stalactites and rock curtains.

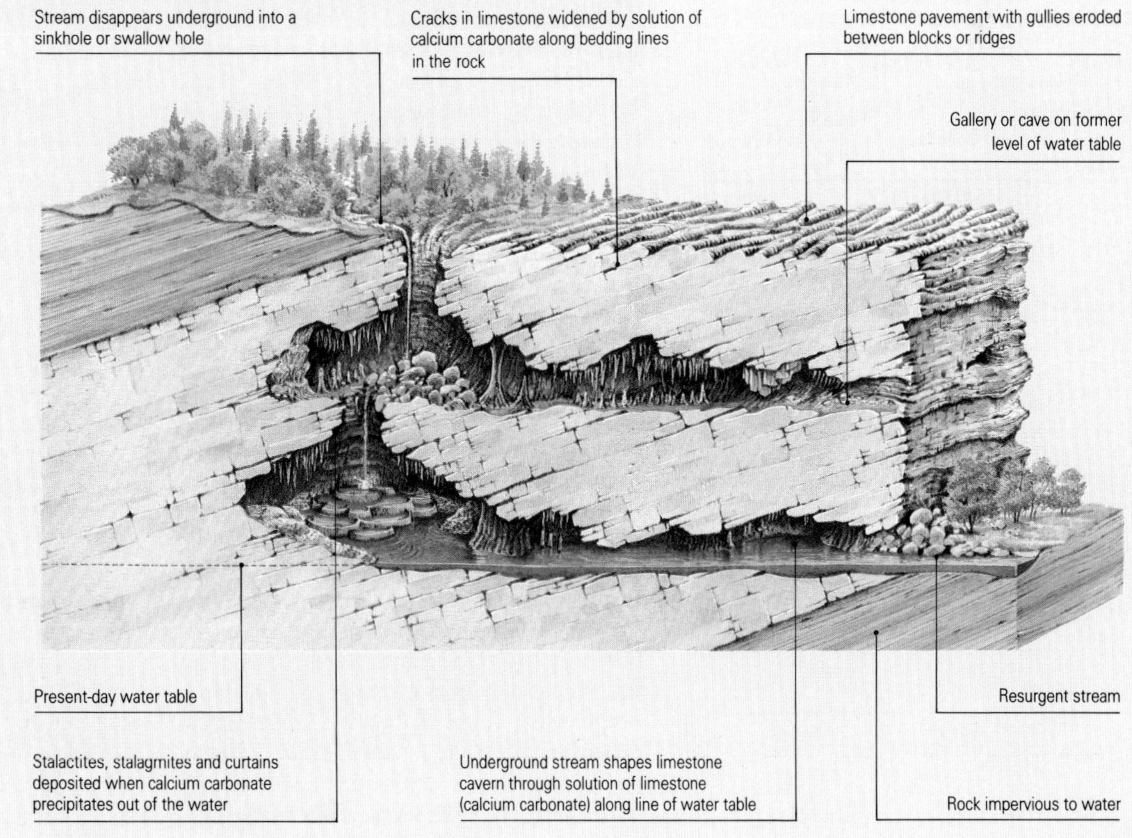

Underground landforms In limestone areas there is little surface water, as it disappears down cracks and swallow holes in the rock. As the water works its way through the rocks it gradually dissolves the limestone, and the passages get larger. Eventually caves can form; gorges result when passage roofs collapse. When the calcite is precipitated out of the water, stalagmites and stalactites are gradually able to develop.

LANDSCAPE ON A GRAND SCALE

THE ANCIENT FOUNDATIONS · A SUCCESSION OF CLIMATES · THE ARID SOUTH

The Soviet Union displays geology on a grand scale. Extending a quarter of the way around the globe, the country contains within its boundaries an immense variety of landforms and climates – high peaks and low plains, flooding rivers and sandy deserts, subtropical beaches and permanently frozen ground. Over much of the region the sheer scale of the gradual change over great distances from one kind of environment to another gives rise to rather monotonous landscapes. High mountains ring the eastern and southern perimeter, forming a hugh amphitheater whose auditorium stretches from the Baltic to the banks of the Yenisei, looking out over vast plains and low flat plateaus to the Arctic Ocean. To the southeast, beyond the mountains, lie grasslands and the desert expanses of Mongolia.

COUNTRIES IN THE REGION

Mongolia, Union of Soviet Socialist Republics

LAND

Area 23,967,200 sq km (9,251,339 sq mi), including largest country on Earth
Highest point Communism Peak,7,495 m, (24,590 ft)
Lowest point Mangyshlak Peninsula, –132 m (–433 ft)
Major features plains and plateaus of north, Ural Mountains, Caucasus Mountains, Pamirs and Altai ranges, Kara Kum, Kyzyl Kum and Gobi deserts, Arctic islands

WATER

Longest river Yenisei, 5,870 km (3,650 mi)
Largest basin Yenisei, 2,619,000 sq km (1,011,000 sq mi)
Highest average flow Yenisei, 18,000 cu m/sec (636,000 cu ft/sec)
Largest lake Caspian Sea, 371,000 sq km (143,240 sq mi), largest area of inland water in world; Lake Baikal is world's greatest in volume, at 22,000 cu km (5,500 cu mi) and depth, at 1,940 m (6,365 ft)

CLIMATE

	Temperature °C (°F) January	July	Altitude m (ft)
Moscow	–10 (14)	19 (66)	156 (512)
Sochi	7 (45)	23 (73)	31 (102)
Krasnovodsk	2 (36)	28 (82)	89 (292)
Ulan Bator	–26 (–15)	16 (61)	1,337 (4,385)
Verkhoyansk	–47 (–53)	16 (61)	137 (449)
Vladivostok	–14 (7)	17 (63)	138 (453)

	Precipitation mm (in) January	July	Year
Moscow	31 (1.2)	74 (2.9)	575 (22.6)
Sochi	201 (7.9)	60 (2.4)	1,451 (57.1)
Krasnovodsk	11 (0.4)	2 (0.1)	92 (3.6)
Ulan Bator	1 (0.04)	76 (3.0)	209 (8.2)
Verkhoyansk	7 (0.3)	33 (1.3)	155 (6.1)
Vladivostok	19 (0.7)	116 (4.6)	824 (32.4)

NATURAL HAZARDS

Earthquakes in Mongolia, southwest and northeast Soviet Union, drought, extreme cold, landslides and avalanches in mountains.

THE ANCIENT FOUNDATIONS

The Soviet Union is an immense plain encircled by high mountains along its borders to the east, south and southwest, and divided in two by the Ural Mountains. The plain has a foundation of ancient Precambrian crystalline rocks more than 600 million years old, which were worn down to a near-flat surface before being covered by sedimentary rocks. Lying on the Eurasian tectonic plate, they provide a stable base for the sedimentary rocks, which have retained their horizontal strata over large areas.

The basement rocks form two platforms. The Russian platform underlies the whole of European Russia – that is, the Soviet Union as far east as the Ural Mountains and Ural river. The old rocks are seen at the surface only in the northeast and southeast. The Siberian platform covers a large part of Siberia, the land east of the Urals between the Arctic Ocean, the Kirgiz Steppe and the mountains along the southern border. It outcrops over a small area in the north and a much larger area in the south. Between these two platforms the basement rocks form a vast, shallow basin (geosyncline) that filled with sedimentary rocks during the Paleozoic era, 590–248 million years ago. These were raised up to form the Ural Mountains when the platforms moved closer together about 280 million years ago during the Hercynian mountain-building period. Movement between the Eurasian and North American plates at that time also raised up fold mountains, such as the Verkhoyansk Range to the east of the Lena river and the Chukot Range in the far northeast of the country.

A turbulent past

The Alpine mountain-building period, which began about 30 million years ago, affected only the southern fringes of the region. The southerly movement of the Eurasian plate against the Indian and Arabian plates folded the deep sediments that had accumulated between them to form mountain ranges that include the Caucasus Mountains and the Pamirs. Some of the older mountains formed in the Hercynian period, such as the Altai range, were lifted up again. Lake Baikal occupies a trench formed by these upheavals. The movements are still taking place, and in 1988 were responsible for the tragic earthquake in Armenia, south of the Caucasus Mountains, which killed 25,000 people.

Along the southeastern and eastern boundaries of the Eurasian plate the movements caused magma to rise to the surface, forming volcanoes. The Kamchatka Peninsula contains Siberia's highest mountain, the volcano Klyuchevskaya Sopka, 4,750 m (15,580 ft). The Kuril Islands are a continuation of the mountains of Kamchatka.

Plains and plateaus

The great plains have nearly always been low-lying, and the sea has advanced and retreated regularly over them. Only the northwest has remained permanently above sea level. With each marine invasion sedimentary rocks were deposited under the sea. The younger rocks occur to the south, with Tertiary rocks 65–2 million years old in the Crimea, the middle Volga plateau and elsewhere.

To the east of the Yenisei river, beyond the West Siberian Plain, the Central Siberian Plateau begins. The land becomes higher, lying mainly between 450 and 900 m (1,500–3,000 ft) above sea level. Here the rocks have not been much disturbed, but there are gentle undulations where they have sagged following small depressions in the platform below.

GREAT INLAND WATERS

The Caspian Sea and the Aral Sea are both saltwater remnants of a former ocean. Lake Superior and the other freshwater lakes of North America are a legacy of ice sheets that covered the land. The greatest freshwater lake by volume, Lake Baikal, contains some 22,000 cu km (5,500 cu mi) of water. It lies in a rift valley formed as a result of tectonic activity in the Earth's crust.

Lake	Area sq km	sq mi
Caspian Sea, Soviet Union/Iran	371,000	143,240
Lake Superior, United States/Canada	83,270	32,150
Aral Sea, Soviet Union	66,500	25,700
Lake Victoria, Tanzania/Uganda/ Kenya	62,940	24,300
Lake Huron, Canada/ United States	59,600	23,000
Lake Michigan, United States	58,000	22,400
Lake Tanganyika, Tanzania/Zaire	32,900	12,700
Great Bear Lake, Canada	31,800	12,100
Lake Baikal, Soviet Union	30,500	11,800
Lake Malawi, Malawi/ Mozambique/Tanzania	29,600	11,400

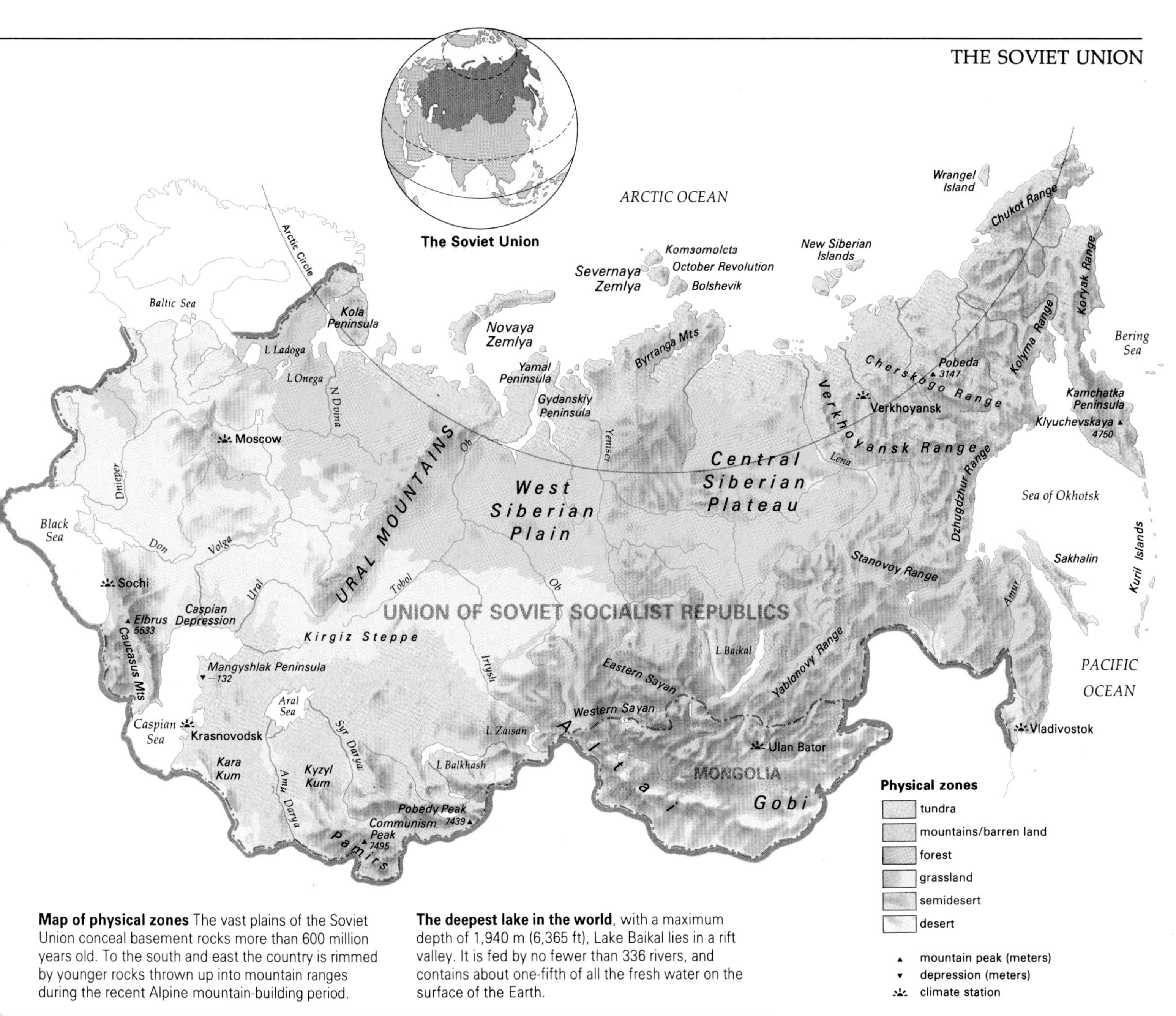

Map of physical zones The vast plains of the Soviet Union conceal basement rocks more than 600 million years old. To the south and east the country is rimmed by younger rocks thrown up into mountain ranges during the recent Alpine mountain-building period.

The deepest lake in the world, with a maximum depth of 1,940 m (6,365 ft), Lake Baikal lies in a rift valley. It is fed by no fewer than 336 rivers, and contains about one-fifth of all the fresh water on the surface of the Earth.

A SUCCESSION OF CLIMATES

The great scale and low relief of the Soviet Union means that large areas have similar landscapes, climates, vegetation and soils. With few natural boundaries formed by bold changes in topography, one zone almost imperceptibly into the next.

The continental interior

The huge interior expanses, which are cut off from the ameliorating effects of any warm ocean except in the extreme northwest, have a continental climate. This is characterized by large differences in temperature between winter and summer, very short spring and fall seasons, low rainfall and strong winds. The country lies farther north than many people imagine, and this makes the continental climate very severe. Moscow lies on the same latitude as Labrador in northeast Canada, and large stretches of the country are north of the Arctic Circle. Verhoyansk

in eastern Siberia has been called the "cold pole" because winter temperatures there can fall as low as −50°C (−58°F). During its short summer, temperatures may reach 37°C (100°F).

Three areas have a more favorable climate. The east coast, particularly the basin of the Amur, has onshore winds in summer that bring rainfall of about 750 mm (30 in). South of the Caucasus Mountains the valleys are well sheltered from the cold north winds, and winters are mild. Summers are hot because of Transcaucasia's southerly latitude, and the climate may best be described as subtropical. The third area with mild climates is the south coast of the Crimean Peninsula in the Black Sea, which is also protected by mountains from cold north winds in winter. With wet, warm winters and hot, dry summers it has a Mediterranean climate.

The plains can be divided into four main zones, running roughly east to west, that reflect gradual climatic transitions from tundra to arid conditions. From north to south they are: tundra, where conditions are too cold for trees; forest; grassland (steppe); and desert, where there is too little moisture available for a rich plant cover. One vegetation zone slowly changes to another over several hundred kilometers in a transitional zone, so it is never possible to state exactly where one starts and the other finishes. The wooded steppe is an example.

The sluggish rivers of Siberia form large deltas. Water readily lies on the surface of the tundra, prevented from draining away by the frozen subsoil.

Dark coniferous forests or taiga, a quarter of all the world's forests, extend over nearly half the country, covering the infertile podzol soils of the northern plains.

The great forest

The dominant impression of the northern landscape is a vast plain, but in fact glaciation has created some relief. Shallow valleys that run east–west mark where meltwater flowed along the front of the ice sheets that advanced over the plains during the ice ages of the past 2 million years. In the northeast material deposited by ice sheets has blocked old valleys and created lakes such as Lakes Ladoga and Onega. Rivers have cut shallow valleys as the land has risen since the great weight of the ice was removed.

Forest is the natural vegetation over the greater part of the Soviet Union. Coniferous forest (taiga) is the most common, covering over a third of the country. It spreads across the land from the Baltic Sea to the Pacific Ocean, and stretches through 2,000 km (1,240 m) of latitude in east Siberia. Forests present a monotonous landscape with only the density of the trees changing – tall, dark thickets on the higher, well-drained ground, thinner forest on the swampy lowlands. Few plants live on the forest floor, but on the trees there are various species of moss and lichen. In the west the main tree species are spruce, pine, fir and cedar, whereas in the east the Siberian and Dahurian larches prevail.

To the south the coniferous forest merges into mixed and then deciduous broadleaf forest. This change reflects the longer growing season and higher temperatures. This is good agricultural land; only isolated patches of the forest remain in a landscape dominated by farming. The largest area of deciduous forest is found in the southeast, where both rainfall and temperatures are higher in the basin drained by the Amur.

Steppe grasslands

Farther south steppe grasslands become more widespread. Flat plains dissected by gullies and ravines are the most typical scenery. The forest steppe is a broad

transitional zone in which patches of forest on the cooler north-facing slopes alternate with steppe on the warmer south-facing slopes.

The steppe is too dry for trees to grow except along the side of river valleys. There is little to break the uniformity of the landscape apart from some small basins formed by erosion, and a few ancient burial mounds raised by human inhabitants. The steppe extends southeast in a broad sweep from the Sea of Azov through the lands between the Ural Mountains and the Caspian and Aral Seas, to the mountain fringe north of Lake Balkhash. The rich black chernozem soils, formerly an expanse of grasses and herbaceous plants adapted to a dry climate, have been plowed and cultivated to form the grainfields of the Soviet Union.

THE GREAT RIVERS

In such a huge landmass, where there are few indentations in the coastline and where the sources of the rivers are such a long way from their mouths, the main rivers are immense. The drainage basins of the Siberian rivers such as the Amur, Lena, Ob and Yenisei could each contain Texas at least three times and France four times over.

As well as their size, the rivers have many other features in common. Their gradients are very shallow because they flow over extensive low-lying areas. The steepest gradient is that of the Volga at 1 : 1,000, about six times less than the Seine at Paris. Still about 2,000 km (1,200 mi) from the Arctic Ocean, the Yenisei is already only 152 m (500 ft) above sea level. This low gradient is associated with relatively slow river flow, but because the volume of flow is so large the rivers and the valleys they occupy are wide: the estuary of the Yenisei, for example is almost 50 km (30 mi) wide.

The flow of the rivers is seasonal, and generally at their lowest level in February, when the catchment areas are frozen. The highest volumes are reached in early summer, when the snow melts and there are summer storms. At this time the rivers flowing into the Arctic overflow their banks, because their mouths are still blocked by ice long after the southern courses have melted. At the mouths of the rivers huge deltas have been deposited. The Volga enters the Caspian Sea through a fan of river channels 200 km (124 mi) wide.

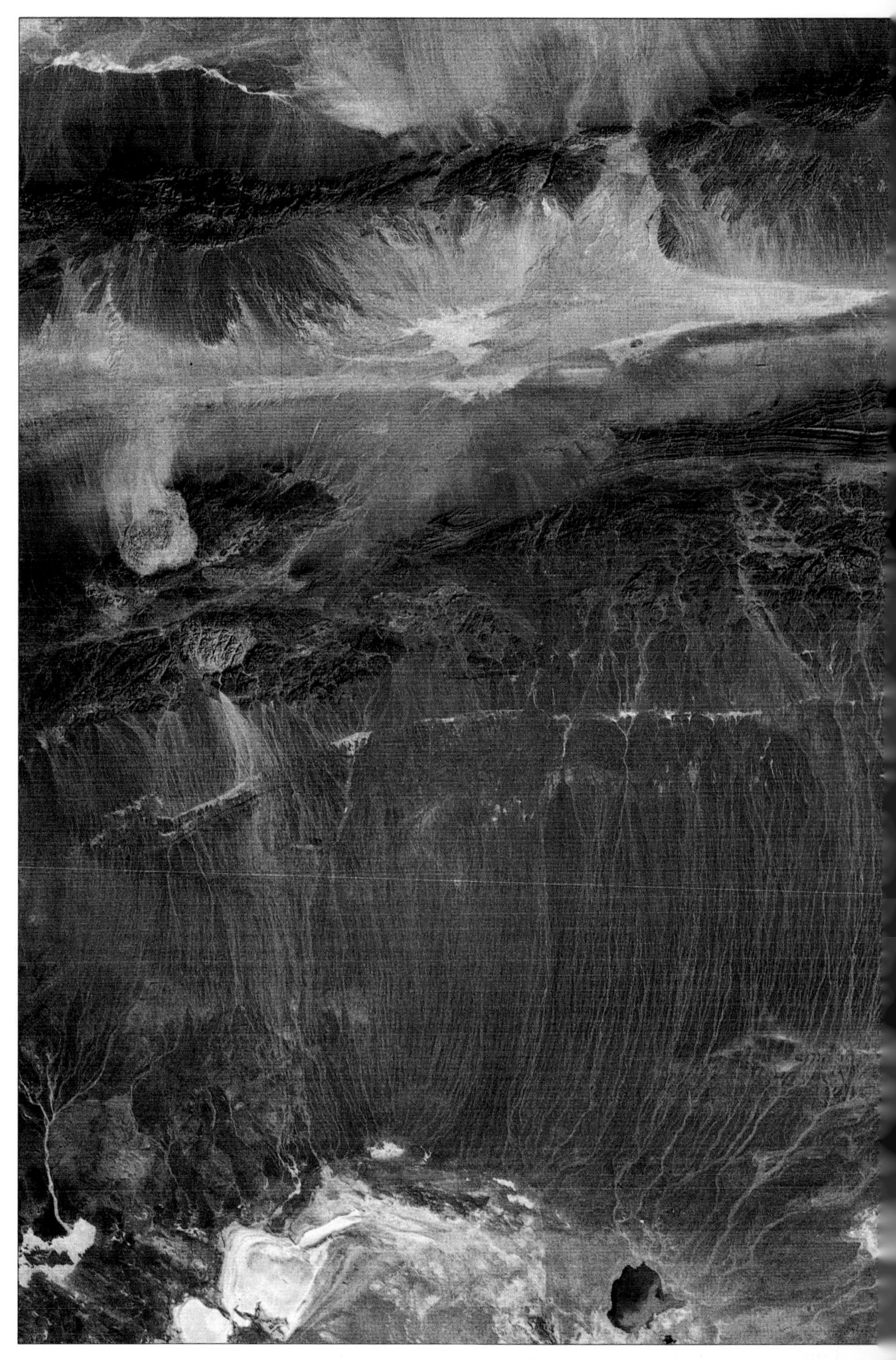

THE ARID SOUTH

Mountains to the south and east, the Caspian Sea to the west and the Kirgiz Steppe to the north mark the boundaries of the Soviet Union's main desert area in central Asia. It covers a huge plain with slopes generally so gentle that they are barely visible. Several types of desert, including clay, gravel and pebble deserts, have developed here but sandy deserts are the most common.

In the central Asian desert large areas of dry land have very little vegetation except in occasional small depressions that collect moisture. In the north there are rich loess soils that can be cultivated if irrigated, but elsewhere wind erosion has left rocks with very thin soils, and over large areas the sand is blown into lines of shifting dunes.

Three main areas within the desert can be identified. The Ustyurt plateau, on the northeastern border of the Caspian Sea, is a block of land about 200 m (650 ft) above seal level surrounded by lines of fractures. To the south lies the mainly sandy Kara Kum (*kum* means desert); between the Syr Darya and Amu Darya rivers is the Kyzyl Kum, which has a thick mantle of sand covering a broken relief, with flat-topped ridges up to about 910 m (3,000 ft).

Beyond the mountains of the Altai range in the southeast of the region lies another great desert area. The Gobi desert extends across half of Mongolia. An area of great extremes of temperature, the Gobi ranges from sandy in the east to stony in the west.

Inland seas

Although precipitation is normally less than 200 mm (8 in) a year in the deserts of central Asia and the potential for evaporation in the intense summer heat is more than 1,000 mm (40 in), the large desert basin of the Soviet Union has two major lakes, the Aral Sea and Caspian Sea.

The Aral Sea is the smaller of the two and has an average depth of about 15 m (50 ft). It is fed by two major rivers, the Amu Darya and Syr Darya, which flow across the desert from the mountains in the southeast. The Caspian Sea, the world's largest inland expanse of water, is 1,945 m (6,384 ft) deep at its southern end. It is fed by the Volga (Europe's longest river), the Ural and the Emba, which flow in from the plains of the north.

Satellite view of the Gobi desert in southern Mongolia. Streambeds formed by flash floods thread the barren slopes of rocky material washed down from mountain ranges. A dry salt pan and a lake occupy the inland basin to the south.

The desert tableland of Ustyurt, near the Aral Sea, a striped landscape of horizontally bedded sandstone. At some time in the past the sandstone was eroded; new, coarser sediments were deposited on the old land surface. In the harsh desert climate these younger rocks are now weathering, and rock fragments are tumbling down the sandstone slopes.

The mountain fringe

Mountains occupy almost a third of the Soviet Union. Many of them, including the Urals, the mountains south of Lake Baikal and the far eastern mountains, do not have very conspicuous relief. The most common landscape is of flattish, rounded hills, broken here and there by higher summits. They represent old mountain systems that have been worn down, pushed up again to form elevated plateaus and then eroded once again.

The principal mountain ranges form a ring around the south and east. The exception is the Ural Mountains, which rise in isolation to separate the European or Russian plains from the Siberian plains. Erosion has reduced their height, and the traveler arriving from the west is hardly aware that the Urals have been reached. For much of their 2,000 km (1,200 mi) length the peaks do not even reach 600 m (2,000 ft).

The highest and most spectacular mountains are found around the edge of the Soviet Union, where the earth movements of the Tertiary period have had most effect. In the east the Kamchatka Peninsula, lying on the edge of the Pacific fault line, is marked by intense volcanic activity. The highest peaks are found in the Pamirs and other ranges to the south of the Siberian plain and Soviet Central Asia. Where the old plain has been pushed up and broken into blocks the mountains look like a giant staircase. Elsewhere earth movements have caused intense folding and faulting and deep rift valleys. High-level deserts have formed in the sheltered basins between the ranges.

The Caucasus Mountains extend for about 1,200 km (750 mi) between the Black and Caspian Seas. They were formed at the same time as the Alps in Europe. Although glaciers still exist, many of the landscape features associated with the ice ages of the past 2 million years, such as U-shaped valleys, hanging valleys and glacial lakes, have been removed by subsequent erosion and deposition. Within the mountains there are small basins separated from one another by high ridges and passes. Most of the mountain areas have been eroded to expose the ancient crystalline rocks lifted up by earth movements. In contrast, the hills in the south of the Crimea and in parts of the Caucasus are limestone, and contain the gorges, pavements and caves typical of karst scenery.

History in the landscape The military road to Georgia passes through the Caucasus Mountains, which began to rise some 30 million years ago and are still rising today. After each uplift the river, faced with an increased gradient, cuts deeply into its old bed, leaving the former valley floor to form a terrace. The many terraces in this valley are evidence of successive periods of uplift.

THE FERGANA VALLEY

Isolated basins surrounded by high mountains are a feature of many of the Soviet Union's mountain areas. One of the largest is the Fergana Valley, a depression running east–west that covers 22,000 sq km (7,720 sq mi) between two mountain ranges north of the Pamirs. The eastern end of the valley is blocked by high mountains that form part of the Tien Shan range, which extends east into China; the western end opens up along the course of the Syr Darya into the Kyzyl Kum. This has made Fergana more accessible than most basins.

Unlike basins in lowland areas, which are formed when the great weight of sedimentary rocks warps the crust to form a geosyncline, the Fergana Valley was created as a result of faulting. The mountains around it rise very steeply from the valley floor, which is covered by a thick layer of sediment.

The area has a very low rainfall, of some 100–300 mm (4–12 in) a year. However, the fertile soils, irrigated by water from the Syr Darya, from oases and from underground sources, have been used to transform the desert into a major agricultural area that produces fruit and a quarter of the Soviet Union's cotton crop. Only small patches of desert remain in the center of the area and along ridges on the periphery.

Permafrost and thermokarst

Traditional houses half sunken into the earth, and more modern buildings abandoned because of cracks and gaps in their walls, dot the city of Yakutsk on the Lena river in Siberia. These are not the effects of an earthquake, but evidence that heat from human habitations has melted what used to be permanently frozen ground beneath them. The ground is as firm as solid rock while it remains frozen, but once the water in it melts it can move as easily as mud.

The northerly position and climate of the Soviet Union are such that half the country is underlain by a layer of permanently frozen ground, or permafrost, between 3 and 1,400 m (10–4,600 ft) deep. The thickness of the permafrost reflects the relationship between the extreme cold of the climate on the surface and the heat from the interior of the Earth. Water occupying the spaces in the soil or rock becomes frozen, making it as impermeable as solid granite.

During the brief summer, higher temperatures cause thawing from the surface downward. The thawed layer is known as the active layer, the permanently frozen ground below it as the permafrost table. The depth of the active layer depends both on how high and for how long surface temperatures remain above freezing and on local characteristics. The layer is deeper in sands and gravels than in finer silts and clays. Vegetation at the surface acts as insulation, and can reduce the depth to which the summer warmth penetrates. Under lakes and in river valleys there may be no permafrost at all.

At the end of the summer, freezing begins from the surface downward, with the result that an unfrozen layer with water under pressure may be trapped between frozen layers. Under these conditions any slight change in temperature, such as that caused by heat from an uninsulated home, for example, can thaw the surface frozen layer.

A pingo can be up to 90 m (300 ft) high and 360 m (1,200 ft) across. At the end of summer the soil water freezes from the surface down, and a layer of water may be trapped against the permafrost. If it freezes, expanding to form ice, a pingo may develop over many years as the surface layers of soil are pushed up.

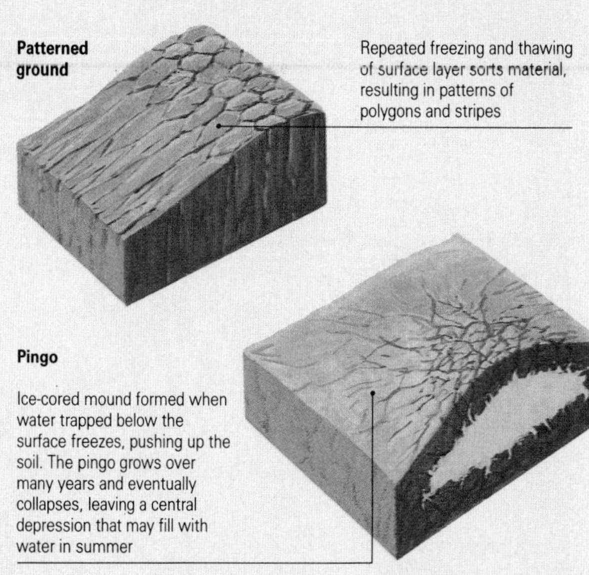

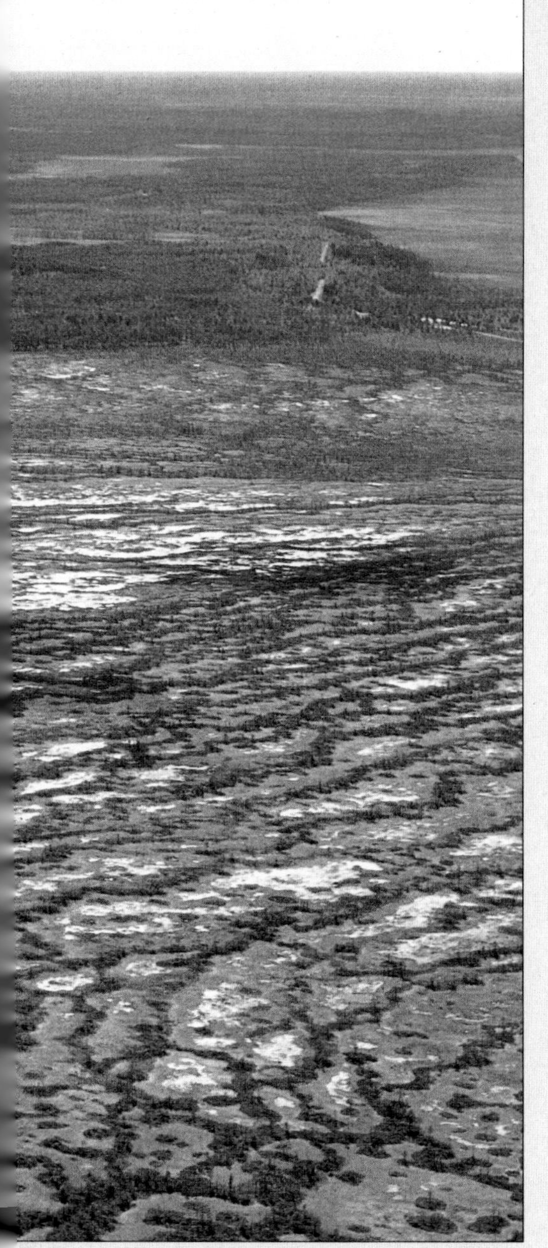

Spotted tundra, with its striking patterns of hummocks and furrows marked by distinct bands of vegetation, is widespread on peaty soils. It may be the result of thousands of years of alternate freezing and thawing of the upper layers of soil.

A distinctive landscape

Melted ice in the active layer cannot soak into the permanently frozen layer below, so water accumulates as the summer progresses. The active layer can become so moist and unstable that just a slight gradient or a small amount of pressure, such as that exerted by a building, is enough to make the whole layer move or start to flow, a process known as solifluction. This can produce ridges, terraces and small steps in the landscape, and can cause buildings to subside.

Another feature of the water-laden soils that undergo repeated freeze–thaw cycles is patterned ground. The name is given to patterns of stones, vegetation or topographical features with a natural regularity, ranging in scale from 1 sq m (11 sq ft) to several square kilometers. In permafrost areas ridges of stones are common that on flat ground form interlocking polygons, but on slopes become elongated until on the steepest slopes they become stone stripes. Mounds covered in vegetation follow a similar arrangment.

There is still much speculation about how such features are formed, but the development of convection cells in the meltwater within the ground seems to play a part. Water is at its most dense at 4°C (39°F). During the spring and summer the meltwater at the surface reaches this temperature sooner than the water lower down. The warmer, heavier water sinks and is replaced by cooler water from below, setting up a convection cell. Materials are moved in the process; the size and shape of the resulting feature will depend on the size and shape of the convection cell, which itself becomes elongated on sloping ground.

Thermokarst is the name given to the distinctive scenery in permafrost areas that is caused by changes in temperature. Its name comes from the karst scenery of limestone areas, which has similar landforms including collapsed features such as depressions. Thermokarst is characterized by subsidence-related features. The permafrost is supersaturated with water. If part of the permafrost layer melts, the excess water is lost by seepage, runoff or evaporation. This decreases the overall volume of the soil, and eventually causes the land surface to collapse.

A spectacular feature of permafrost areas is the pingo. Pingos are domes that rise up where water has collected under the ground and has frozen. As the water freezes it expands, forcing the surface upward into a dome. Pingos are often found in the center of depressions called alases, a feature of thermokarst scenery.

Permafrost areas are among the world's most sensitive environments, and are easily damaged by development projects, such as the construction of buildings, pipelines and roads, or simply the removal of the insulating layer of forest. The development projects themselves may be damaged, through flooding or subsidence, by the changes they cause to the permafrost layer. For both conservation and development to succeed, it is necessary to understand how changes to the environment can be minimized.

Predicting volcanic eruptions

Volcanic eruptions often take place unexpectedly, and as a result may cause great loss of life. Attempts to predict eruptions center on two techniques: the monitoring of the volcano's shape, and the detection of earth tremors associated with the volcano.

As the volcano's magma chamber fills with molten lava the mountain swells, especially where the overlying lava crust is weak. This swelling accelerates just before an eruption, as the intense pressure below forces molten lava toward the surface.

One of the sites where volcanoes are being studied closely is the Kamchatka Peninsula in the Far East of the Soviet Union. Lying on the Pacific "Ring of Fire", where two of the Earth's crustal plates are in constant slow collision, Kamchatka has many volcanoes.

A typical example of the use of earth tremors to predict eruptions came in 1955. Earth tremors began to take place frequently along the peninsula. The focus of these tremors was traced to the volcano Bezymianny, which was thought to be extinct. The number of tremors built up to over two hundred a day. Then suddenly Bezymianny erupted. Earthquakes continued for over a month as the volcano spewed out ash and smoke. When it went quiet, the earthquakes subsided. Four months later a tremendous earthquake shook the peninsula. The volcano exploded so violently that the top 200 m (650 ft) were blown off.

Research into the patterns of earthquake and volcanic activity could help to prevent a major disaster in the future, but very few volcanic areas are monitored regularly enough for accurate predictions to be made.

Trouble brews in a Soviet volcano There are about 80 volcanoes on the Kamchatka Peninsula, some rising to 4,750 m (15,675 ft) above sea level. About 33 of these are active.

MOUNTAIN, PLATEAU AND DESERT LAND

MEETING PLACE OF CONTINENTS · LANDFORMS AND THE CLIMATE · SOILS AND VEGETATION

The long dry season in the Middle East is responsible for the parched appearance of the land away from the Mediterranean, Black and Caspian Sea coasts. Some of the world's hottest and driest deserts are found here, including the Rub al-Khali in Arabia and Iran's Dasht-e-Kavir. Landforms include sand dunes, dry riverbeds (wadis) and salt pans. A harsh climate is also found in the mountains and plateaus of the north, which stretch from Anatolia to the Hindu Kush. The spring snowmelt in Turkey's eastern highlands and Iran's Zagros Mountains swells the great Tigris and Euphrates rivers, which water dry but fertile plains. The Red Sea and the Dead Sea are both an extension of Africa's great Rift Valley. Earthquakes shake the region, and dust storms are a hazard. Around the Gulf are the world's greatest petroleum reserves.

COUNTRIES IN THE REGION

Afghanistan, Bahrain, Iran, Iraq, Israel, Jordan, Kuwait, Lebanon, Oman, Qatar, Saudi Arabia, Southern Yemen, Syria, Turkey, United Arab Emirates, Yemen

LAND

Area 6,595,735 sq km (2,545,954 sq mi)
High point Concord Peak, Hindu Kush, 5,407 m (17,740 ft)
Lowest point Dead Sea, −400 m (−1,312 ft), lowest point on land surface on Earth
Major features plateaus of Anatolia and Iran, Hindu Kush, Zagros and Elburz Mountains, deserts of Arabia and Iran

WATER

Longest river Euphrates, 2,720 km (1,700 mi)
Largest basin Euphrates, 1,105,000 sq km (427,000 sq mi)
Highest average flow Euphrates, 2,860 cu m/sec (101, 000 cu ft/sec)
Largest lake Caspian Sea, 371,000 sq km (143,240 sq mi), largest area of inland water in the world

CLIMATE

	Temperature °C (°F) January	July	Altitude m (ft)
Samsun	7 (45)	23 (73)	44 (144)
Haifa	14 (57)	27 (81)	5 (16)
Amman	8 (46)	25 (77)	771 (2,529)
Basra	12 (54)	34 (93)	2 (7)
Riyadh	16 (61)	40 (104)	609 (1,998)
Kandahar	5 (41)	32 (90)	1,010 (3,313)

	Precipitation mm (in) January	July	Year
Samsun	81 (3.2)	39 (1.5)	731 (28.8)
Haifa	129 (5.1)	1 (0.04)	499 (19.6)
Amman	68 (2.7)	0 (0)	273 (10.7)
Basra	26 (1.0)	0 (0)	164 (6.5)
Riyadh	24 (0.9)	0 (0)	82 (3.2)
Kandahar	22 (0.9)	0 (0)	225 (8.9)

NATURAL HAZARDS

Chiefly earthquakes, landslides, dust storms

MEETING PLACE OF CONTINENTS

The mountains and the basins of the Middle East have both been formed by the collision of continents. North Africa and Arabia contain remnants of Gondwanaland, the great landmass of the southern hemisphere that began to split up and slowly drifted northward during the Mesozoic era (250–65 million years ago). Eventually these continental plates of the Earth's crust made contact with Laurasia, a similar landmass in the northern hemisphere.

The movement of plates

Between the two landmasses sediments were laid down in the Sea of Tethys. These were folded to form mountains in the north of the region during the collision of the plates in the Tertiary period, 65–2 million years ago. The mountains are part of a system stretching from the Atlas Mountains of North Africa, through the European Alps to the Himalayas. In northern Turkey they run parallel with the Black Sea shores; in the south lie the Taurus Mountains. The two link in the vicinity of Ararat (5,165 m/16,945 ft) before diverging again in Iran to form the Elburz Mountains (highest peak Damavand, 5,671 m/18,600 ft) in the north and the slightly lower Zagros Mountains in the south. Farther east the ranges then merge again in Afghanistan to form the massive Hindu Kush.

Smaller plates within the area also play an important role. The Turkish plate is moving almost due west in relation to the Eurasian and African plates; its northern boundary is marked by a fault in north Anatolia. The Red Sea and the Gulf of Aden have been formed as the result of Arabia moving away from Africa at about 1 cm (0.4 in) a year. The rift that separates them is part of the great Rift Valley system, which extends farther north to the Dead Sea lowlands. Here the Arabian plate appears to have moved north about 100 km (62 mi) in the past 5 million years relative to the small and apparently fixed Sinai plate in northeast Africa.

These movements of the Earth's crust continue today. As in other parts of the world, there are many violent earthquakes along the major plate boundaries. When these take place close to cities, such as Tehran, situated to the south of the Elburz Mountains, they cause considerable damage and loss of life.

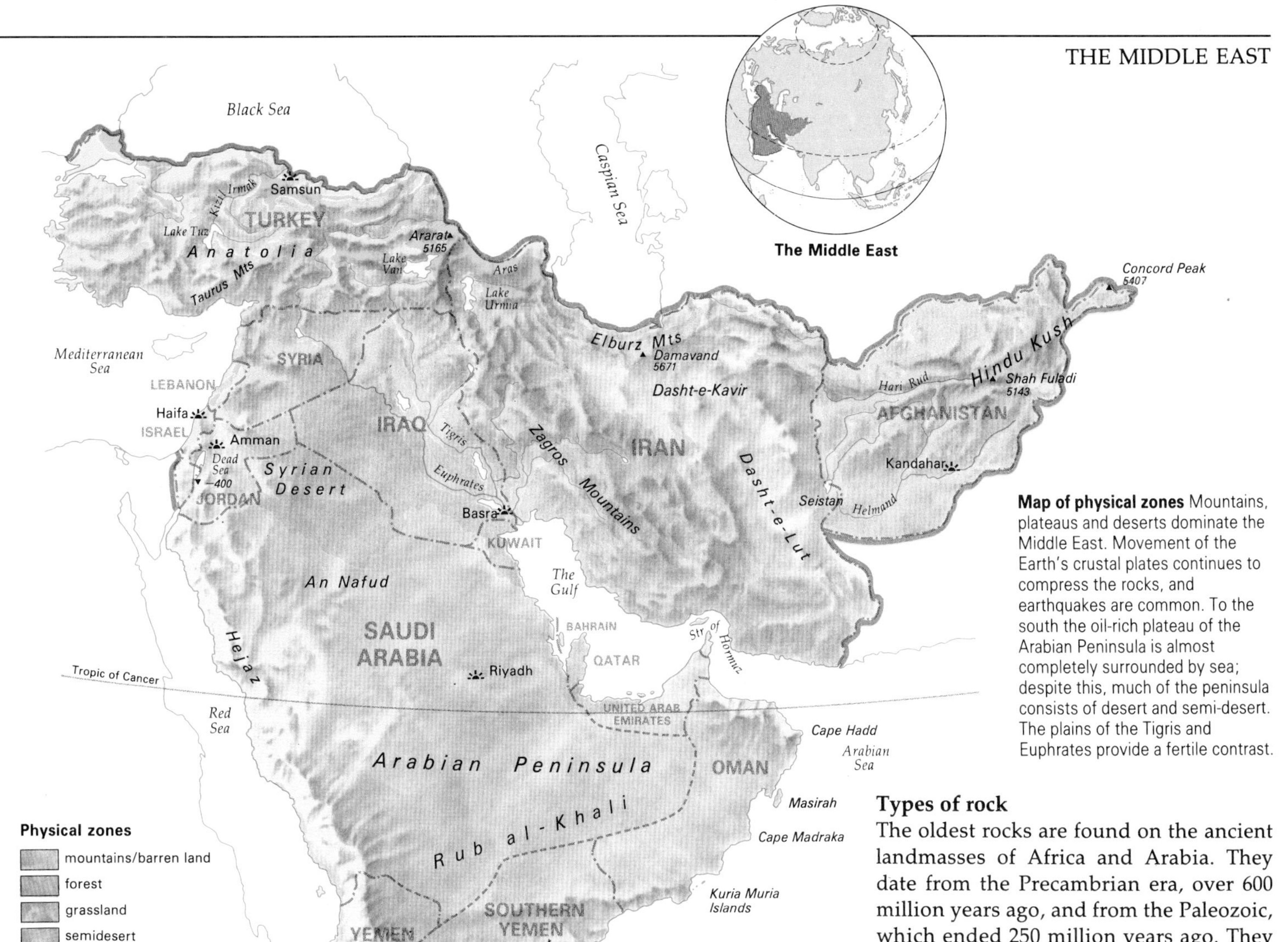

Map of physical zones Mountains, plateaus and deserts dominate the Middle East. Movement of the Earth's crustal plates continues to compress the rocks, and earthquakes are common. To the south the oil-rich plateau of the Arabian Peninsula is almost completely surrounded by sea; despite this, much of the peninsula consists of desert and semi-desert. The plains of the Tigris and Euphrates provide a fertile contrast.

Types of rock

The oldest rocks are found on the ancient landmasses of Africa and Arabia. They date from the Precambrian era, over 600 million years ago, and from the Paleozoic, which ended 250 million years ago. They are exposed in western Saudi Arabia. Northward the rocks become progressively younger, and they are mostly of sedimentary origin. In eastern Mediterranean lands and in southern Iran marine sediments, especially limestones and marls of the Mesozoic era, are common.

Rocks of Tertiary age, mostly marine sands, clays, marls and limestones, are generally confined to the lowest parts of the region, including a wide zone parallel to the Gulf. They also occur in inland basins of Turkey and Iran. Thick layers of unconsolidated sediments from the Quaternary (the last 2 million years) are found in upland basins and along the major valleys.

Igneous rocks, chiefly basalts, are found in many areas that are associated with weaknesses in the Earth's crust in Turkey, in Iran and in the major fault zones along the Dead Sea lowlands and the Red Sea. The best-known volcanoes in the region, Ararat and Damavand, are extinct. Elsewhere lavas have welled up along major fissures, producing widespread flows, such as those of the mountainous Jebel Druze in southern Syria.

The Hindu Kush at Bamian in Afghanistan forms part of the great mountain system that stretches from North Africa to Indonesia. Rocks originally laid down under the sea have been pushed up high enough for their peaks to be permanently covered with snow.

LANDFORMS AND THE CLIMATE

Broadly speaking there are four types of landscape in the Middle East: mountains or uplands, and three different zones in which eroded rock material has been deposited – alluvial fans, alluvial plains and salt lakes or salt deserts.

Shaping the landscape

Most erosion takes place in the mountains and uplands, where the altitude ranges through many hundreds or even thousands of meters and slopes are steep. Freeze-thaw is the major type of weathering and produces angular debris (scree). Large quantities of sediment have filled in the river valleys, many of which consequently have flat floors.

Alluvial fans are common where eroded material is deposited by streams and rivers as they slow down on leaving the highlands. The land slopes at an angle of about 15 degrees, and there are few hills or other relief features. The coarsest material is deposited near the top of the fan, where mud flows also occur. Major cities such as Tehran, the capital of Iran, and Damascus, to the west of the Syrian Desert, are sited on alluvial fans.

Alluvial plains with very gentle slope angles are the chief landform in many parts of the Middle East, notably on the plateau of Anatolia in Turkey and in central Iran. Most are made of finer sediments that are carried beyond the alluvial fans by running water. Fine-grained material on the surface is lifted and carried away by the wind, to be deposited again elsewhere, sometimes in the form of sand dunes, leaving bare desert pavement behind.

Water flows into these areas only during the spring snowmelt, when the rivers are in flood. Evaporation in summer exceeds the inflow of water and the lakes become increasingly salty. Some dry out completely, and the minerals in the water are precipitated as a salty crust. The Caspian, the world's greatest inland sea, is a salt lake. It does not dry up during the summer months because it is fed by the Volga, which drains part of the Soviet Union to the north.

Patterns of rainfall

The most characteristic feature of climate in the Middle East is the dry season. In winter areas of low atmospheric pressure (depressions) track eastward across the northern part of the region, bringing precipitation almost everywhere in the north. In summer the depressions pass mostly over northern parts of Turkey and the Elburz Mountains, so precipitation is limited to the northern fringe. Elsewhere sinking air causes high pressure and stable conditions and there is often no rain for months. Precipitation is greatest in winter almost everywhere, with the exception of the shores of the Black and Caspian seas and southern Arabia. In southern Arabia the summer monsoon winds bring rains that vary in quantity from year to year and in how far they penetrate north.

Most of the Middle East receives less than 600 mm (24 in) of rain a year. The heaviest precipitation is on mountain ranges. There is less rain toward the east and south away from the Mediterranean coast. There are some well-watered areas, however, generally confined to narrow coastal belts backed by mountain ranges. Parts of the Black Sea and Mediterranean coasts of Turkey, for example, receive over 1,000 mm (40 in) a year. The eastern shoreline of the Mediterranean and the high mountains behind it also receive

Effects of chemical weathering In the area of Cappadocia in central Turkey limestone has been chemically weathered by rainwater to form narrow gullies and pinnacles.

Cold winters and hot summers characterize the climate in the north of the region. Shallow valleys are carved by streams fed by the snow that in winter covers the hills near Gaziantep in south-central Turkey. This is one of the wetter areas of the Middle East.

substantial rainfall. Similarly, in western parts of the Caspian Sea lowlands more than 1,000 mm (40 in) a year may fall. The final zone of high precipitation is a long continuous belt running southeast from eastern Turkey parallel with the crest of the Zagros Mountains. Only occasionally do totals rise above 800 mm (31.5 in) a year, but the sheer size of the area means that it has tremendous importance for the water balance of the region.

In the uplands and mountains, and particularly those of Turkey and Iran, much of the precipitation falls as snow. This is released in spring when the snow begins to melt, feeding the Tigris and Euphrates, and the waters of the Helmand river in the Hindu Kush.

Extremes of temperature

Summers are hot in almost the whole region. Even on high plateaus temperatures can rise to over 35°C (95°F). Only in the highest parts of Turkey and Iran and near the Mediterranean coast do they normally remain below 30°C (86°F). Temperature contrasts are much greater in winter. In the south temperatures are mostly 10–20°C (50–68°F); to the north they decline rapidly across the highlands of Turkey and Iran, with lows of –10°C (14°F). The greatest annual temperature range, of more than 30°C (54°F), is found in the mountain areas of eastern Turkey. In most of the northern highlands of Turkey and Iran an annual range of over 25°C (45°F) is common. The southern shores of the Red Sea and the coastal area of Southern Yemen have an annual range of less than 10°C (18°F). The lowest range of all, less than 5°C (9°F), is found on the southern coast of Oman.

WHERE THE WATER GOES

Water is in short supply in the Middle East. Huge amounts are lost in dry areas by evapotranspiration (evaporation, and transpiration from plants). Even in Turkey, one of the region's wetter countries, two-thirds of the precipitation is lost in this way. Only if precipitation exceeds evapotranspiration can rivers flow and water-bearing rocks (aquifers) be replenished. In the Middle East such surpluses are restricted to small areas in the north, parallel to the mountains. The two largest rivers, the Tigris and Euphrates, receive most of their flows from melting snows. The water surpluses are transported hundreds of kilometers to the arid lowlands of Iraq, where they make cultivation possible.

Surplus water often moves underground. On the margin of the highlands water percolates through the coarse debris of alluvial fans. Once in the aquifer, the water moves slowly, often no more than a few meters a year, but the amount of water that is stored, safe from evaporation, can be enormous. These groundwater reserves supply the region's oasis settlements.

At least 80 percent of the water used by people in the Middle East is needed for irrigation. In the drier parts, traditional methods can require up to each year 10,000 cu m of water for every hectare (142,000 cu ft per acre). Much of it evaporates from the fields and channels. In very arid countries, such as Kuwait and Saudi Arabia, water is obtained by desalination of sea water, but even oil-rich countries can find such methods expensive.

SOILS AND VEGETATION

Soils and vegetation are formed by the interaction of relief, geology, climate and human activity. In the Middle East soils are of two main types: desert soils where calcium or sodium salts have accumulated because they have not been washed out (leached) by percolating water, and the thin soils of mountain areas.

Desert soils

The most widespread is the red desert soil found over much of the Arabian Peninsula and southern Iran. There is little organic content, so the soil lacks nutrients and is not very fertile.

Gray desert soils are found in cooler but still arid areas such as eastern Iran. Calcium carbonate may build up just below the surface, cementing together a layer of rock fragments (calcrete or caliche).

In slightly wetter conditions, such as on the desert margins of southern Iran, Iraq, Syria and Jordan, reddish prairie or reddish brown soils are found. They are transitional between soils of arid areas, in which the dominant direction of water movement is upward through evaporation, and soils with at least moderate rainfall, in which water moves downward, percolating through the soil.

THE GULF PETROLEUM DEPOSITS

The oilfields of the Middle East are located in deep sedimentary basins along the margins of the Zagros Mountains. For oil to accumulate in rock formations certain conditions must be met. There has to be a porous reservoir rock in which the oil can accumulate, and a geological structure such as an upfold (anticline) in the rock strata to trap the oil. An impermeable cap rock above the reservoir is needed to prevent the oil from moving upward through the rocks; this is usually provided by salt deposits. Last but not least, there must be rocks close by in which the oil originally formed.

In the Middle East shales deposited during the Jurassic and Cretaceous periods (213–65 million years ago) are thought to be important source rocks for petroleum. Shales are formed when mud is compressed. Most petroleum in the Middle East has been formed from these marine deposits, which were rich in organic material derived from the remains of extremely small floating plants and plankton.

In the Gulf region there are five main geological settings in which oil has accumulated between the folds of the Zagros Mountains and the ancient shield rocks of the Arabian Peninsula. The most important of these is the limestone of Iran, which was laid down some 38–15 million years ago, during the Oligocene and early Miocene epochs, in deposits that are now between 30 and 500 m (100–1,600 ft) thick.

The oilfields are found on both sides of the Gulf and extend northward into Iraq. All the countries in this area are important oil producers. The largest oil reserves yet to be tapped are believed to be in Iran, Iraq and Saudi Arabia.

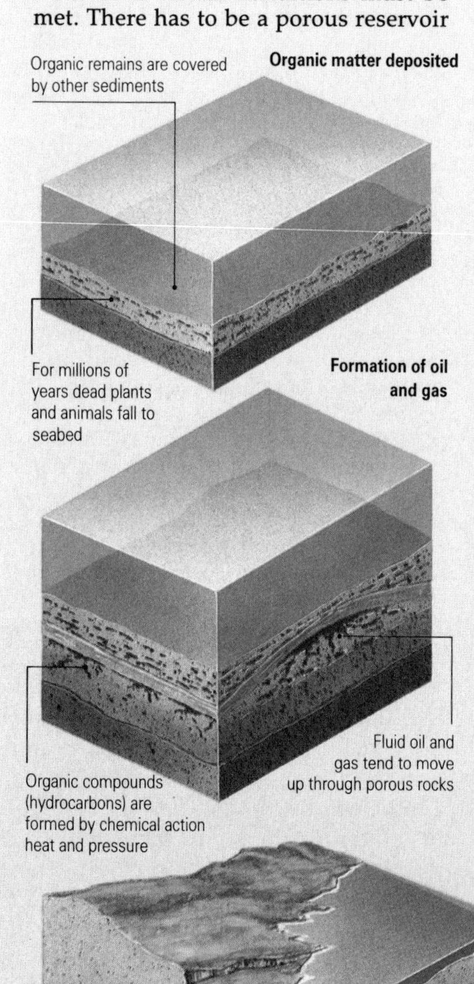

The geological conditions needed for oil and natural gas to collect under the ground are limited. The most common is an anticline of pervious or porous rock capped by an impervious layer.

Oil reservoirs

Oil and gas collect in porous rocks beneath a domed cap of impermeable rock in an upfold (anticline)

Oil and gas trapped by a fault where impermeable rocks block further movement

Soils of mountains and valleys

In the highlands of the north there are a variety of soil types. The abrupt contrasts in the environment make it difficult to generalize, but most soils are thin and stony, and unsuitable for cultivation.

The deeper alluvial soils found along all the major valley systems are much better for growing crops. Such soils are found in the Tigris–Euphrates lowlands of Iraq, part of the Fertile Crescent, one of the earliest centers for the development of agriculture some 10,000 years ago.

Remnants of natural vegetation

The long history of cultivation in the region makes it difficult to reconstruct the original pattern of vegetation. In the highland areas of the eastern Mediterranean, Turkey and Iran, extensive forests were probably dominant, while on the foothills and lowlands forests gave way to grassland. Remnants of the original forests are still found on the slopes of the northern Taurus ranges of Turkey, and also on the northern slopes of Iran's Elburz Mountains. In central Iran and most of the Arabian Peninsula semidesert scrub would have prevailed.

Since farming was first established the activities of humans have modified the vegetation cover. Forests have been cut down for building and for fuel, with the result that today very few dense forests remain in the region.

As the human population has increased, particularly in the 20th century, severe grazing pressure has damaged pastures, leading to soil erosion and desertification. The vegetation that exists today is only a degraded remnant of that of former times.

Salt in the land

During summer in the arid Middle East little water remains on or near the surface of the ground. The topsoil is baked to a dry dust under the hot sun. Any moisture in the soil is drawn upward to the surface,

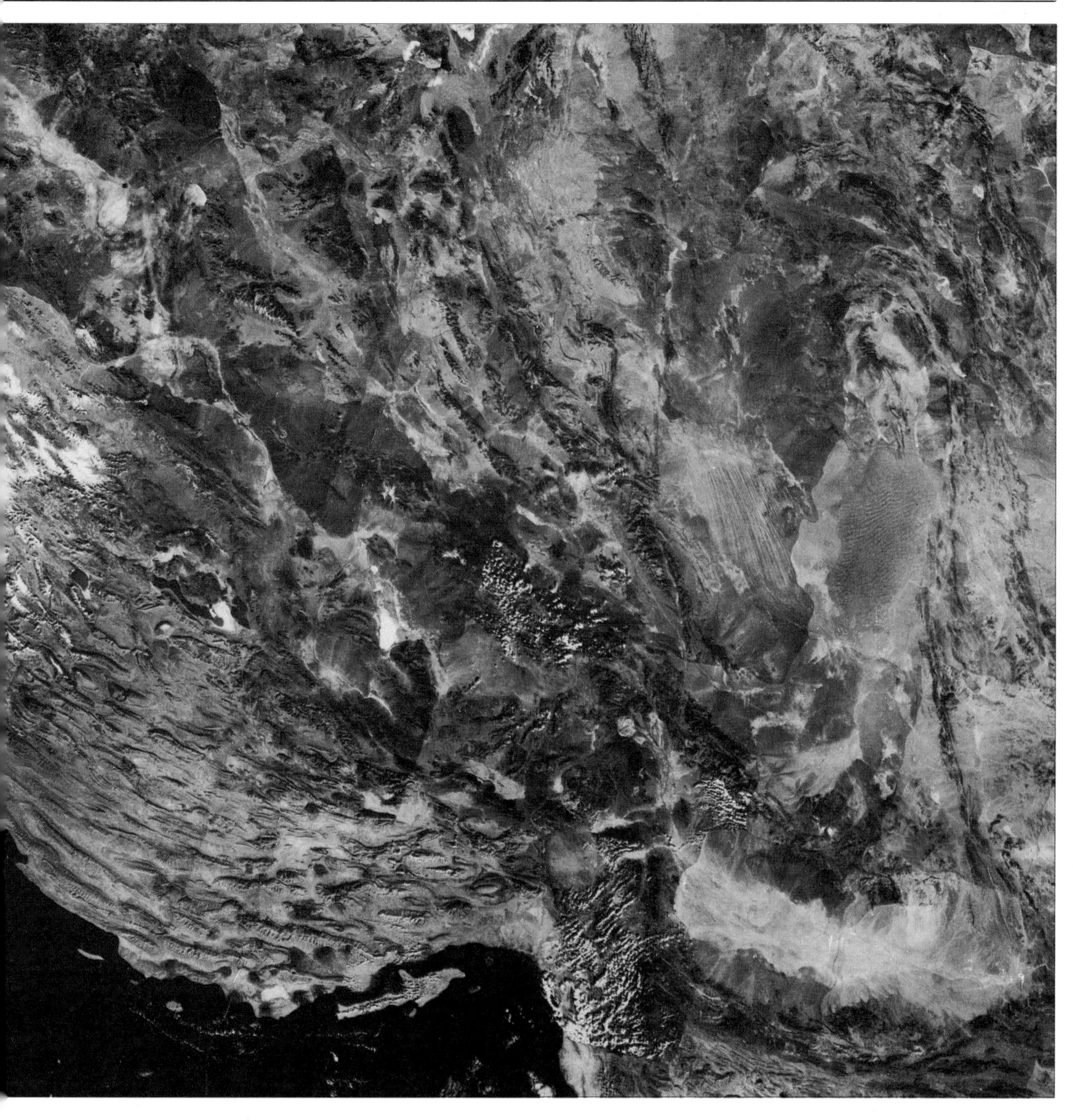

The structure of southern Iran This satellite image covers the area north of the Strait of Hormuz. In the center are the Zagros Mountains. The lines at top right are yardangs, ridges up to 15 m (50 ft) high formed by erosion of alternate soft and hard layers of rock.

where it rapidly evaporates.

The salts that are dissolved in the water are precipitated out in the soil. Over time this upward movement can lead to the topsoil becoming so salty that arable cultivation is impossible. The salts form a hard layer on the surface, impeding drainage. Irrigation can worsen the problem. Salination has been affecting the Fertile Crescent for more than 4,000 years, and modern irrigation methods have increased the proportion of salinated soils in much of the region. Half the irrigated areas in Iraq and in the Euphrates valley in Syria are now affected, and one-seventh in Iran. Reclamation of such soils is at best difficult and often impossible.

On shallow slopes and on plains in the Middle East groundwater seeps out at points where the water table reaches the surface. It quickly evaporates, leaving behind a highly concentrated brine solution. Seepage points can be from 1 m (3 ft) to over 40 m (130 ft) wide and many hundreds of meters in length.

Salt-weathering is associated with seepage in many of the inland basins of the Middle East, and helps to explain the breakdown of resistant rocks. In these hot, dry conditions salt crystals grow quickly, breaking down cobbles and pebbles into very small particles. This fine-grained material can be transported much more easily by wind or water than can heavier, coarse-grained fragments. Salt weathering can also attack the foundations of buildings and other structures, and therefore presents a hazard in these dry areas.

Dust storms and wind erosion

In the dry Seistan depression between Iran and Afghanistan the houses have no windows in their north walls. This traditional feature of house design helps to prevent the dust from entering during the dust storms that blow in from the north – on 80 days in a typical summer.

Air travelers are also affected by this phenomenon of the Middle East. Every summer international airports along the Gulf to the southwest have to close for several days when the *Shamal* wind blows in dust from the floodplains of the Tigris and Euphrates. At Kuwait airport dust storms (defined internationally as dust-raising events that reduce visibility to below 1 km/3,300 ft), are recorded on average 27 days a year.

In the north of the region dust storms are associated with eastward-moving cold fronts, and are more frequent in winter. In parts of eastern Syria there are dust storms on average 15 days a year.

Land moved by the wind

High winds are essential for dust storms to blow up. As the dry wind passes over areas of sparse vegetation it picks up the small particles, which are not bound together as they generally are in more humid areas.

The size of wind-eroded material varies from clay (less than four microns/0.004 mm across), through silt (0.004–0.06 mm) to fine sand (0.06–0.2 mm). The heavier fine sand is blown along, bouncing up to a few centimeters above the surface. Silt can be carried to heights of many meters, while fine clay particles can be carried in the air for days – and have even been found in the atmosphere above Alaska after making the 10,000 km (over 6,000 mi) journey from Mongolia's Gobi desert.

Desert rocks and dry farming

Sand-sized and smaller particles blown by the wind can scrape away and polish the surface of rocks like sandpaper, fashioning the detail of rock pillars, rock islands (inselbergs) and other features.

Very fine particles are removed by the wind from the slopes of the alluvial fans and the plains at the foot of the uplands. Heavier particles that have moved down the slopes remain, leaving behind an "armored" surface of pebbles and stones that protects the ground from further erosion. Such processes produce stony deserts (reg) and rock deserts (hamada). In sandy deserts (erg) such as the Rub al-Khali,the Empty Quarter of Arabia, the wind shapes the loose sand into dunes of various shapes: crescent-shaped, star-shaped and long, narrow dunes.

In the dry Middle East dust storms remove valuable soil from agricultural areas, and young crops can be destroyed when blasted by sand. Plowing and other disturbance of the land surface in dry areas increases wind erosion, as it breaks down the soil structure and buries vegetation that would otherwise help to hold the particles and prevent their removal by the wind. These effects are particularly marked in dry farming areas in Jordan, Syria and Iraq.

In low-lying areas wind-blown silt, known as loess, may accumulate in deposits several meters thick. Loess is widespread in the central plateau of Iran and in southern Afghanistan, forming layers up to 5 m (17 ft) thick.

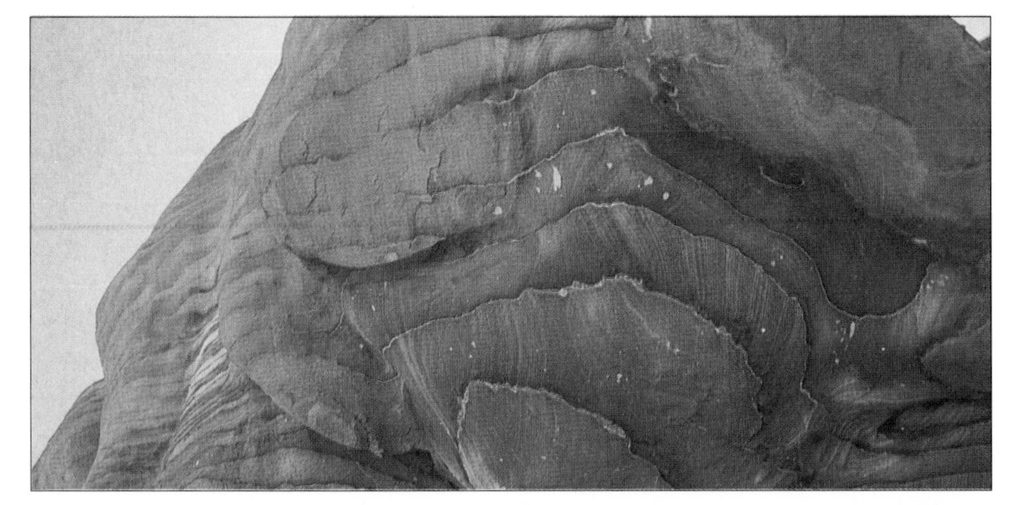

Patterns of erosion In arid areas there is little or no vegetation to protect rocks from the processes of weathering and erosion. The exposed rocks are broken down as a result of intense heating during the day and rapid cooling at night. Moisture condensing on the cool surface of the rock hastens exfoliation, in which the outer layers of the rock peel off. The fragments are worn down by wind-borne particles until they too can be carried by the wind, and erode other rocks in their turn. On this rock outcrop at Petra in southwest Jordan the different layers have been etched by particles of sand blasted against the rock by the wind. The effect is greatest near the base where larger, more abrasive grains of sand are bounced along the ground.

Rock pillars are a common feature of arid landscapes, such as this one in the Negev desert in Israel. The rock is undercut in several ways. Weathering is more rapid around the base of rocks because any moisture condensing on the cool rock at night tends to collect there; any flash floods will also erode the base, where wind erosion is also more powerful. The top of the pillar may consist of a more resistant layer of rock.

Desert tableland of the Dasht-e-Lut in eastern Iran. The wind has worn away layers of sandstone to form a plain of almost bare rock. Some solitary pillars remain, but they are being undercut by wind erosion and the debris will eventually be blown out of the area.

THE WORLD'S GREATEST DESERT ZONE

A CHANGING LANDSCAPE · CLIMATIC INFLUENCES · SAND AND WATER

In the plateau northeast of the Sahara's Ahaggar mountains rock paintings dating from 6,000 BC show a land very different from the desert landscapes of today. Hunters pursued their prey on savanna grasslands, lakes and rivers, and pastoralists grazed their cattle on the abundant vegetation. Northern Africa has undergone several changes of climate, and old river valleys and dry lake beds in the desert are evidence of wetter periods. Today, arid and semiarid land stretches from the Atlantic Ocean to the Red Sea. It is the largest hot, dry region on Earth. Flat expanses of sandy and rocky desert are broken by isolated rocky outcrops. There are also some dramatic mountain ranges, including the High Atlas and the Ethiopian Highlands. Only two major rivers, the Nile and the Niger, flow throughout the year.

COUNTRIES IN THE REGION

Algeria, Chad, Djibouti, Egypt, Ethiopia, Libya, Mali, Mauritania, Morocco, Niger, Somalia, Sudan, Tunisia

LAND

Area 14,887,110 sq km (5,747,919 sq mi)
Highest point Ras Dashan, 4,620 m (15,158 ft)
Lowest point Lake Assal, Djibouti, −150 m (−492 ft)
Major features Atlas ranges, Ethiopian Highlands, Sahara, world's greatest desert, northern part of East African Rift Valley

WATER

Longest river most of the Nile's 6,690 km (4,160 mi) length, the world's greatest for a river, and 2,802,000 sq km (1,082,000 sq mi) basin is in the region
Highest average flow Niger, 5,700 cu m/sec (201,000 cu ft/sec) on lower section
Largest lake Chad, 25,900 sq km (10,000 sq mi)

CLIMATE

	Temperature °C (°F) January	July	Altitude m (ft)
Ouarzazate	9 (48)	30 (86)	1,136 (3,726)
Timbuktu	23 (73)	32 (90)	273 (895)
Tripoli	11 (52)	27 (81)	84 (276)
Alexandria	15 (59)	26 (79)	7 (23)
Wadi Halfa	13 (55)	32 (90)	155 (508)
Addis Ababa	16 (61)	15 (59)	2,360 (7,741)

	Precipitation mm (in) January	July	Year
Ouarzazate	6 (0.2)	2 (0.1)	123 (4.8)
Timbuktu	0 (0)	65 (2.6)	225 (8.9)
Tripoli	46 (1.8)	0 (0)	253 (10.0)
Alexandria	44 (1.7)	0 (0)	169 (6.7)
Wadi Halfa	0 (0)	1 (0.04)	3 (0.1)
Addis Ababa	24 (0.9)	228 (9.0)	1,089 (42.9)

World's highest recorded temperature, 58°C (136.4°F), Al Aziziyah, Libya; Wadi Halfa is one of the world's driest places

NATURAL HAZARDS

Drought, earthquakes in mountains of northwest

A CHANGING LANDSCAPE

The ancient land of Africa was once the center of a vast continent called Gondwanaland, which also included Antarctica, Australia, India, New Guinea and South America. It comprised Precambrian granites and sandstones, rocks over 600 million years old, overlaid by sedimentary deposits. Gondwanaland broke up in stages between about 250 and 100 million years ago.

Deformations of the Earth's crust raised the land surface and warped it into a series of saucer-like depressions such as the Sudan and Chad basins. The basins were surrounded by high rims or swells, many of which now act as watersheds for major drainage systems.

Another feature of the massive changes in the land was a tilting of the continent. It rose in the south and subsided in the north, dividing Africa into high and low areas. About 70 million years ago, a lowering of the crust in the north allowed the sea to flood the land, and later to deposit sediments that became sandstones, shales and limestones. These now provide the main water-bearing rocks (aquifers), and also contain mineral deposits such as oil and natural gas.

In the east of the region tectonic movements in the Earth's crust around the middle of the Cenozoic era some 30 million years ago ruptured the lowest strata of rocks. Associated volcanic activity together with rifting and faulting accounted for the formation of the Rift Valley, the world's greatest geological depression, which cuts across the Ethiopian Highlands from north to south and divides Africa from Arabia along the line of the Red Sea.

Tectonic movement also explains the most dramatic relief feature in northern Africa, the Atlas ranges in the northwest. They were formed by the folding of sedimentary rocks, principally limestone, and some older underlying rocks, that took place when the northward-drifting African continent ruptured as it moved into the Eurasian landmass.

Present-day climates

The climates of northern Africa are of four major types: Mediterranean, arid, semiarid and humid tropical. These climate zones are arranged in east–west belts running parallel with the Equator. The pattern is complicated by the Ethiopian Highlands and the Indian Ocean to the southeast, whose influences give Somalia and Djibouti a semiarid rather than the expected humid–tropical climate of other areas in the same latitudes.

More than half the region lies within the tropics, and few locations south of the Tropic of Cancer have daytime average temperatures below 20°C (68°F) in even the coolest months of December and January. Variations between day and night temperatures can be considerable because clear skies at night allow more radiation from the land surface to escape, allowing temperatures to drop.

Toward the north and the Mediterranean Sea seasonal extremes are more marked. Temperatures in August can exceed 30°C (86°F), especially in the desert interior. The world's highest recorded temperature was at Al Aziziyah in northern Libya, but winter temperatures in December and January can be low. Frost

Sulfur springs at the Danakil Depression in Ethiopia on the coast of the Red Sea, which lies on the volcanic faultline of the East African Rift Valley. Water laden with sulfur bubbles up from fissures in the ground. At the surface the water evaporates, leaving encrusted yellow sulfur deposits.

An inhospitable landscape This desert scene in northern Niger is typical of much of the region. Desert climates are dry and hot; they occur in latitudes where global air circulation brings dry air down from the lower atmosphere, in inland continental areas, or in the rain shadow of mountains.

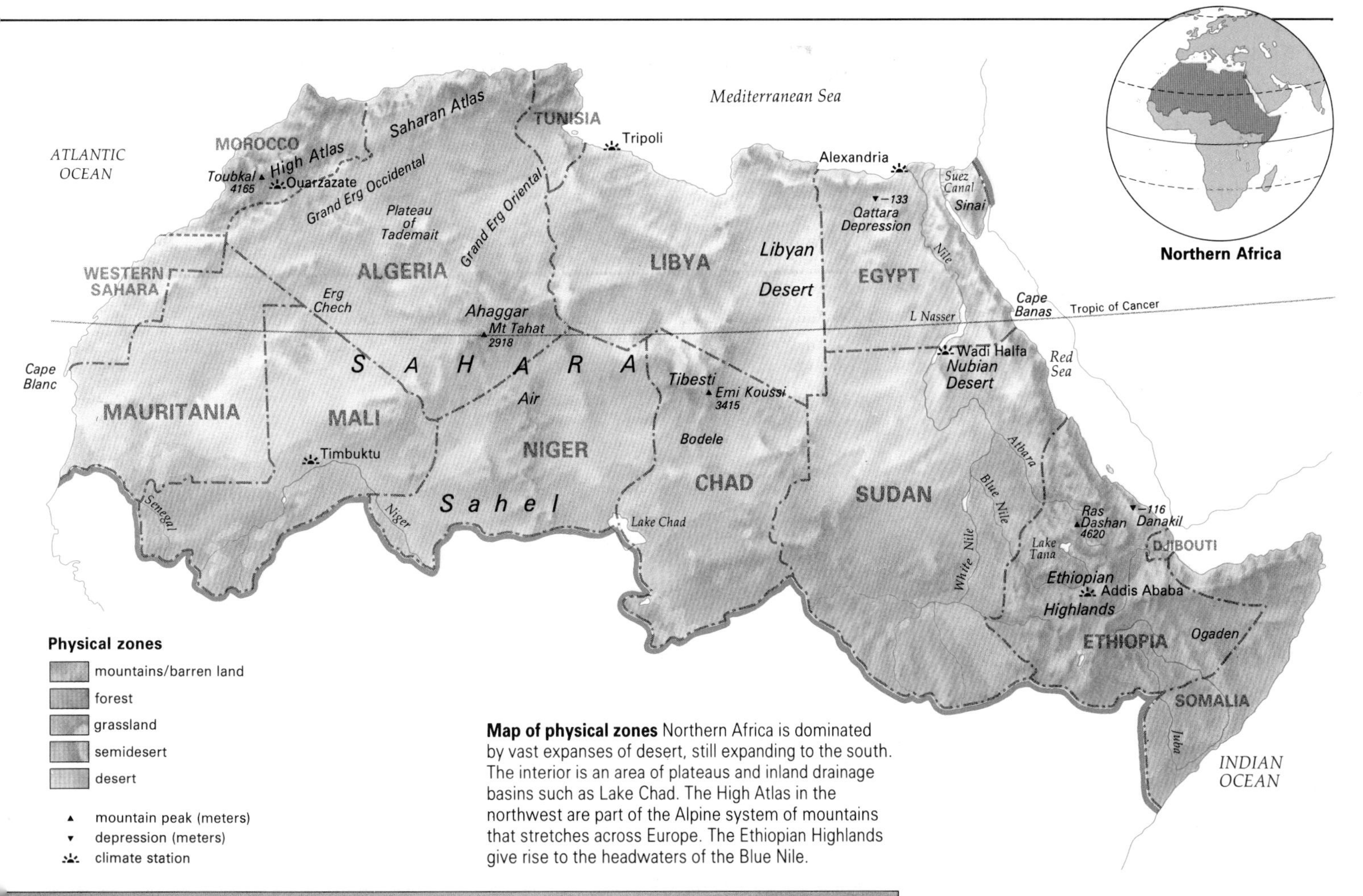

Map of physical zones Northern Africa is dominated by vast expanses of desert, still expanding to the south. The interior is an area of plateaus and inland drainage basins such as Lake Chad. The High Atlas in the northwest are part of the Alpine system of mountains that stretches across Europe. The Ethiopian Highlands give rise to the headwaters of the Blue Nile.

and snow are not unknown along the Mediterranean fringe, and in the mountains of the High Atlas there is a short skiing season.

Despite the generally hot climate, precipitation – or the lack of it – is the most important climatic feature. Nowhere is there rainfall throughout the year, and in parts of the Sahara one or two showers a year may be above average. However, southern Chad, the southern half of Sudan and most of Ethiopia, and some coastal fringes of Morocco, Algeria, Tunisia and Libya expect to receive more than 400 mm (16 in) in any one year. Very high annual rainfalls of over 1,000 mm (39 in) are recorded in parts of upland Morocco in the northwest, Ethiopia in the southeast and in the more southerly regions of Sudan, Chad and Mali. Along the Mediterranean coast rain usually falls in the cooler months, so evaporation losses are not great. In the semiarid open savanna country of the Sahel, along the southern fringes of the arid Sahara, nearly all rainfall comes in the hot season as the intertropical convergence zone (ITCZ) crosses the region. This trough of low atmospheric pressure moves north and south of the Equator with the overhead Sun, and brings rain-bearing clouds.

Climate and vegetation

The four climate zones can be identified by their vegetation. The Mediterranean

Chimneys of lava Plugs were formed when lava solidified in the vents of volcanoes in the Ahaggar mountains of southern Algeria. The cones eroded, leaving the harder plugs standing clear above the surrounding rocks.

zone has much scrubland, with luxuriant trees in sheltered, moist locations. Close to the sea and near salt flats the vegetation is salt-tolerant (halophytic). The wide arid zone of the Sahara is devoid of vegetation apart from occasional shrubs and grasses. North and south of the desert is a narrow semiarid zone where rainfall is higher, and the shrubby vegetation is supplemented by grassland and drought-tolerant (xerophytic) thorny trees such as acacia. The most abundant vegetation is found in the narrow humid–tropical zone in the south, forested with hardwords.

CLIMATIC INFLUENCES

While ancient geological processes explain many of the large-scale physical features of northern Africa, more recent environmental changes have dramatically transformed the surface landscape. Virtually all areas can be divided into two categories: those in which the rock type and complex local history are the dominant influences in scenery formation (azonal landforms), and others where climatic factors have been predominant (giving zonal landforms).

Landscape-forming processes

Local factors have been crucial in shaping landforms throughout the region, where outcrops of relatively hard, resistant rock became isolated as softer rocks were eroded away. Landscapes containing steep, dome-shaped hills of crystalline rock (inselbergs) are typical. Perhaps the most spectacular of these landforms are the vast escarpments of the central Sahara. The strata around the Ahaggar mountains dip northward. The less resistant rocks have been eroded and the resistant rocks have remained, forming two parallel south-facing scarps. Other azonal features are the lakes and marshes that have filled old sedimentary basins, such as Lake Chad, the sudd (papyrus) marshes of southern Sudan and the inland delta of the Niger in Mali.

Climate-induced zonal landforms are more widespread. Although the Sahara

DAMMING THE NILE

The river Nile provides a lifeline for the people of Sudan and Egypt; attempts have been made throughout history to control its annual flood. In the 1960s the Aswan High Dam was constructed in Egypt to guarantee regular supplies of water, irrigate crops and generate electricity. The waters of the world's longest river were backed up some 480 km (300 mi) to form a vast reservoir, Lake Nasser. Two other dams were built in Sudan, one across the Nile's last tributary, the Atbara, at Khashm al-Girba and the other on the Blue Nile at Roseires, to increase the country's share of the Nile water.

Recently, the advantages of the High Dam, which was completed in 1970, have been questioned. Vast amounts of water are lost by evaporation from Lake Nasser, and the lake is silting up. Silt has historically provided a renewable source of fertile soil around the Nile further downstream, but much of this is now lost, trapped by the reservoir. The resulting reduction in the Nile's load means that downstream of Aswan it cuts a deeper channel, affecting the efficiency of irrigation systems. More seriously, so much water is now extracted that the discharge into the sea is negligible and the Nile delta has been gradually eroded by longshore drift, as waves break at an angle to the shore and shift the delta deposits along it. The loss of freshwater flowing into the Mediterranean has affected fisheries, particularly around Cyprus.

Attempt were made to increase the usable water of the Nile by building the Jonglei Canal to prevent the loss through evaporation and seepage of half the waters of the White Nile as it passes through the sudd swamps in southern Sudan. However, the project has now been abandoned owing to civil war and controversy over the canal's ecological and social implications.

receives little rain, water is an important agent of erosion and deposition. When the rain comes it tends to fall in violent downpours. The huge volumes of rain create flash floods that sweep through steep valleys, or wadis. These floods can move heavy loads of sediment, erode and undercut valley sides.

In the semiarid savanna regions there is a short wet season when surface runoff of water creates sheet floods, which wash away topsoil and other fine surface materials. The humid, wetter climate of the savannas also encourages chemical weathering of the rocks, producing clays and ironstone/laterite soils.

Other zonal landforms include the glaciated scenery found in some parts of Ethiopia and in the High Atlas range. The glaciers that were formed in the ice ages of the Pleistocene (2 million to 10,000 years ago) created features such as cirques at valley heads, and deposited moraines of eroded material at the glacier's snout.

Wet and dry phases

Although the dunes and other landscape features of the Sahara show signs of continuing development, some desert landforms suggest that dramatic changes took place when the climate was very different. There is geological evidence for a series of climate epochs since the breakup of Gondwanaland, but the past 20,000 years account for most of the present-day surfaces and landforms in northern Africa. Global changes in climate that caused advances and retreats of ice sheets and glaciers in the northern hemisphere also affected landscape-forming conditions in Africa. Temperatures, rainfall and sea level fluctuated, thus creating sequences of wet (pluvial) and dry (interpluvial) phases. The work of archaeologists investigating traces of human settlements has made a useful contribution, as it has helped to date landscape features and the stages of their development.

The wet phases increased the flow and extent of rivers and affected drainage patterns. Luxuriant vegetation grew in the now-arid Ahaggar and Tibesti mountains. River valleys and lakes were formed, including one huge lake over 1,000 km (600 mi) in length, of which Lake Chad is a remnant. It was in the dry phases that arid landforms such as sand dunes originated, and the many changes of climate led to the shifting of the edges of the Sahara as dune formations expanded and contracted.

About 5,000 years ago, global warming led to greater aridity. The levels of the Nile and other rivers dropped, leaving abandoned river valleys, channels, lake shores and basins. These features can still be seen today in the desert. For example, wadis are desert valleys originally carved by running water but now dry and frequently filled with sand. In the northwest lakes formed in wetter climates dried up, leaving flats, called chotts and sebkhas, that are often salt-encrusted.

During this period people, animals and their crops were forced to move to wetter areas. A new phase of Saharan sand-dune formation began, with material that was eroded in earlier rainy periods being reworked by the wind. In areas near major sand sources, the wind blew the sand into dunes and vast sand seas.

Slopes terraced for agriculture in the Ouzirhimt valley of the High Atlas on the Sahara's northern fringe, which consist of several ranges. Those near the Mediterranean Sea are wetter than those farther inland, with rainfall mainly in the winter. Folded during the Alpine mountain-building period, they are deeply incised with steep-sided river valleys.

SAND AND WATER

Northern Africa is dominated by the Sahara. This vast desert stretches from the Atlantic Ocean to the Red Sea, making much of the region a difficult place for humans to live in. Rivers and wells provide vital supplies of water, and maximum use is made of these limited resources. In some areas there are also valuable deposits of oil and gas beneath the surface, allowing some countries in the region, for example Algeria and Libya, to export a surplus overseas.

The Sahara desert covers over 8.6 million sq km (3.3 million sq mi), nearly a third of the area of Africa. Its base rock is a shield of highly eroded crystalline rocks broken in the center by the volcanic mountains of the Tibesti, rising to 3,415 m (11,204 ft) at Emi Koussi, and of the Ahaggar, whose Mount Tahat is 2,918 m (9,573 ft) high. The lowest point is the Qattara Depression, 133 m (436 ft) below sea level. Stretching across the interior are stony plains (reg), broken only by flat, bare, rocky uplands called hamadas.

About a quarter of the Sahara is made up of sand swept into dunes by the wind. Such areas include the Libyan desert and parts of northern Algeria. The lack of rain and surface water, along with extreme day and night temperatures and the constant Saharan wind, make this an inhospitable environment. Few animals and plants live here, but those that do have developed strategies for conserving and utilizing water. The Tuareg nomads wander the desert with their animals, following the pattern of the rains, but return to wells in the dry season. Desert plants may lie dormant for years, but are able to grow and reproduce rapidly when rain does come. The very arid desert interior is surrounded by semiarid desert margins with a savanna zone.

Rivers and wells

In this hot, dry region there are four sources of water supply for humans: rivers, some of which originate outside the region; underground rocks (aquifers); the sea; and rainfall, directly or indirectly.

Where they are present, rivers are the most important water source, but apart from a number of streams in the area of the Atlas ranges, the only perennial rivers are the Niger and the Nile. The 4,200 km (2,600 mi) long Niger in the west yields water for drinking and irrigation, and is an important fishery for the riverside communities of Mali and Niger. It provides a similar lifeline for the people living beside it as the Nile does along most of its length in the east of the region.

The Nile is the world's longest river, at 6,690 km (4,180 mi). Its remotest headwaters rise in the lake plateau of eastern Africa and flow north as the White Nile before joining the Blue Nile, which rises

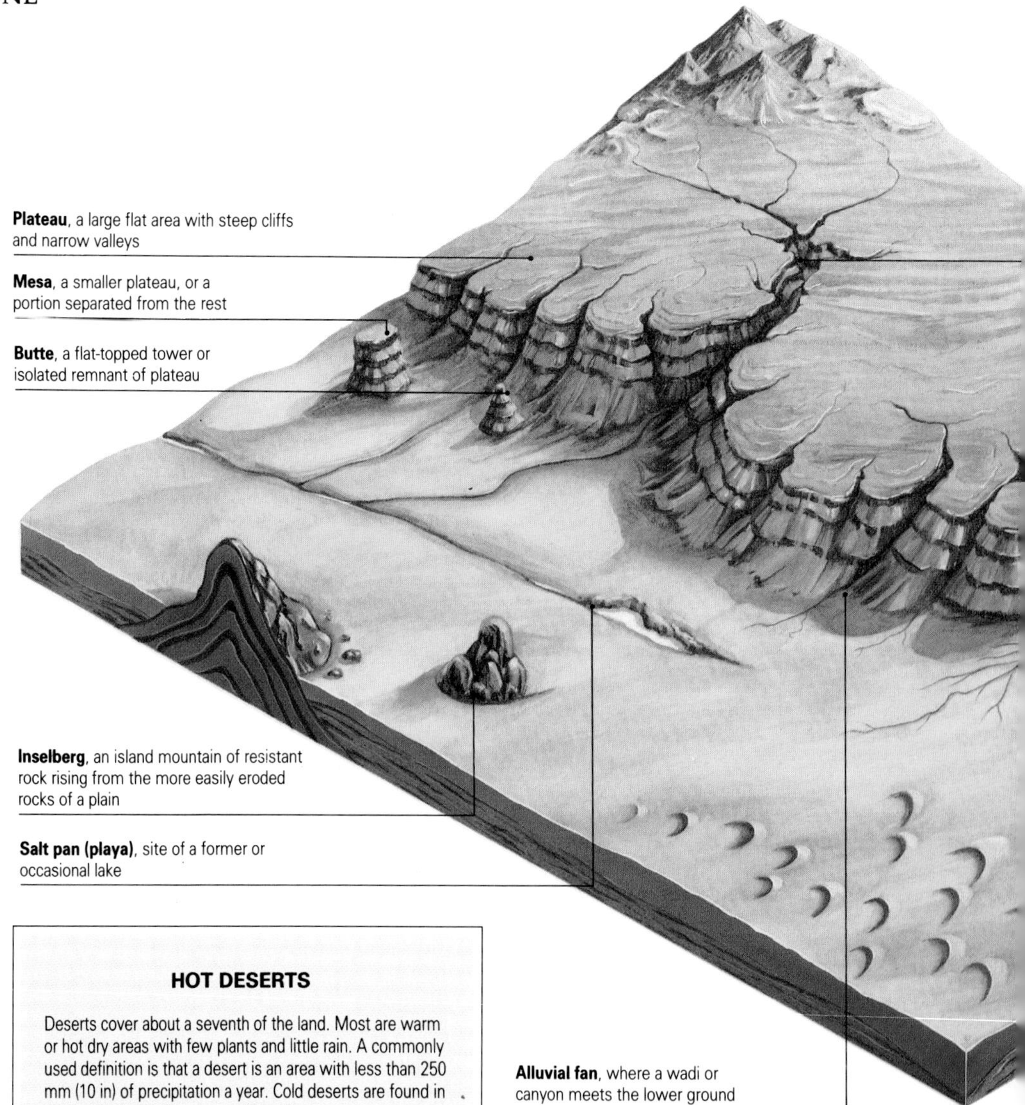

HOT DESERTS

Deserts cover about a seventh of the land. Most are warm or hot dry areas with few plants and little rain. A commonly used definition is that a desert is an area with less than 250 mm (10 in) of precipitation a year. Cold deserts are found in Antarctica, North America and Eurasia.

Continent and principal deserts	sq km	sq mi
Africa		
Sahara	8,600,000	3,320,000
Kalahari	260,000	100,000
Namib	135,000	52,000
Eurasia		
Gobi	1,300,000	500,000
Rub al-Khali	650,000	250,000
Kara Kum	350,000	135,000
Kyzyl Kum	300,000	115,000
Takla Makan	270,000	105,000
Dasht-e-Kavir	260,000	100,000
Syrian	260,000	100,000
Thar	200,000	77,000
Australasia		
Great Victoria	647,000	250,000
Great Sandy	400,000	150,000
Simpson	145,000	56,000
North America		
Great Basin	492,000	190,000
North Mexico	450,000	175,000
Sonoran	310,000	120,000
Mojave	65,000	25,000
South America		
Patagonian	673,000	260,000
Atacama	140,000	54,000

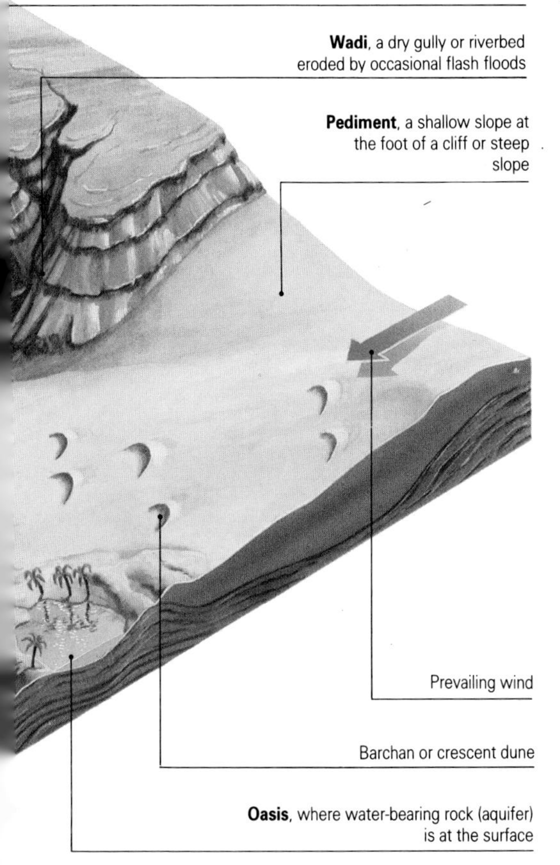

Desert landforms Because they are not softened by vegetation cover, desert landscapes are dramatic. They are sculpted by wind and water. Water flow is periodic, with flash floods created by storms rushing through wadis. Alluvial fans are formed as the eroded material is deposited, and surface flow can lead to temporary lakes in inland drainage basins. Wind-blown deposits form dunes; they can also erode by undercutting rocks. There are different types of desert: hamada are rocky upland deserts with little sand; erg are sandy deserts; and reg are stony plains.

An oasis in the great western sand sea of the Sahara. Oases form when a depression in the desert reaches down as far as the water table.

TYPES OF DUNES

The Sahara desert contains some of the largest areas of sand dunes in the world. These vast, patterned expanses are among the most awesome and regular landforms on our planet.

The sand that makes up the dunes is blown from alluvial plains, lake and sea shores and from weathered sandstone and granite rock. Dunes can be found in a variety of geometric forms depending on the supply of sand, the wind direction, seasonal variations, the vegetation cover and the shape of the ground.

A tail dune forms when an obstacle such as a small bush or hill checks the velocity of the wind, causing it to deposit its sandy load. Small dunes, less than 30 cm (12 in) in height, that form in the lee of small shrubs are called nebkha. Probably the best-known dune type is the barchan, a crescent-shaped dune with horns pointing in the same direction as the wind. Barchans are often found in groups that migrate downwind over the years .

Seifs (from the Arabic word meaning sand) are also known as linear, or longitudinal, dunes. They form long straightish ridges, with slip-faces on both sides, that run parallel to the wind. Seifs can form from barchans if there is a slight seasonal shift in wind direction. Ridges of seif dunes can be over 100 m (330 ft) high and can stretch for hundreds of kilometers. In areas where winds blow from many different directions rhourds, or star dunes, form. They have arms extending from a central peak and can be up to 150 m (500 ft) high and up to 2 km (1.2 mi) across.

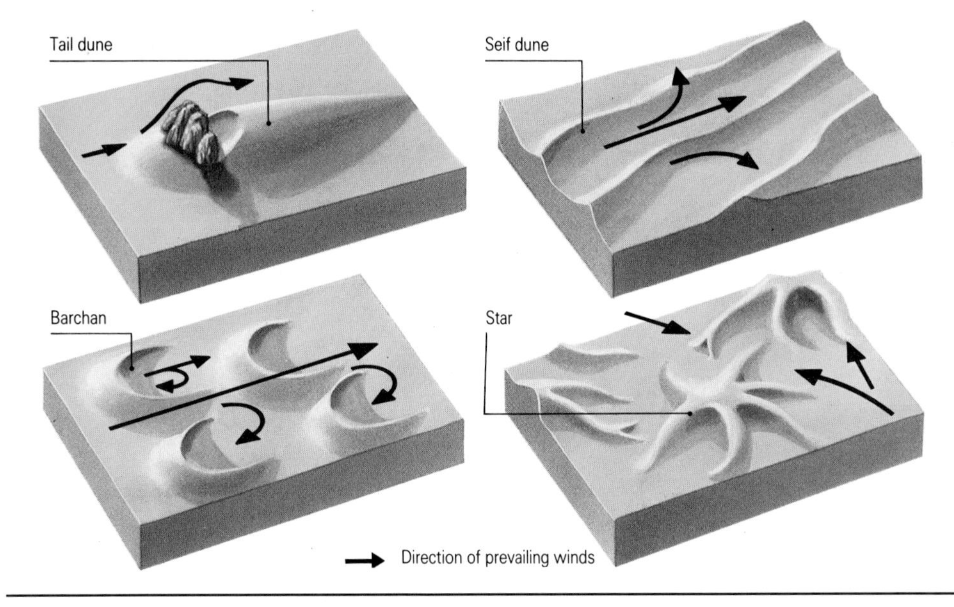

in the Ethiopian Highlands to the east. The Blue Nile provides four-fifths of the annual flow, which is at its greater in August. In the dry season the White Nile provides most of the flow. Nearly all the basin is desert apart from occasional oases. Below Lake Nasser a narrow floodplain is cultivated by irrigation.

The most widespread source of water are underground aquifers. Some contain water which accumulated during wetter periods a long time ago. Aquifer sources are finite, and this is being starkly highlighted by the drier climates of today. Water stores are being used more quickly than they are being replenished, an imbalance that causes the water tables to fall. Some springs and aquifers have dried up completely or are becoming difficult to exploit, particularly in coastal Morocco and Tunisia.

Tunisia and Libya have both experimented with purification and desalination of water from the sea and from salt lakes, but so far large-scale extraction has proved very expensive. Another water-gathering technique, the collection of rainwater from pools, is widely used in countries of the semiarid Sahel south of the Sahara.

Oil and gas

As well as underground aquifers, northern Africa has large underground reservoirs of oil and gas. The formation of deposits depends on certain geological conditions. Oil and gas (hydrocarbons of organic origin) are found in sedimentary basins, usually of limestone and sandstone, where there is normally an anticline (upfold) in the rocks to act as a trap, and where there are strata above and below non porous rocks, which are impervious to seepage. In these underground reservoirs the oil gradually separates and floats on top of the water, and natural gas vapors form above it. Commercial exploitation takes place by drilling down through the rock layers to tap the reserves.

The spreading desert

The problem of desertification became well known in the mid-1970s, when many countries of the Sahel, the semiarid zone on the southern fringe of the Sahara desert (notably Mauritania, Mali, Niger, Chad and Somalia), experienced a succession of drought years. The effect of these droughts was an increase in soil erosion as well as the reactivation of fossil dunes, which extended the edges of the desert. A United Nations conference popularized the use of the term "desertification" to describe this deterioration of the landscape close to, but not necessarily on, the margins of the desert.

Explanations of desertification linked increased surface erosion and gullying, declining crop productivity, deforestation and other factors with a combination of over-exploitative use of the land and climatic deterioration. More recently, the role of governments in the desertification process has been highlighted. In poor and politically unstable countries money and other resources cannot often be spent on environmental schemes, and environmentally sensitive legislation has a low priority, as it still does elsewhere.

Climatic change and spreading deserts

Analyses of annual rainfall trends across northern Africa and the Middle East in the 1960s and 1970s seemed to show a correlation between sequences of dry years and accelerated desertification. More recent research, however, has demonstrated that the links are more complicated, and relate to changes in the distribution of rainfall throughout the year as well as to absolute amounts.

It is now recognized that the countries of the Sahel have experienced almost continuous dry years since the early 1960s. More rain fell in 1978 and also in 1988, when flood disasters threatened parts of Chad, Niger, Mali and Sudan. Rainfall has tended to concentrate in fewer but heavier storms, loosening the topsoil before it is washed away under heavy rainfall and in flash floods. Extreme rainfalls are also becoming more evident. Khartoum, at the confluence of the Blue Nile and White Nile in Sudan, for example, received only 5 mm (0.2 in) of rain in the whole of 1984, but over 200 mm (8 in) in one storm in 1988.

Another form of climatic change is the apparent shortening of the wet season. First showers throughout the Sahel now tend to come later in the year than they did in the period before the 1960s. This phenomenon is often coupled with a break in mid-season rainfall or a premature end to the season. An early end to the wet season can often lead to widespread losses of staple foods such as millet and sorghum, which require water right up to the ripening stage.

The combination of these factors – less total rain, rain falling in fewer, heavier showers, and a short rainy season – has serious implications for the landscape. As well as greater erosion, gullying and other losses of topsoil, there is greater pressure to increase crop and livestock production in order to have security against the unpredictable conditions, and this stimulates more intensive land use. This in turn sets in motion other damaging environmental forces such as overgrazing and deforestation.

Climate alone is not the cause of desertification; the problem exists in other parts of the world where climatic changes have been less dramatic. It has been demonstrated in the Sahel, however, that climatic changes of such magnitude can trigger a whole series of environmentally disastrous consequences.

The landscape of northern Africa has been shaped by a succession of different climates in the past; each climatic period has added its own particular ingredient to the scenery. Desertification can be seen as only the latest stage in this progression. Human activity has simply accelerated the process.

Desertification As the Sahara becomes even drier the vegetation cover declines, and there is little to stop the onward progress of the moving sand. This once prosperous oasis has been abandoned to the desert.

Resisting the sands A depression in the desert is deep enough for the roots of the date palms to reach the water table. The hollow is protected from the drifting sand by walls, themselves in danger of being buried.

AFRICA'S CENTRAL PLATEAU

EVOLUTION OF THE LAND · LANDSCAPES OF CENTRAL AFRICA · AN EQUATORIAL CLIMATE ZONE

Size for size, Africa has fewer mountains and less low-lying land than other continents. The ancient land of Africa south of the Sahara was not subject to intense folding of sedimentary rocks, and much of the region has remained stable for so long that erosion has worn the land into a series of level or very gently undulating plateaus where ancient rocks regularly outcrop at the surface. The highest land is in the east of the region, where the world's greatest geological depression, the Rift Valley, is a deep scar running north–south through eastern Africa. This faulting of the Earth's crust was accompanied by volcanic activity. Rainforest covers much of the basin of the river Congo and the coasts of west Africa. Savanna grassland in other parts gives way in the north to the southern fringe of the Sahara desert.

COUNTRIES IN THE REGION

Benin, Burkina, Burundi, Cameroon, Cape Verde, Central African Republic, Congo, Equatorial Guinea, Gabon, Gambia, Ghana, Guinea, Guinea–Bissau, Ivory Coast, Kenya, Liberia, Nigeria, Rwanda, São Tomé and Príncipe, Senegal, Seychelles, Sierra Leone, Tanzania, Togo, Uganda, Zaire

LAND

Area 8,979,034 sq km (3,465,907 sq mi)
Highest point Kilimanjaro, 5,895 m (19,340 ft)
Lowest point sea level
Major features Jos Plateau and Adamawa Highlands in west, Congo basin, Ruwenzori Range, mountains and Rift Valley in east

WATER

Longest river Congo (Zaire), 4,630 km (2,880 mi)
Largest basin Congo, 3,822,000 sq km (1,476,000 sq mi)
Highest average flow Congo, 39,000 cu m/sec (1,377,000 cu ft/sec)
Largest lake Victoria, 62,940 sq km (24,300 sq mi)

CLIMATE

	Temperature °C (°F) January	July	Altitude m (ft)
Dakar	21 (70)	27 (81)	23 (75)
Ngaoundéré	22 (72)	21 (70)	1,119 (3,670)
Lisala	25 (77)	24 (75)	460 (1,509)
Bukoba	21 (70)	20 (68)	1,137 (3,729)
Lodwar	29 (84)	28 (82)	506 (1,660)
Mombasa	28 (82)	24 (75)	58 (190)

	Precipitation mm (in) January	July	Year
Dakar	0 (0)	88 (3.5)	578 (22.8)
Ngaoundéré	2 (0.1)	256 (10.1)	1,511 (59.5)
Lisala	63 (2.5)	190 (7.5)	1,626 (64.0)
Bukoba	151 (5.9)	49 (1.9)	2,043 (80.4)
Lodwar	8 (0.3)	15 (0.6)	162 (6.4)
Mombasa	30 (1.2)	72 (2.8)	1,163 (45.8)

NATURAL HAZARDS

Drought, floods, earthquakes

EVOLUTION OF THE LAND

The central African region straddles the Equator, and stretches for more than 6,000 km (3,700 mi) from east to west. It is part of the vast plateau of hard, ancient rocks that outcrop on over half of Africa's surface. Elsewhere, they are covered with sedimentary or volcanic rocks. These reached the surface when the continent was cracked by huge faults 35–25 million years ago.

The ancient blocks of Africa

The landmass of this part of Africa is built around huge masses of the Earth's crust (cratons) that became stable 2,000 million years ago. The western African craton occupies parts of Guinea, Liberia, the Ivory Coast and Ghana. The Congolese and Kalahari craton occupies part of southern Zaire. Diamonds and gold are often associated with these blocks, which have become anchors for the continent.

Around the cratons are the Precambrian granites and gneisses of old mountain chains dating from more than 600 million years ago. They contain minerals, including copper, lead, tin and zinc. Superimposed on these ancient foundations there are sedimentary rocks, mainly sandstones and shales but occasionally limestone. They are found primarily in the shallow basins such as the Congo (Zaire) basin and the middle Niger basin, which were formed during the Hercynian mountain-building period about 250 million years ago. Pressure on the rocks by movements in the north caused some areas to swell up and others to subside in gentle waves of up to 1,000 km (620 mi). Other sediments are found at the bottom of rift valleys, in faulted coastal basins and along the narrow coastal plains.

By about 120 million years ago the continent of Gondwanaland was splitting up, creating new continents and smaller remnants of granite rock masses, such as the Seychelles in the Indian Ocean 1,800 km (1,120 mi) off the Kenyan coast.

Vast basins

The basin covering parts of Congo, Zaire and the Central African Republic is surrounded by Precambrian rocks, but contains sandy sediments deposited by the Congo river over many millions of years. Much of the basin is now covered with tropical rainforest, but the forest covers old dunes that were formed by the wind, implying that parts of the basin were arid in the past. Such major changes of climate were associated with the cooling of the Earth during glacial periods, the most recent of which ended about 10,000 years ago. The soil is poor and contains few mineral resources. Cut off by high mountains in the east and rapids in the west where the river flows over the rim of the basin, this area remains one of the least

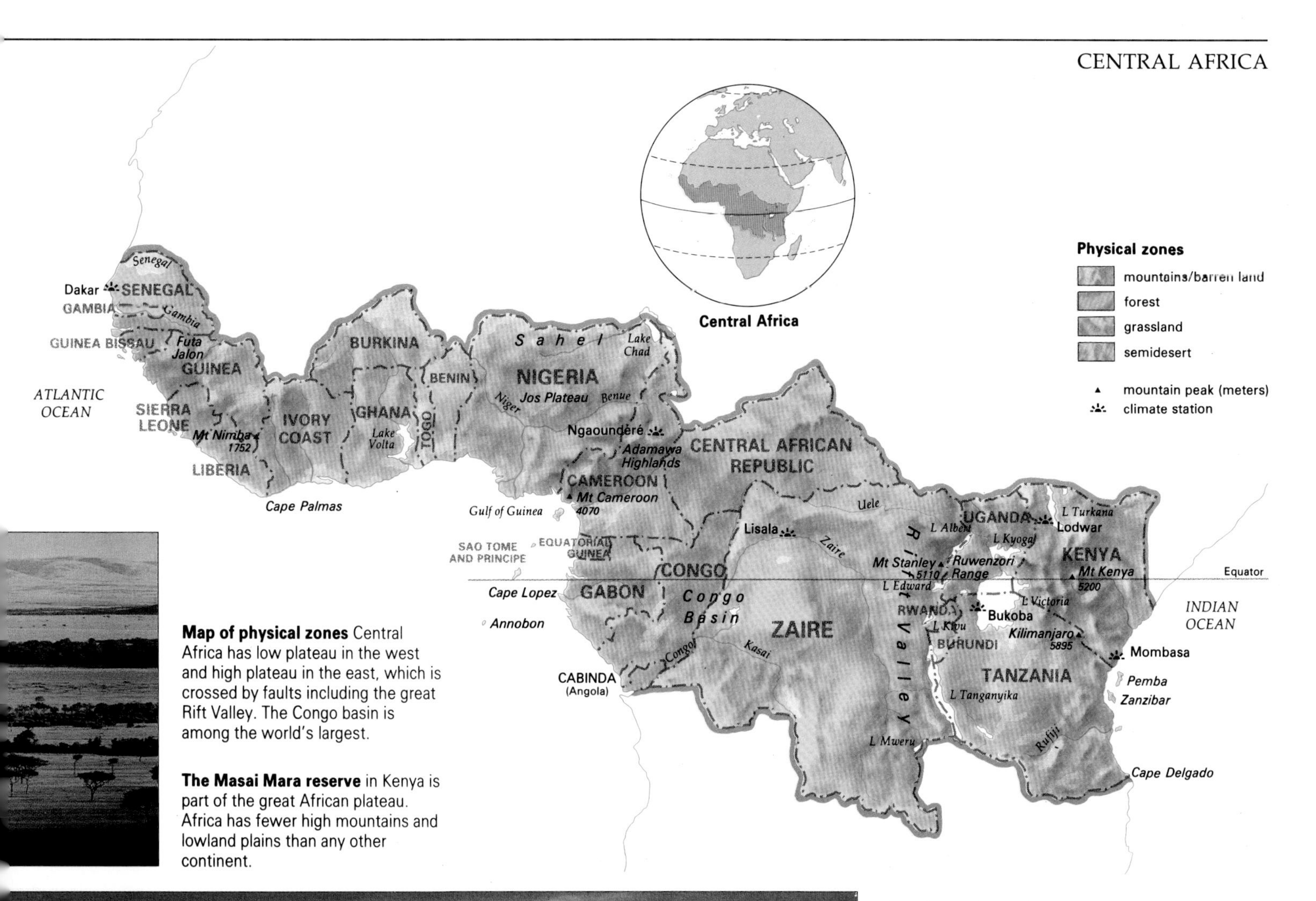

Map of physical zones Central Africa has low plateau in the west and high plateau in the east, which is crossed by faults including the great Rift Valley. The Congo basin is among the world's largest.

The Masai Mara reserve in Kenya is part of the great African plateau. Africa has fewer high mountains and lowland plains than any other continent.

developed in Africa. Elsewhere, in Kenya, Tanzania, Uganda and Zaire, the sedimentary rocks have deposits of coal.

Volcanoes, mountains and valleys

During the Miocene and Pliocene periods between 25 and 5 million years ago, movement of the Earth's tectonic plates broke parts of Africa's stable ancient blocks, producing rift valleys and mountain ranges. One arm of the Rift Valley runs east of Lake Victoria; the other stretches in a great arc that separates Burundi, Rwanda, Uganda and Tanzania from Zaire. In the west the lower Niger and Benue rivers occupy another rift system. This probably originated as a split formed when the old continent of Gondwanaland broke up. Sediments have accumulated in it to a depth of 10,000 m (33,000 ft), and huge quantities of oil are found in the marine shales. On the eastern edge are the faulted mountains of the Cameroon highlands and islands, including São Tomé and Príncipe. Volcanic activity has accompanied the movements of the land throughout the region; Kilimanjaro, the highest mountain in Africa is a dormant volcano.

Dawn over Kilimanjaro The summit of this massive dormant volcanic cone, which lies nearly on the Equator, is permanently covered with snow. It is associated with the earth movements that created the African Rift Valley between 25 and 5 million years ago.

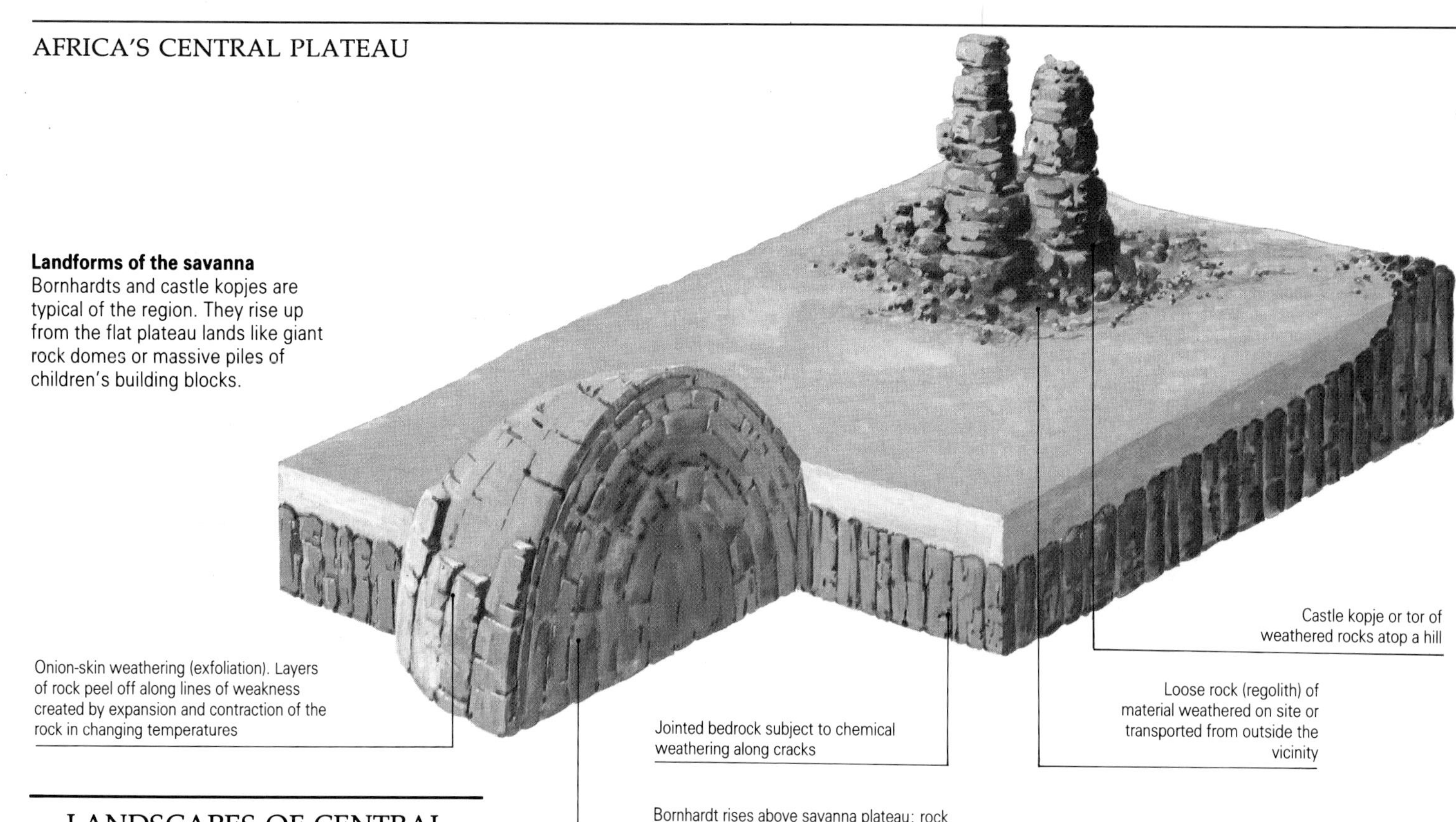

Landforms of the savanna
Bornhardts and castle kopjes are typical of the region. They rise up from the flat plateau lands like giant rock domes or massive piles of children's building blocks.

Bedrock and regolith

Magma, formed within the Earth's crust, is forced up toward the surface through weaknesses in the rocks. As it slowly cools and solidifies it forms a hard crystalline rock, such as granite, lying just beneath the land surface.

Weathering processes

The overlying rocks are removed by erosion, lessening pressure on the rocks underneath. They develop cracks, which are attacked by weathering processes while still under the ground. In one such process – hydration – the minerals absorb water and expand.

Rock features emerge

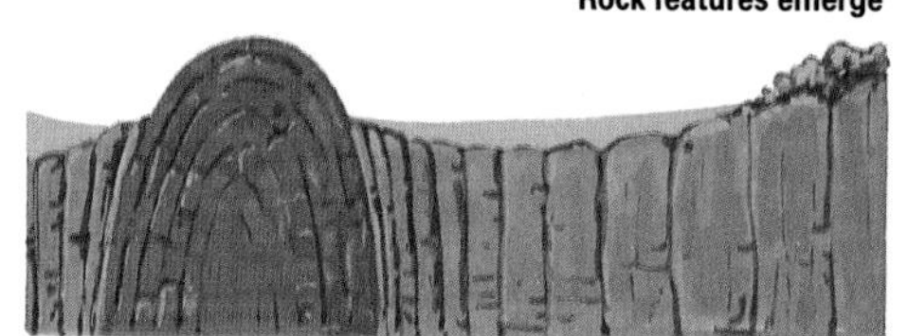

Castle kopjes develop where there are joints and cracks in the rock. A solid rock dome becomes a bornhardt as the overlying rocks are removed by erosion. Onion-skin weathering caused by temperature changes peels off thin sheets of rock on the surface.

LANDSCAPES OF CENTRAL AFRICA

The landscapes of the region show very clearly the influences of the rock types below the surface and the geological events that have taken place over the past hundred million years and more. It is part of the huge block of level or slightly undulating land that makes the vast plateau of the continent of Africa, broken occasionally by mountain blocks, volcanoes and deep fault valleys. The plateau is tilted. To the north and west of the Congo river the land is part of Low Africa, where the plateau rarely exceeds 300 m (1,000 ft) in altitude. To the south and east is High Africa, where the plateau is above 900 m (3,000 ft).

The low plateaus

West Africa stretches westward from the highlands of Cameroon. The ancient land surface has been eroded into expanses of level, low-lying land extending for hundreds of kilometers. Where old, resistant rocks have been exposed at the surface, they form areas of higher land such as the Fouta Djallon of Guinea where the headwaters of the river Niger rise, less than 250 km (150 mi) from the Atlantic. Farther east, where the river Niger cuts through the hard rocks of a group of hills on the Jos Plateau, deep gorges provide suitable sites for dams. As it approaches the sea the Niger forms Africa's largest delta – 250 km (150 mi) long and 320 km (200 mi) wide.

The highlands of Cameroon, separating the Congo basin from western Africa, extend into the Central African Republic.

LATERITES

The term laterite is used mainly to describe tropical soils containing hard layers with a high content of iron and aluminum oxides. These oxides are formed by chemical weathering of rocks under tropical conditions. The iron oxides are red or reddish brown, while the aluminum oxides are pale buff or yellow in color.

In seasonally humid climates and under forest or woodland, the iron is moved down through the soil by water (leaching). The process is assisted by organic acids absorbed by the moisture. The iron is deposited within the soil when it dries, often forming small pea-sized nodules. With time these concretions increase in number and may form a compact, nodular layer within the soil. If a valley is eroded, the water table falls and the iron-rich layer becomes increasingly hard and crystalline. The hardened layer (duricrust) can resist erosion sufficiently to form benches around valleys or even flat cappings on isolated hills.

Most duricrust cappings are ancient and very restricted in extent. However, laterites found near the water table in valleys and on the lower slopes of many hills (groundwater laterites) are more recent. When the soil is eroded they are exposed, and will then harden to form pavements of laterite.

Above the treeline near Mount Kenya Like most high mountains in Africa, Mount Kenya – 5,200 m (17,058 ft) – is a volcanic peak. Eruptions, accompanied by lava outflows and faulting, occurred throughout eastern Africa in the Miocene and Pliocene epochs.

Mount Cameroon is volcanic in origin and forms the highest peak, at 4,070 m (13,354 ft). The area is still active. On 21 August 1986 a cloud of poisonous gas burst from Lake Nyos and enshrouded the sleeping inhabitants of nearby villages. At least 1,700 people and hundreds of animals died.

The huge shallow depression of the Congo basin is generally flat, but rises by a series of giant steps to an almost circular rim, which is broken into a series of plateaus fringed by faults. They rise to less than 1,500 m (5,000 ft) in the west and north, but reach 1,750 m (5,750 ft) in the southeast and between 2,500 and 5,000 m (8,200–16,400 ft) in the mountains of eastern Zaire. Where the rivers flow from one level to the next there are rapids and waterfalls. On the Congo, falls near the coast prevent ocean-going ships from reaching the main navigable parts of the river inland.

The high plateaus

The high mountains on the eastern edge of the Congo basin form the western arm of the Rift Valley system. The seismic and volcanic activity that was responsible for the formation of the Rift Valley is still recorded every 10 to 12 years in the volcanic Virunga Mountains northeast of Lake Kivu. Lake Kivu itself has formed behind a lava flow that dammed the Rift Valley. A number of lakes are found along the bottom of the valley. The largest is Lake Tanganyika, which at 1,470 m (4,825 ft) is the second deepest lake in the world. Along the eastern side of the valley is found the Ruwenzori Range.

The vast, flat grassland plains of the high plateau of eastern Africa stretch away toward the eastern arm of the Rift Valley. It forms a shallow depression that is partly occupied by Lake Victoria, whose shores are bordered by Tanzania, Uganda and Kenya. It is Africa's largest lake, but has a depth of only 81 m (265 ft). The plateau is highest in Kenya, where the eastern Rift Valley is at its most dramatic. The high mountains along the edge are old volcanoes, and several are high enough to be capped with snow. Kilimanjaro, in northern Tanzania, is a huge conical mountain whose snow-covered heights can be seen hundreds of kilometers away across the plains.

From the high lava plateaus of the east, the ancient continental surface slopes down toward the Indian Ocean. Isolated high blocks stand above the surface, which is notched into large steps, or escarpments, as it descends.

Over the region as a whole there are only narrow coastal plains surrounding the plateau. They are at their widest along the coast of Tanzania and, in the west, in the Niger delta and Gabon.

The effects of heat and moisture

Landforms throughout the region show the influence of a tropical climate that causes rapid chemical weathering of rocks and creates a deep mantle of weathered material. Some rock landforms appear to have been excavated as this mantle has been worn away, leaving a scattering of boulders and large rocky domes (inselbergs) of granite or gneiss, which protrude above the undulating plains and plateaus that are so typical of the region. Where there is high rainfall, deep weathering can leave soils with few available nutrients. However, the picture is rarely as simple as this, as it overlooks the great complexity of tropical environments. There are highly productive soils on lavas and other basic rocks as well as barren sandy sediments in the river areas, where rainforests recycle their nutrients in a self-sustaining system.

AN EQUATORIAL CLIMATE ZONE

The whole of Central Africa lies within the tropics, so the climate and vegetation are influenced by the burning strength of the overhead Sun and its seasonal passage between the northern and southern hemispheres. There are nevertheless considerable climate variations within the region.

Temperatures are generally high and seasonal variations small because the Sun maintains a high angle throughout the year. Most temperature differences are the result of variations in altitude, distance from the sea and the amount of rainfall. In the humid forests along the coasts of Low Africa in the west, daily and seasonal temperatures remain relatively even. Maximum daytime temperatures rarely exceed 33°C (91°F), and at nighttime they rarely fall below 18°C (64°F). On the drier savanna grasslands away from the cooling effects of the sea, temperatures range between 13° and 38°C (55°–100°F).

In High Africa temperatures are lower because of the increase in altitude. The range of temperatures at Nairobi in Kenya's central highlands, 1,820 m (5,970 ft) above sea level, is between 11° and 26°C (52°–78°F). On the slopes of Kilimanjaro there are vertical climate zones, which change from equatorial to tundra within a few kilometers.

A fragment of Gondwanaland When Gondwanaland split up it formed new continents and smaller rock masses such as the Seychelles, lying off the Kenyan coast. These granite rock islands are today a popular tourist resort.

A bird's-eye view of mangrove forest in the Niger delta. The trees thrive on the rich silts deposited by the river, and their aerial roots form an effective barrier against coastal erosion.

Extremes of rainfall

The circulation of air between the tropics

is the major factor affecting rainfall in the region. Intense heating over the equatorial zone causes air to rise, and pulls in air from the high pressure (anticyclonic) cells where cooler air is sinking over the deserts of the Sahara to the north and the Kalahari to the south. Where the northerly and southerly air masses meet is known as the Intertropical Convergence Zone (ITCZ). This zone forms an irregular boundary across the region, and shifts north and south with the movement of the overhead Sun.

Across western Africa and into the Congo basin the ITCZ migrates from a few degrees south of the Equator in January to about 21°N in August. Dry air from the Sahara meets the moist air from the Atlantic Ocean. Convection currents are set off by waves in the dry air masses of the easterly trade winds; clusters of convective cells move slowly west–southwest, bringing short, heavy storms. Much of the rain falls in the highlands of Cameroon, inland from the Gulf of Guinea, where moist air moves in off the sea all year and rises over the mountains.

The rain is heaviest along the west coast, where it averages between 1,400 and 3,000 mm (55–118 in) a year. It falls in two peak periods that coincide with the passage of the ITCZ as it moves north and then south again. Farther north, the wet seasons become closer together and eventually merge into one. Because the distance the ITCZ penetrates inland and the strength of the convection currents varies, the amount of rain received can fluctuate by more than half the annual average from one year to the next. As the total is only around 800 mm (31 in) a year in northern Nigeria, this can cause serious problems for farmers.

Rainfall in eastern Africa is also influenced by the passage of the ITCZ, but the pattern is modified by mountainous relief in some areas, by the presence of lakes and by the influence of the Indian Ocean monsoon, which brings rain when the ITCZ moves over the lands of southern Africa. Around Lake Victoria, which straddles the Equator, there are two rainy seasons when the ITCZ passes overhead, bringing about 1,200 to 2,000 mm (47–79 in) of rain a year, ample for productive farming. In northern Kenya there is often less than 400 mm (16 in) a year, and across much of Tanzania the rains are less than 800 mm (31 in), barely enough for farming.

TROPICAL RAIN

The heaviness or intensity of tropical rainfall is measured as the amount of rain falling in a given period, usually expressed as millimeters per hour. The heavier the rain, the greater is its velocity, size and potential energy. At a rate of about 10 mm an hour (0.4 in) raindrops are 1 mm (0.04 in) in diameter, while at 100 mm an hour (4 in) they can be up to 3 mm (0.1 in) across. Falls of 400 mm an hour (16 in) have been recorded in Liberia, but this rate is exceptional, and seldom lasts more than a few minutes. Rates of 25 mm an hour (1 in) are common.

Soils have a maximum rate at which they can absorb water, called their infiltration capacity. When rainfall exceeds this, the water either lies on the ground in puddles or, if it falls onto a slope, runs off and is not available to plants and crops. In areas where rainfall is low and arid conditions prevail, any loss of water can ruin a harvest.

When rainfall intensity is high the energy of the raindrops may dislodge soil particles, which are then transported downslope by runoff. In Uganda more than 40 percent of rainfall can be lost to runoff, which in places removes 80 tonnes per hectare (32 tonnes per acre) of soil in one rainy season. This is how serious problems of soil erosion can begin, and is the main reason why soil conservation measures are so important in the tropics.

Adaptable vegetation

The natural vegetation reflects both the total amount of rainfall available and its seasonality. The rainforests grow where there is abundant rainfall of more than 1,400 mm (55 in) a year and the dry season is very short, usually less than three months. Such rainforests are the richest land-based ecosystems on Earth, and are capable of sustaining themselves even on poor soils.

Away from the rainforest areas, the number of trees declines and the amount of savanna grassland increases in direct relation to the decreasing rainfall. On the fringes of the Sahara (the Sahel) and in northern Kenya so little rain falls that only plants able to withstand arid conditions can grow.

Along much of the coast fringing the Indian Ocean mangrove forests grow on the tidal flats. Their tight mass of aerial roots form an almost impenetrable barrier that protects the coast.

Africa's Rift Valley

Lava once poured from vents in the center of these volcanoes at Oldonyo Lengai valley in Tanzania. Violent volcanic activity accompanied the formation of the East African Rift Valley, which took place about 1.9 million years ago. Many old volcanoes are now found along the valley edge, and about 30 volcanoes in the area are believed to be active.

Home of early man Olduvai Gorge in Tanzania contains the oldest remains of our human ancestors. This steep-sided ravine is 48 km (30 mi) long and 90 m (295 ft) deep. The deposits exposed on its sides cover a timespan from just over 2 million to 15,000 years ago. The Olduvai Beds were laid down in a lake basin on top of Pliocene volcanic and Precambrian metamorphic rocks.

From space the great fissure of the Rift Valley system that runs through eastern Africa is one of the most visible features of the Earth's land surface. Beginning between 35 and 25 million years ago, gigantic movements of the continental crust threatened to tear Africa apart along a line stretching from the Red Sea to Mozambique. These were further movements of the kind that had caused the breakup of the great continent of Gondwanaland, which had taken place 100 million years earlier.

The rift runs from Turkey, through the Dead and Red Seas and across Ethiopia to Lake Turkana in northern Kenya. From here one arm of the system runs south to the east of Lake Victoria and another curves around to the west; they link up again to the southeast of Lake Tanganyika. From here the rift fault runs south to reach the coast near the mouth of the Zambezi river – a total length of almost 10,000 km (6,200 mi).

The western arm of the Rift Valley began to form in the middle of the Tertiary period between 35 and 25 million years ago when lava that started to pour out at the surface was fractured by faulting. About 13 million years ago activity began to the east, and this was followed by the formation of the boundary fault 3.3 million years ago. The East African Rift Valley itself is about 1.9 million years old.

The Earth's fractured surface

The scenery produced by these fractures is dramatic and impressive. Typically, parallel cliffs between 35 and 60 km (20–37 mi) apart stand 600 to 800 m (1,970–2,800 ft) above a flat-bottomed trench. However, the floor of the valley varies considerably in altitude. At Nakuru in Kenya on the eastern section the floor is 1,800 m (5,900 ft) above sea level, but the bottom of Lake Tanganyika on the western section is 650 m (2,140 ft) below sea level. The undulations show the way in which the fault systems have been superimposed on a series of great swells and depressions that form part of the underlying structure of the whole continent.

Evidence of the volcanic activity that accompanied the faulting can be found throughout eastern Africa. The Aberdare Range of western Kenya is the eroded base of a 3 million-year-old volcano. To the east, Mount Kenya (5,200 m/17,058 ft) was formed about 1 million years ago and

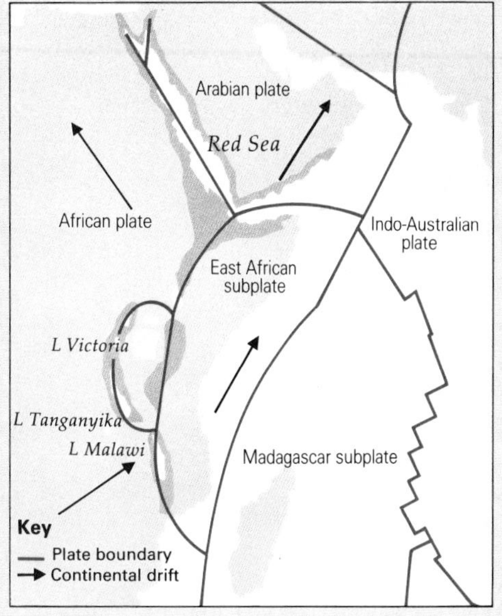

The Earth's greatest depression
The Rift Valley cuts down through Africa. The Red Sea forms one end of the split, and it reaches the coast near the mouth of the Zambezi. The East African subplate may eventually split off from the mainland, as Madagascar did 50 million years ago.

The formation of the Rift Valley
Between 35 and 25 million years ago the upward movement of material beneath the Earth's crust caused gigantic faulting of the rocks in the area, forming a rift valley. This activity was accompanied by volcanoes, and molten material forced its way up through the faults. The movements threatened to pull the continent apart. As the two parts moved away the valley between them widened, and filled with lava and sediment.

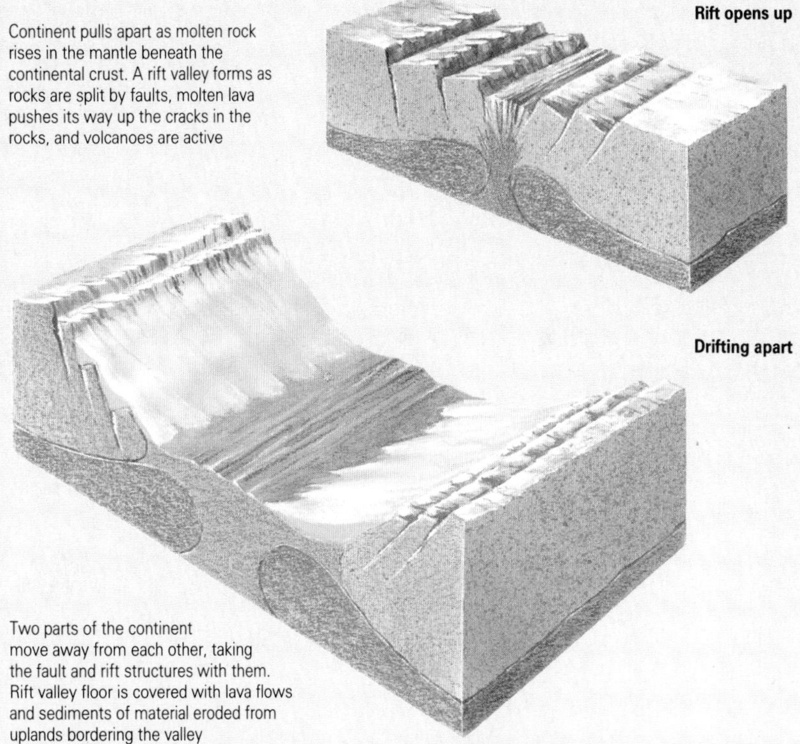

Continent pulls apart as molten rock rises in the mantle beneath the continental crust. A rift valley forms as rocks are split by faults, molten lava pushes its way up the cracks in the rocks, and volcanoes are active

Two parts of the continent move away from each other, taking the fault and rift structures with them. Rift valley floor is covered with lava flows and sediments of material eroded from uplands bordering the valley

has been carved by ice and rain into great pinnacles of rock.

The Ngorongoro crater in northern Tanzania is a caldera formed when a huge volcano exploded and collapsed; it is a flat plain 30 km (19 mi) across surrounded by walls that are 600 m (1,970 ft) high. Today it supports a variety of wildlife second to none and attracts thousands of tourists every year. To the east the dormant Kilimanjaro still preserves its volcanic cone and could erupt again. In all about 30 volcanoes are still active along the course of the Rift Valley.

Although the rift valleys were "let down" between normal faults arranged like a series of giant steps on either side, there are a number of lesser faults extending from the valley that have resulted in local uplift and volcanic activity. The valleys have also been partially filled with lavas and sediments as they have become deeper. They would otherwise be about twice as deep. The deposits have preserved many important archeological sites, the best known of which is Olduvai Gorge in Tanzania. Here the upright hominid *Australopithecus bosei* evolved.

A new island in the making?

The movements that have formed the Rift Valley system still continue. As recently as 1 million years ago, the sides of the Rift Valley were lifted still more as faulting took place along the floor. The subsidence of the floor caused molten rock to be squeezed out. Continuing volcanic activity is evidence today that the Earth's tectonic plates continue to move. If these movements continue, millions of years hence eastern Africa may well be an island separated by an ocean from the rest of the continent.

FROM LAKE MALAWI TO THE CAPE

THE ROCKS OF HIGH AFRICA · PATTERNS OF RAIN AND DROUGHT · ROCKS, SOILS AND PLANTS

The saucer-shaped plateau of southern Africa rises sharply from narrow coastal plains, rimmed by mountains that reach their highest point in the Drakensberg. These slope down inland to the ancient, eroded land of the savanna plains and the arid desert of the Kalahari. The Namib Desert extends along the western edge; the region's northern limit is marked by the southern extremities of the vast Congo basin and the East Africa Rift Valley as it passes through Lake Malawi. There is a varied climate and vegetation, ranging from the distinctive scrub cover, known as fynbos, of the Cape ranges to deciduous woodland and semidesert succulents. Fossil dunes, dry lake beds and salt pans in the western interior supply evidence that there were both drier and more humid climates in the past.

THE ROCKS OF HIGH AFRICA

Southern Africa is at the raised end of Africa's tilted plateau, often called High Africa; much of the land behind the narrow coastal plains lies above 1,000 m (3,300 ft). The region is composed of a Precambrian shield over 600 million years old, whose hard rocks outcrop over about one-third of the land surface. Younger sedimentary rocks edge the region. These have been extensively folded and faulted, especially in the Cape ranges in the extreme southwest.

In the western interior windblown sediments of red sand have been deposited that reach thicknesses of 300 m (1,000 ft) on the Namibian–Angolan border. These are the beds of the Kalahari, a vast desert tract of sand.

The Drakensberg range of mountains, on the saucer-shaped plateau's high southeastern rim, arc capped with volcanic lava formed in the Mesozoic era 248–65 million years ago. Rivers have cut through the rim, creating deep gorges in the southeast and in the valleys of the Zambezi and Limpopo. The Victoria Falls on the Zambezi are among the most spectacular waterfalls in the world; those on the Tugela river in the Drakensberg are the highest in Africa. Rising in the Drakensberg, the Orange river flows west to the Atlantic Ocean.

Most of the central plateau is covered by plains broken in places by isolated steep-sided hills called inselbergs. In the center of the saucer is a vast depression, which includes the inland drainage

COUNTRIES IN THE REGION

Angola, Botswana, Comoros, Lesotho, Madagascar, Malawi, Mauritius, Mozambique, Namibia, South Africa, Swaziland, Zambia, Zimbabwe

LAND

Area 5,751,800 sq km (2,220,194 sq mi)
Highest point Thabana Ntlenyana, 3,482 m (11,425 ft)
Lowest point sea level
Major features interior plateau, salt pans and deltas, Kalahari and Namib Deserts, Karoo tableland, Cape ranges in southwest, Drakensberg range, Madagascar

WATER

Longest river Zambezi, 2,650 km (1,650 mi)
Largest basin Zambezi, 1,331,000 sq km (514,000 sq mi)
Highest average flow Zambezi, 16,000 cu m/sec (565,000 cu ft/sec)
Largest lake Malawi, 29,600 sq km (11,400 sq mi)

CLIMATE

	Temperature °C (°F) January	July	Altitude m (ft)
Lusaka	22 (72)	16 (61)	1,279 (4,195)
Bulawayo	21 (70)	14 (57)	1,345 (4,412)
Cape Town	22 (72)	13 (55)	12 (39)
Toliara	27 (81)	20 (68)	9 (30)
Antananarivo	19 (66)	13 (55)	1,310 (4,297)

	Precipitation mm (in) January	July	Year
Lusaka	224 (8.8)	0 (0)	829 (32.6)
Bulawayo	134 (5.3)	0 (0)	589 (23.2)
Cape Town	9 (0.4)	96 (3.8)	652 (25.7)
Toliara	71 (2.8)	4 (0.2)	342 (13.5)
Antananarivo	286 (11.3)	10 (0.4)	1,270 (50.0)

World's greatest recorded 24-hour rainfall, 1,870 mm (73.6 in), Réunion island

NATURAL HAZARDS

Drought

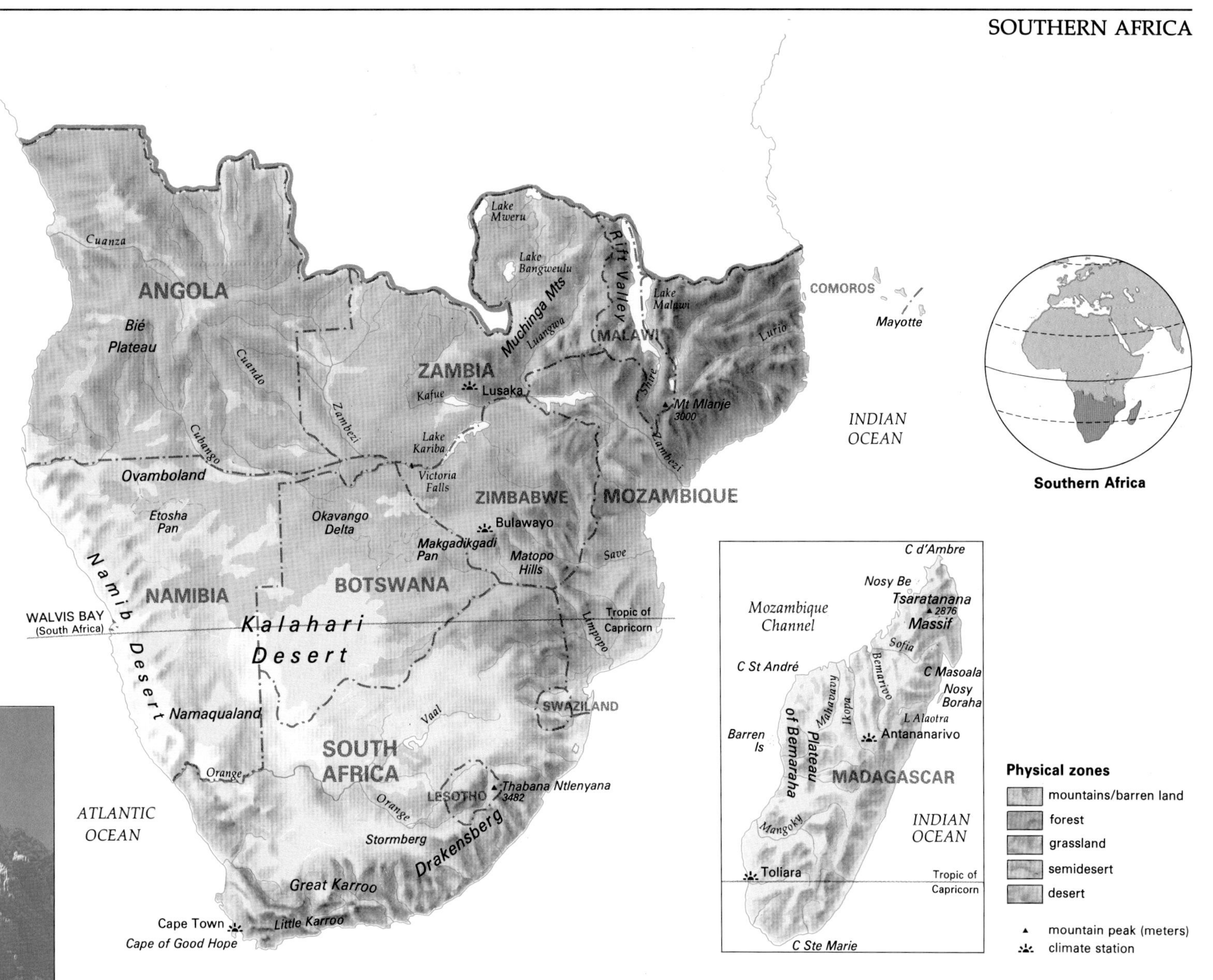

Map of physical zones Southern Africa's elevated plateau is formed of ancient metamorphosed rock. Rich mineral deposits are found in its southeast mountain rim.

Snowcapped peaks The mountain ranges of the southwest Cape are composed of sedimentary rocks.

Islands off the coast of Africa Craters on the coastline of Pamanzi island, Mayotte, clearly illustrate the volcanic origins of the Comoros Islands. Other islands such as Ascension and St Helena in the Atlantic and Mauritius in the Indian Ocean were created by volcanic eruptions. Madagascar, the fourth largest island in the world, rises from the submarine plateau that broke away from the African continent 50 million years ago.

systems of the Okavango Delta and the Makgadikgadi Pan in Botswana. Between the high rim and the ocean is a narrow coastal strip. This broadens in the northeast into Mozambique's coastal plain, through which the Zambezi flows to the Indian Ocean.

The huge island of Madagascar became separated from the continental landmass during the breakup of Gondwanaland about 135 million years ago. Its structure is similar to that of the mainland.

Evolution of the landscape

About 180 million years ago southern Africa occupied a central, elevated position within the ancient supercontinent of Gondwanaland. The breakup of Gondwanaland created new base levels (the level down to which running water can erode), and this initiated widespread and rapid erosion by rivers. For some 160 million years, valleys were widened by the leveling of slopes through a combination of surface wash erosion and weathering. The same process occurs in semiarid areas today. The resulting plains, called the "African erosion surface", are best developed in the interior of the region.

Toward the end of the early Miocene epoch, about 20 million years ago, gigantic movements lifted up the continent by about 200–300 m (650–1,000 ft) and also tilted it down toward the west. This set off a new phase of river erosion when a lower-level landscape, the "Post-African I" erosion surface, was formed.

During this phase the surface was worn away on rock that had already been weathered. The depth of weathering varied according to the jointing patterns of the bedrock; for example, in granitic rocks greater weathering occurred in well-jointed bedrock, while more massive blocks were less affected because water could not penetrate them so easily. Renewed erosion exposed features of the rock beneath, creating landforms in the shape of steep-sided piles of boulders called tors or castle kopjes, and rock domes and rock knobs (bornhardts). Good examples of bornhardts can be found in eastern Zimbabwe and in the Matopo Hills south of Bulawayo.

Large accumulations of river and wind-blown sediments were deposited in the Kalahari basin during this period. In the late Pliocene, about 2.5 million years ago, further lifting and westward tilting of the continent took place, possibly triggered by movements of semi-molten rocks deep beneath the surface. In a new erosion phase deep gorges and waterfalls developed along river valleys.

Evidence of climatic change

Southern Africa experienced alternating phases of wetter and drier climates during the late Quaternary period. Evidence of the arid phases is provided by fossil sand dunes near the Okavango Delta. Active dunes – those still subject to movement – are restricted to areas where rainfall is under 150 mm (6 in), such as the Namib Desert coast. The extensive river and lake sediments and dry valleys in the area of Okavango and Makgadikgadi are evidence of more humid climates in the past. There are large salt deposits at the once water-filled Makgadikgadi Pan. In this area of northern Botswana raised shorelines show that there has been progressive shrinking of what was once a huge lake.

PATTERNS OF RAIN AND DROUGHT

Southern Africa's seasonal climates affect the amount of water available for plant growth, regulate the flow of rivers, and have a major influence on the nature of the landscape and on the activities of the people who live there.

During the southern summer (October to March) intense heating results in the development of low-pressure cells over the region. These draw in moist winds from the Atlantic and Indian Oceans, which meet in a broad convergence zone stretching across the continent. The warm, moist air masses rise, cool, and form towering cumulus clouds whose tops may reach 12,000 m (39,000 ft). Heavy downpours of rain follow. When the upper atmosphere is dry, as occurs occasionally when high pressure is dominant over southern Africa, then the buildup of clouds is inhibited and there may be a period of drought. In the southwest the offshore cold Benguela Current cools the air moving toward the Namib Desert so that it cannot absorb much moisture. When the air is warmed over the land its humidity falls yet again, giving very low rainfall in the area.

In the winter (April to September) the low pressure moves north with the overhead Sun. The cooling of the air brings high-pressure systems over the region. Stable, dry atmospheric conditions prevail and the sky is normally free of cloud

"The smoke that thunders" The Victoria Falls on the Zambezi get their local name from the roar of the river as it cascades over a resistant band of basalt rock and from the wall of spray that reaches high into the air.

The Karroo landscape showing "Kraal Bos" (*Galena africana*), a species of drought-tolerant shrub that is characteristic of this arid tableland area in South Africa's Cape Province. The vegetation is in its dormant winter phase.

during the day and at night, giving frost in exposed valleys and in the mountains, especially in the south. Subzero temperatures and low humidity in late winter may cause "black frosts", which are damaging to broadleaved plants. The westerly winds, positioned well to the south of the continent in January, move north during the winter, bringing rains to the southernmost Cape area.

From Mediterranean to desert climate

There are four main climatic zones in southern Africa. In the southern tip of the region, the western Cape has a warm, temperate Mediterranean-type climate with cool, wet winters and dry, warm summers. The high windward slopes receive up to 2,000 mm (80 in) of rainfall a year, but the leeward slopes and intervening valleys receive very little.

The interior of the region, distant from the sea and lying in the rain shadow of the Drakensberg, has a tropical continental climate with semiarid conditions. Rainfall is low (under 400 mm/16 a year) and falls in the summer. The winters are dry, with low humidity, clear skies and high temperatures. Conditions are similar in western Madagascar, which lies in the rain shadow of the island's mountains and is also dry with seasonal rainfall. The extreme southwest of the island is semiarid too, despite its proximity to the sea.

Along most of the eastern edge of the region there is a tropical marine climate. Most rain falls in the summer months; there is no true dry season because the easterly airflow is affected by the highland along the coast, giving winter rainfall in higher areas. Mist is common on the lower slopes of the Drakensberg. Annual average temperatures in the far north are about 26.6°C (80°F), with a range of 3.3°C (6°F); in the southeast they average 21°C (70°F), with a range of 7.2°C (13°F). Similarly, the mountainous area of eastern Madagascar receives over 1,400 mm (55 in) of rainfall in an average year, though with local extremes of as much as 3,400 mm (134 in).

Along the western edge, the Namib Desert has a hot desert climate. Rainfall is very low; no month has an average temperature below 6.1°C (43°F).

RAINFALL CYCLES

In the early 20th century rainfall was believed to be decreasing throughout southern Africa. The rapid expansion of drought-tolerant karroo plants into more humid areas reinforced the idea that the spreading of semiarid areas was creating a "Great Southern African Desert". Subsequent research has shown that there is no overall trend of decreasing rainfall, and that the spread of these plants is more likely to be related to overgrazing by livestock.

Analysis of rainfall records shows that there is an 18-year cycle in the summer rainfall area of northeast South Africa. The 1960s, for example, were generally dry, whereas the 1970s were wetter; trends indicated that the late 1980s would be drier. Awareness of the variability of supply means that water must be stored in large reservoirs and wise use made of water for domestic, industrial and agricultural purposes.

Rainfall cycles are found in other parts of the world, such as Australia, the southwestern United States and Mexico. It is thought that they may be linked to periodic changes in circulation patterns in the atmosphere.

ROCKS, SOILS AND PLANTS

The nature and distribution of soils in southern Africa are influenced by the bedrock, climate, vegetation, how long the soil has had to develop, and recent human activity. Non-lime soils are found on the old shield rocks and Kalahari sands in the north and northeast, where higher rainfall promotes weathering and leaching. Deeply weathered acidic soils also develop where conditions are similar in eastern Madagascar.

Lime (calcareous) soils are found on a variety of rock types in lower rainfall areas in the southwest and central parts of the region, including the southern Kalahari. Here weathering is less intense and base elements are not leached out of the soil. Poorly developed soils (regosols), associated with recent river or wind-blown deposits like those in the Namib Desert, are more localized. Local differences in soil produce repetitive sequences or patterns known as catenas. They are found in areas of gently undulating terrain where there are also differences in drainage conditions between hill crests and valley floors, but usually derive from the same bedrock; for example, granitic rocks in Zimbabwe.

The differentiation of soils on slopes commonly found in savanna areas occurs where the seasonal rainfall flows over the land surface and down into the ground. This causes leaching of the hill crest soils and waterlogging of the basal slopes. In deeply eroded areas such as the middle Zambezi valley and the Lesotho uplands, the rates of erosion are greater than rates of soil formation, so soils are shallow, stony and patchy.

Pressures on the land

Erosion threatens agricultural and grazing land in many areas as they come under increasing pressure from population growth. The problem is severe in the northeast and in eastern Madagascar, where gullies are easily formed in the sandy soils after a sudden downpour, particularly where vegetation cover is sparse. As the sides of the gully deepen more land is made useless for farming.

Erosion problems are serious in areas of bornhardt terrain, where clearance of the land surrounding the hills has exposed sandy soils and where there are high rates of runoff from the rock outcrops. Not all gullying, however, is due to human activity. Sodium-rich soils, for example, are naturally prone to tunnel erosion and gullying.

Land bordering deserts is threatened by desertification – a natural process caused principally by climate change. Drier climates with fewer, heavier bouts of rain that give rise to flash floods, increase erosion, which rapidly extends the margins of the desert. Human activity intensifies the problem – the use of wood for fuel, and grazing by livestock, strips the land of vegetation, making the soil susceptible to wind erosion. The United Nations Environment Program has estimated that 3.5 percent of southern Africa is natural desert; a further 16.5 percent is at very high risk of desertification, 24.3 percent at high risk and 59.2 percent at moderate risk.

A diverse vegetation

The plant cover of southern Africa is extremely diverse. The distribution of plants is affected mainly by rainfall and the length of the dry season. Savanna vegetation, for example, occurs where seasonal rainfall is between 500 mm and 1,400 mm per year (20–55 in). On well-drained slopes in a number of areas, such as the Zambian plateau, sparse, deciduous woodland called miombo is found. In drier regions woodland gives way to scrub and thickets. Toward the semiarid interior the vegetation becomes sparse, and succulent plants dominate. In the far south the Cape area, with its variety of

FEATURES OF AN ERODED LANDSCAPE

Pediments, gently sloping rock platforms that lie at the foot of steep escarpments or isolated hills (inselbergs) are typical landforms of southern Africa's arid plains. They are usually considered to be the result of erosion because they often cut across different layers of rock. On most pediments the bedrock is overlain by debris. Some of this is transported material, and some is formed by weathering on the spot. They represent a late stage in the erosion of a landscape.

Southern Africa's flat landscapes scattered with inselbergs are sometimes said to have been worn down by a process known as prolonged pediplanation. One explanation suggests that as the steep faces of scarps and the debris slopes at their feet were worn back in parallel, gently inclined pediments or replacement slopes developed at their foot. An alternative view is that pediments were formed, in the first instance, by deep weathering of bedrock. This was subsequently followed by redistribution of the debris by processes such as sheetwash, when water flows over the land surface.

Pediplanation is particularly conspicuous in areas of crystalline rocks such as granite, as these produce crumbly, coarse material when they are chemically weathered. As a result irregularities in relief are more easily reduced by weathering and erosion.

The Blyde river has cut down through soft sedimentary rock and basaltic lavas to form a deep gorge.

An inland delta The Okavango river drains into a vast basin in the Kalahari. It is subject to drastic changes of flow. Densely growing papyrus reeds block channels in the sand, altering their course.

soil conditions and microclimates, has a rich, natural evergreen shrub vegetation known as the fynbos.

Vegetation on Madagascar, formerly part of the continental landmass, resembles that of the mainland, but subsequent isolation has resulted in the evolution of many unique species. The vegetation of the volcanic islands is derived mainly from long-distance seed dispersal, though endemic species have evolved on older islands such as Réunion.

History in the landscape

Southern Africa's arid landscapes contain a number of features that enable us to unravel climatic changes that took place in the long distant past. This can be helpful in attempting to identify cyclical patterns of climate today, and to predict changes in the future. At times during the last 50,000 years wind and water action has sculpted the landscape and reworked the surface materials much more actively than at present. Fossil, or fixed, dunes cover much of the Kalahari desert in the western interior. They are the most recent of the sediments accumulated in the region's central depression during the Cenozoic era of the last 65 million years; the deposits can reach 300 m (1,000 ft) deep in places.

Three main groups of fossil dunes have been identified. The dunes north of the Makgadikgadi Pan are about 18,000 years old; the northern dunes between the Okavango Delta and the Etosha Pan are about 30,000 years old; the southern dunes are older than these, but have not been dated. These dune systems indicate that much drier conditions prevailed in the past, and the different ages of the systems show that there have been several arid climate phases within the last 50,000 years. The drier periods coincided with an increase in the size of polar ice sheets and mountain glaciers elsewhere.

Today active dunes are found only in the Namib Desert. Here the dune field is 100–120 km (60–75 mi) wide and extends about for nearly 400 km (250 mi) – about one-third of the desert's entire north—south extent – along the southwest coast south of Walvis Bay.

In the central Namib massive linear dunes are found, up to 250 m (820 ft) high and 16 km (10 mi) or more long. Plant cover on dunes is very sparse. It survives mainly as a result of fog moisture in the coastal areas. Farther inland, where rainfall is higher, herbacous and woody vegetation grows on the eroded and now stabilized remnants of former dunes. Satellite images can be used to detect and map the distinct lines of the vegetation growing on these ancient dunes.

It is possible that a longterm cycle of arid and humid phases is still in progress. Reduction in rainfall and prolonged drought today may be part of this sequence, and a contributory factor to the problem of desertification now confronting large areas of southern Africa.

Soil evidence of change

Elsewhere in the region, particularly in the northern interior and the east, shallow, less conspicuous deposits of eroded rock particles are found. Rainfall in these areas is about 600–900 mm (24–35 in) per year. As the weathered material is carried as sheetwash by water flowing across the surface of the ground, it accumulates on low-angled footslopes or valley floors to form deposits known as colluvia. These colluvial deposits vary in texture depending on the source bedrock and on the process of sorting that takes place as the fragments gradually move downslope or downstream.

Sections through a colluvial deposit show distinct layers, including earlier soils or paleosols. These appear to have formed at times in the past when organic matter accumulated in the topsoil or when higher rainfall allowed the growth of denser plant cover. Layers containing organic material are separated by mineral layers of sands and gravels, formed by soil erosion on adjacent hillslopes in periods of low rainfall.

The exact nature and timing of these changes have yet to be pieced together over the whole of southern Africa, but the presence of Stone Age human artefacts in the colluvia show that most of the sediments were deposited between 30,000 and 12,000 years ago. The alternating layers are also a record of the long history of climate variation in the region.

Riches in the sand The active dunes of the Namib Desert stretching along the coast from the exit of the Orange river to Walvis Bay, contain the largest diamond deposits in the world. They are believed to come from rocks found farther inland. Over several millions of years erosion debris from these rocks has been carried toward the coast. As the softer particles were ground up, the harder diamonds were left beneath windblown layers of sand.

Trees and scrub in the desert grow at the foot of gigantic dunes. The vegetation has established itself on the shallow mounds of earlier dunes that have now become stable.

A COLLISION OF CONTINENTS

THE MAKING OF A SUBCONTINENT · LAND OF THE MONSOON · THE LIFELINE OF INDIA

The pendant shape of India projects south from Asia into the Indian Ocean. The subcontinent extends some 5,600 km (3,500 mi) between the snowcapped Himalayas in the north and the island of Sri Lanka in the south, and from the barren deserts in the west to the wet jungles in the east. Some of the oldest rocks in the world are found on the tapering peninsula, while the Himalayas, the world's highest mountains, are still being lifted up. Between the two, the fertile plains and deltas of the great Brahmaputra, Ganges and Indus rivers support one of the world's chief concentrations of people. Climates range from glacial to tropical, and the influence of the seasonal wet monsoon winds reaches almost everywhere. Vegetation also varies greatly, from tropical rainforests to alpine plant communities.

COUNTRIES IN THE REGION

Bangladesh, Bhutan, India, Maldives, Nepal, Pakistan, Sri Lanka

LAND

Area 4,476,064 sq km (1,727,276 sq mi)
Highest point Mount Everest, 8,848 m (29,028 ft), highest on Earth
Lowest point sea level
Major features Himalayas, world's highest mountain range, plains and deltas in north, Thar Desert, Deccan plateau

WATER

Longest river Brahmaputra and Indus both 2,900 km (1,800 mi)
Largest basin Ganges, 1,059,000 sq km (409,000 sq mi)
Highest average flow Brahmaputra, 19,200 cu m/sec (678, 000 cu ft/sec)
Largest lake Manchhar, Pakistan, 260 sq km (100 sq mi); reservoirs in India are larger

CLIMATE

	Temperature °C (°F) January	July	Altitude m (ft)
Jacobabad	15 (59)	39 (95)	56 (184)
Simla	19 (66)	5 (41)	2,205 (7,232)
New Delhi	14 (57)	32 (90)	216 (708)
Katmandu	10 (50)	24 (75)	1,334 (4,376)
Chittagong	20 (68)	28 (82)	14 (46)
Trincomalee	26 (79)	30 (86)	3 (10)

	Precipitation mm (in) January	July	Year
Jacobabad	7 (0.3)	27 (1.0)	88 (3.5)
Simla	61 (2.4)	424 (16.7)	1,577 (62.1)
New Delhi	25 (1.0)	211 (8.3)	715 (28.2)
Katmandu	19 (0.8)	378 (14.9)	1,328 (52.3)
Chittagong	10 (0.4)	642 (25.3)	2,858 (112.5)
Trincomalee	211 (8.3)	54 (2.1)	1,727 (68.0)

World's highest recorded annual rainfall, 26,470 mm (1,042.1 in), Cherrapunji, northeast India

NATURAL HAZARDS

Cyclones, storm surges, flooding of great river deltas

High in the western Himalayas, the waters in the upper reaches of the Indus valley flow past the mountains at an altitude of 3505 m (11,500 ft). The river, fed by streams formed by melting snows, erodes the mountains to give steep, V-shaped valleys.

THE MAKING OF A SUBCONTINENT

There are three major structural elements to the subcontinent of India. In the south is the ancient and stable triangular mass of peninsular India, in the north the geologically young folds of the Himalayan mountain chain. Between them lie depressions now filled with sediments deposited by the great Brahmaputra, Ganges and Indus rivers.

The plateau of peninsular India (the Deccan) and Sri Lanka once formed part of the ancient landmass called Gondwanaland, which moved north until it collided with the Asian landmass. Hard, resistant granites over 600 million years old are widespread in this ancient shield area, as are gneisses, granite which has

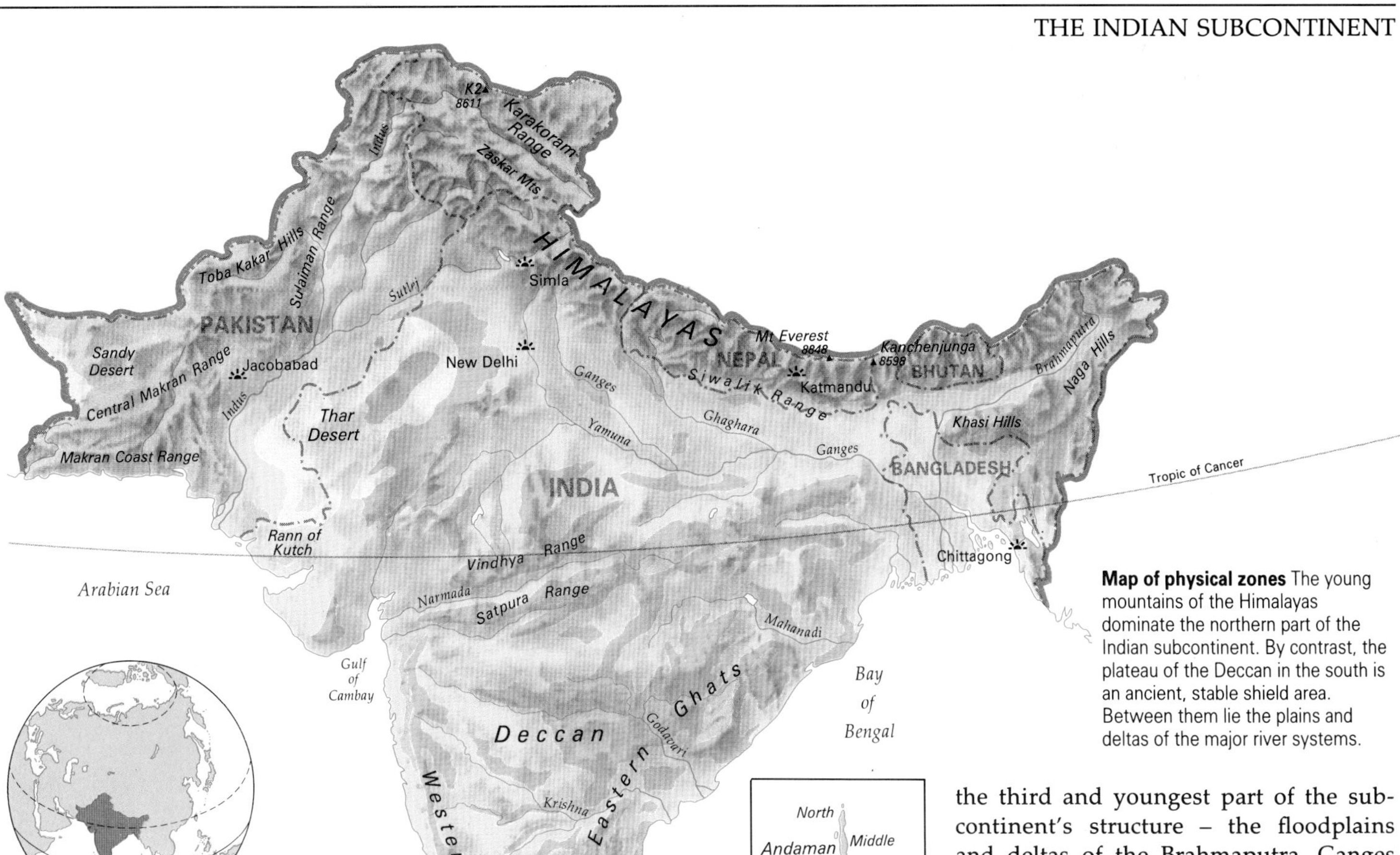

Map of physical zones The young mountains of the Himalayas dominate the northern part of the Indian subcontinent. By contrast, the plateau of the Deccan in the south is an ancient, stable shield area. Between them lie the plains and deltas of the major river systems.

been metamorphosed under great heat and pressure. There are also other less intensively altered but still ancient sedimentary rocks. A typical landscape is one of plateaus, with more resistant blocks rising abruptly above them. An isolated plateau that includes the Khasi Hills in northeastern India is an extension of this.

There are younger, yet still very ancient, sedimentary rocks in dips or basins within the older structures. These form both gently rolling plateaus and spectacular scarplands, particularly along the northern edge of the Deccan, where they are cut by waterfalls. Much younger sediments exist in long depressions, particularly in the east of the peninsula, which is fringed by a coastal plain and a series of deltas formed by sediment that has been washed down from the plateaus.

The west of peninsular India is covered by volcanic lavas that reach an immense thickness of up to 3,000 m (10,000 ft) toward the coast. This lava plateau has been tilted toward the east; the gentle slopes facing east contrast with the steep scarp of the Western Ghats, which faces the Arabian Sea.

The Himalayas

The mountains of the Himalayas began to form about 40 million years ago, when the slowly drifting crustal plate of India collided with that of Eurasia. The Himalayas are the world's highest mountain system, with an average height of about 6,000 m (19,000 ft). Major rivers, such as the Brahmaputra, that originally flowed south from the region of Tibet were diverted by the rising land, then cut down to the south as the land was raised, and now cross the Himalayas in spectacular gorges. River basins and ranges were formed, notably in the far north, as well as remote plateaus areas such as the Ladakh Range.

Plains, deserts, deltas and atolls

Between the mountains of the north and the stable mass of peninsular India lies the third and youngest part of the subcontinent's structure – the floodplains and deltas of the Brahmaputra, Ganges and Indus. The floor of these vast plains is made of sediments brought down from the Himalayas, and to a lesser extent from the north of the Deccan plateau. Once well-watered parts of these alluvial plains have become very arid: in the Thar or Great Indian Desert there are abandoned river channels as well as sand dunes and other desert landforms.

Deltas form where rivers deposit sediment more quickly than it can be removed by the sea. In the northeast of the Indian subcontinent this process takes place on a grander scale than anywhere else on Earth. Almost the entire country of Bangladesh is included in the combined delta of the Brahmaputra, Ganges and other rivers. The delta is the world's largest, covering about 60,000 sq km (23,000 sq mi).

The deposits that form the delta also extend hundreds of kilometers out below sea level into the Bay of Bengal and as far south as Sri Lanka. At times during the past 2 million years, when glaciers covered larger areas of the Himalayas than at present and the sea level was lower, the Brahmaputra and Ganges poured sediment brought from the Himalayas down to positions that are now far out into the ocean.

In contrast to the dramatic relief of most of the subcontinent, the coral atolls of the Indian Ocean – the Laccadive, Maldive, Andaman and Nicobar groups – have very little relief. The highest land on the 2,000 islands of the Maldives is only 2 m (6.6 ft) above the level of the sea.

LAND OF THE MONSOON

Within the subcontinent of India are some of the hottest, coldest, driest and wettest parts of the world. In the north are the cold mountain ranges of the Himalayas, in the south the equable tropical atolls of the ocean, to the northwest the arid Thar Desert, and to the southeast the hot, humid rainforest of Sri Lanka. The pattern of relief dictated by geological history has been greatly modified by these different climates.

Monsoon climates

The seasonal system of winds known as the monsoon dominates the climate of the subcontinent. Each year the change from predominantly offshore northeasterlies in winter to onshore southwesterlies in summer brings rain to most of the area from June to October. In winter air is drawn from the land toward a low pressure belt farther south, where the sun's heat is stronger. The Coriolis effect, caused by the Earth's rotation, swings the winds to the northeast. This gives onshore winds in the extreme southeast, which receives some rain in winter, unlike most of the rest of the subcontinent.

During spring and summer the land heats up more quickly than would be expected because the barrier of the Himalayas prevents cooler air from being drawn south. Moist air drawn in from the sea to replace rising air cools, condenses and falls as rain. The Coriolis effect causes these summer monsoon winds to converge. This effect is accentuated by relief: in northeast India there is very high rainfall – Cherrapunji holds the world record for annual rainfall.

The development of the monsoon is influenced by the strong westerly subtropical jet stream. During the summer this high-altitude wind migrates north of the Himalayas, allowing the moist winds to penetrate farther inland.

Superimposed upon this broad pattern is the effect of altitude, with barriers such as the Himalayas and the Western Ghats causing very high concentrations of rainfall and lower temperatures.

Weathering and erosion

In the tropical south chemical weathering is the chief process by which the land surface is broken down. It is largely caused by the rainwater that percolates

down through the rocks, becoming more acid as it passes through decaying vegetation. In much of peninsular India and on the plains of the Indus and Ganges there is less weathering of this kind because there is much less rain. In the more mountainous north weathering is mainly the result of alternating freeze and thaw (frost shattering), though because of the high relief and continuing rise in the level of the land constant erosion removes the weathered layer.

The processes of erosion also depend upon climate. From the wet upland areas major rivers bring large amounts of sand and silt, the products of soils eroded by some of the heaviest rainfalls in the world. In the most arid parts of the subcontinent sand blows into massive dunes, and erosion by rivers takes place only in short periods immediately following the sporadic but intense rainstorms. In the Himalayas glaciers also play their part in eroding and transporting rock from the mountains.

Reddish laterite soils are found in many parts of southern India. This iron-rich crust is produced by chemical weathering under a moist tropical climate. It is commonly red or brown in color because of the iron oxides that have remained while other minerals, such as silica, have been leached out by rainwater. Although best developed on the iron-rich Deccan lavas, laterite underlies much of the Indian peninsula. It has been formed by up to 60 million years of deep chemical weathering. Climate changes during the last few million years have also contributed to erosion, but in some areas laterite remains as a capping on hills, protecting the less resistant layers beneath.

Sediments washed down from the Deccan by heavy monsoon rains are deposited on the lower land in the state of Tamil Nadu in southern India. To overcome the seasonal variation in rainfall, farmers irrigate the land.

The spectacular scenery of the Deccan plateau. The ancient shield surface has been deeply cut by rivers such as the Krishna, which rises a few kilometers from the west coast and flows across the plateau to the Bay of Bengal in the east.

From forest to desert

The subcontinent's natural vegetation has been substantially modified, especially by deforestation. The Himalayas have alpine vegetation between the snowline and the mountain forests. The wet western margins of peninsular India, much of Sri Lanka, and areas north and east of the Bay of Bengal are covered by dense tropical rainforest; thorn forest and scrub cover the drier inland and eastern areas of the peninsula and the margins of the Thar Desert. In much of the northern part of the region there is monsoon forest.

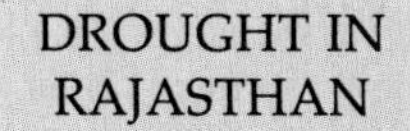

DROUGHT IN RAJASTHAN

During the 1980s a major drought came to parts of northwest India. Rivers flow only occasionally in most of this semiarid area, which includes much of the Thar Desert, and when they do they drain into inland lakes that dry up fast. Rainfall in "normal" years is less than 300 mm (13.5 in). In the Thar Desert sand, scrub and rocky outcrops extend over a large area – some 260,000 sq km (100,000 sq mi).

In 1983 the lakes and reservoirs vital for irrigation in a dry land were filled; in the dry years that followed many became empty. Although there may be water underground it is often too salty for irrigation. Water was rationed, and desperate measures such as bringing in water by train became routine. In recent years the Rajasthan Canal has been built in a mammoth effort to bring waters from the north.

Dry and humid spells have alternated in the area for a long time. From geological and archaeological evidence it is also clear that the Indian part of the Thar Desert has gradually been becoming more arid during the last 10,000 years. The process of desertification is probably the result of long-term climate change brought about by changes in wind direction, but may have been accentuated by human activity. Destruction of forests in semiarid areas can speed up the process of desertification. In the Indian states closest to the Thar Desert about half of the forests were destroyed during the 1970s.

THE LIFELINE OF INDIA

Rivers have an immense significance in the Indian subcontinent. They have carried down material eroded from the Himalayas and the Deccan plateau and deposited it over millions of years to create the great plains of northern India, one of the world's most fertile farming regions. As a populous agricultural area, the plain of the Ganges is second only to the Chang valley of China.

For thousands of years the rivers of the subcontinent have been used to irrigate large areas that would otherwise have been too dry for cultivation, and to provide the water supply for towns and villages. Today their waters also generate hydroelectricity. Culture and religion reflect this enduring importance. The Ganges is Mother of Life to the Hindus of India. Where its waters join those of the Yamuna at Allahabad up to 30 million people may gather to take a holy bath.

River flow and rainfall

The catchment areas drained by these rivers are relatively small compared with those of some of the world's great river systems. From their sources in the Himalayas to their deltas, the Brahmaputra and Indus flow for about 2,900 km (1,800 mi), the Ganges for about 2,505 km (1,557 mi). They are easily surpassed in length by the rivers of other continents. However, their rate of flow matches that of some of the world's greatest rivers. The Brahmaputra has the seventh largest discharge in the world, and the Ganges, Indus, Godavari and Mahanadi (the last two from the plateaus of the peninsula) are all included in the top forty. The joint discharge of the Brahmaputra and Ganges into the Bay of Bengal is exceeded only by the Amazon in South America and the Congo in Central Africa.

Rainfall is concentrated into a fairly short season. As a result water runs off the surface of the land, as opposed to percolating through it, at a much higher rate than in more equable climates. The arrival of the monsoon rains, generally in June, has a dramatic effect on the rivers, with a concentration of flow into a very short time. In the most arid regions rivers may flow only immediately after sporadic rainfall. Rivers that in the dry season are a broad sandy channel with no more than a trickle of water can become an impassable flood of water several kilometers wide during the monsoon season.

The regimes of the Ganges and Brahmaputra are very similar. There is a rapid rise to maximum flow in July–August, at the peak of the monsoon, and a low flow for the rest of the year. More than half the delta of Bangladesh is regularly flooded by rivers, and up to a third of it may be inundated each year to a depth of a meter (3.3 ft) or more.

River landscapes

There are two reasons for the exceptional quantity of sediment carried by the rivers draining from the Himalayas. The height of the land, and the fact that it is still rising, means that rainfall is high and the process of erosion very active. By continuing to cut down into the rocks, rivers have created steep slopes that have frequent landslides. The second reason is

GREAT RIVERS IN FLOOD

The annual flow of the Brahmaputra and Ganges through their combined delta is less than that of the Amazon or Congo. But allowing for their smaller drainage basins, they have a very high rate of flow, and they carry three times as much sediment for every square kilometer of watershed as the Huang, and twenty times as much as the Amazon.

River	Average annual flow cu km	cu mi
Amazon, Brazil	3,768	902
Congo (Zaire), Congo/Zaire	1,256	301
Chang, China	688	165
Mississippi–Missouri, United States	566	133
Orinoco, Venezuela	538	129
Paraná, Argentina	493	118
Brahmaputra, India/Bangladesh	476	114
Indus, Pakistan	443	106
Irrawaddy, Burma	443	106
Ganges, India/Bangladesh	440	105

Cutting through the mountains The Indus river at Skardu, Pakistan. The low level of the water reveals the surrounding floodplain. On the right is a remnant of a higher floodplain, formed before the base level dropped and the Indus cut down again into the valley floor.

LIVING ON A DELTA

Most of Bangladesh is delta, and over half of it is flooded every year during the summer monsoon season. In many countries flooding is regarded as a hazard, yet in delta regions it is part of normal landscape development. Deltas receive both the solid and the dissolved nutrients carried by rivers. They are consequently very fertile and often support a large human population – over 100 million in Bangladesh. Floods are therefore both a hazard and an asset.

A new danger is being caused by deforestation in the catchment areas of the Ganges and Brahmaputra, particularly in Nepal and the Indian Himalayas. In October 1988 three-quarters of Bangladesh was flooded for three weeks. These were the worst floods of the century, killing over 2,000 people. Thousands more died from disease and starvation, and 25 million people were made homeless. The waters destroyed perhaps a billion dollars worth of crops, 3,500 km (2,200 mi) of roads and 250 bridges.

The delta area is also subject to catastrophic storm surges. Cyclone winds up to 200 k/h (120 mph) whip up waves as much as 9 m (30 ft) high. Devastation and an enormous death toll may follow when a storm surge passes over the densely populated delta.

that deforestation, especially in Nepal, leaves the ground unprotected against the intense monsoon rains. Deforestation is now having a similar impact in river catchment areas along the northern side of the Deccan.

The Indian subcontinent has a rich variety of landscapes formed by rivers. Himalayan rivers flow through well-defined valleys, with frequent lake basins, before reaching the plains of northern India and Pakistan. Sandy alluvial fans are built out where rivers emerge from the Himalayan foothills and flow down onto the floodplains. Rivers flowing off the Deccan plateau produce spectacular gorges headed by waterfalls where the waters are cutting down into the basalt. In the dry season they can be completely dry, whereas in the wet season the roar of water can be heard 16 km (10 mi) away. In the desert regions rivers flow infrequently, often ending in inland lake basins such as those in the northwest of India.

The great expanses of the northern plains are covered by a fertile layer of sandy deposits. Levees or mud banks are among the few natural features that break up the flatness of the land. They are formed in times of flood when the river deposits the coarser material in its load of sediment as it slows down on leaving the main channel. In the deltas of the Indus and Ganges–Brahmaputra, the rivers and distributaries into which they divide also flow above the general level of the land, their channels bounded by levees to either side. Beyond the levees the land slopes down into dips (backswamps) where finer-grained material is deposited during floods. The constant shifting of the river channels means that sandy deposits are widespread across the delta.

The floodplain of the Ganges in the state of Bihar, in northeast India. Monsoon rains and melting snow in the Himalayas swell the sacred river to bursting point, periodically causing catastrophic floods that are made worse by deforestation in the catchment area.

Asia's giant mountain wall

The name Himalayas is a Nepalese word meaning "home of the snows". This "home" is by far the greatest mountain system of the world. With an average height of about 6,000 m (19,000 ft), they stretch 2,500 km (1,550 mi) from the far north of the subcontinent through Pakistan, India, China, Nepal, Sikkim and Bhutan to the southern bend of the Brahmaputra river in the east. The system includes the world's highest mountain, Mount Everest (8,848 m/29,028 ft), which is still known by the name of a 19th-century British surveyor but also bears the Tibetan name Chomolungma. The third highest peak, Kanchenjunga (8,598 m/ 28,208 ft), is also to be found in the Himalayas; the second highest, K2, lies nearby to the northwest in the Karakoram Range. There are some thirty peaks higher than 7,620 m (25,000 ft) in the system, and nearly ninety of the hundred highest peaks in the world are to be found in the Himalayas; every one of them is more than 7,315 m (24,000 ft) high.

The Himalayas are in three parallel parts. In the north are the perpetually snow-covered Great Himalayas. This is where the highest mountains are to be found. Flanking them to the south are the Lesser Himalayas, which attain their greatest height at the western end of the range. Further south still, the Siwalik Range or Outer Himalayas are much lower, being below 1,000 m (3,300 ft). They are formed of coarse sandstone laid down as a result of erosion of the main Himalayas and subsequently cut through by rivers. Although much lower they are nonetheless very rugged. Between them and the Lesser Himalayas are scattered river basins, such as that of the Katmandu valley in Nepal.

A wide valley containing the headwaters of two of the subcontinent's great rivers runs along the northern edge of the Greater Himalayas. The Indus flows to the west in this east–west oriented furrow, the Brahmaputra flows to the east.

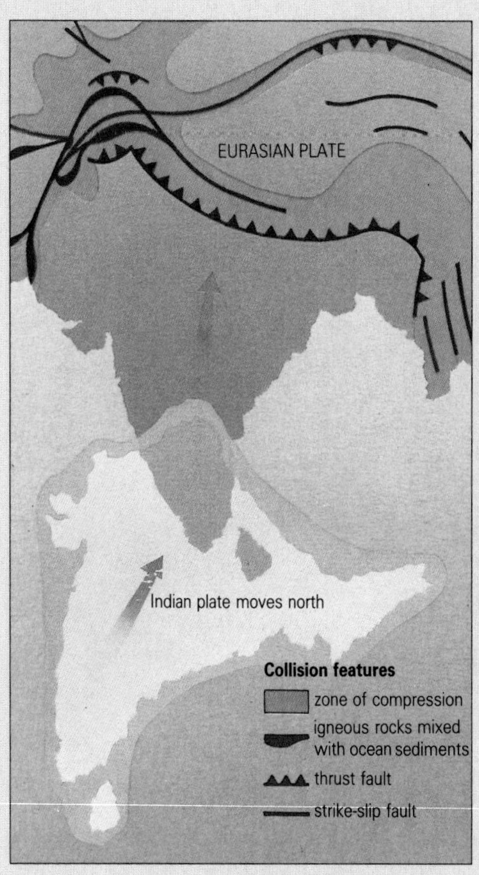

EURASIAN PLATE

Indian plate moves north

Collision features

zone of compression

igneous rocks mixed with ocean sediments

thrust fault

strike-slip fault

Collision course About 40 million years ago the Indian plate collided with the Eurasian plate, buckling the sediments between them. The collision is marked by long thrust faults and shorter strike-slip faults.

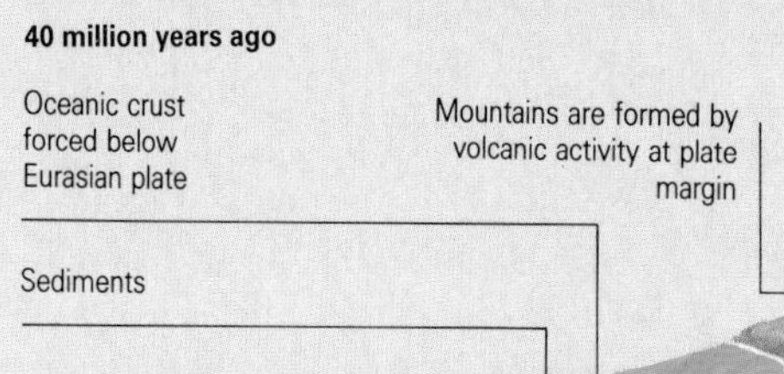

40 million years ago

Oceanic crust forced below Eurasian plate

Mountains are formed by volcanic activity at plate margin

Sediments

Magma

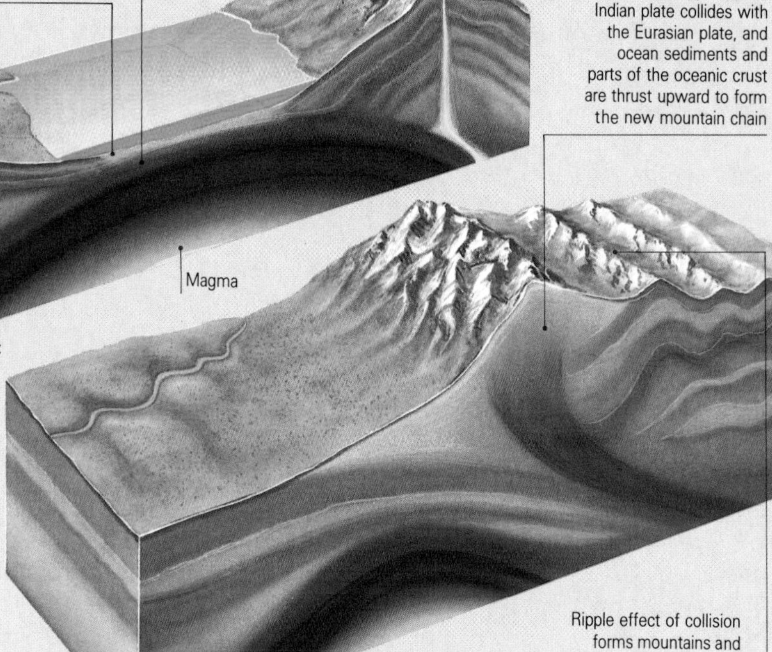

Building the Himalayas As the plate carrying the subcontinent of India moved toward the Eurasian plate, the shallow sea that separated them became smaller, and the oceanic crust was forced under the Eurasian plate to become part of the mantle again. A line of volcanoes was produced above the molten crust. The Indian subcontinent was pushed ever closer to the continent of Asia until the continents themselves collided. The line of collision ismarked by thrust faults in northern India. To the north of this the sedimentary rocks have been folded and faulted, pushing up to form the Himalayas and down to depress the mantle itself, making this the thickest part of the Earth's crust. Evidence of the scale of vertical movement is provided by the marine fossils to be found on the mountain peaks.

The peak of Annapurna stands 8,078 m (26,504 ft) above sea level in the Great Himalayas. The Himalayan mountains are made of sedimentary rocks laid down in the sea and and raised by earth movements; they still gain 1 m (3.3 ft) every 1,000 years.

Growing ever higher

The colossal structure of the Himalayas is the result of a major event in the history of the Earth – the collision between two continents. The part of the Earth's crust that carries peninsular India became detached from the southern supercontinent of Gondwanaland when it broke up in the Cretaceous period (144–65 million years ago). It drifted slowly northeast away from Africa at a rate of about 1 km (0.6 mi) every 10,000 years. About 40 million years ago the Indian plate met the Eurasian plate, the rate of drifting was halved, and the ancient sedimentary rocks on the ocean floor between the two landmasses were pushed up to begin the formation of the Himalayas. As the mountains were thrust up over millions of years, there were also massive dislocations, including faults with angles of less than 45° to the horizontal. Some of these reverse faults pushed great mountain blocks one over the other to distances of 50 km (30 mi) or more.

The oceanic crust beneath the Indian Ocean continued to spread, pushing the Indian plate still farther north. The Indian plate moved under the Eurasian plate, thickening the crust to 70–90 km (40–55 mi) beneath the Himalayas and Tibet to the north, thicker than anywhere else on Earth. In the past few million years the whole region of the Himalayas has risen some 3,000 m (10,000 ft), partly as a result of the deep crust floating upward on the underlying mantle to reach a new equilibrium (isostasy) and partly because India continues to push into Asia, though now at a slower rate of about 1 km (0.6 mi) per 100,000 years. Earthquakes regularly occur, providing proof that mountain-building is still going on.

Averaged over the whole area, every thousand years the land has been lifted some 2 m (6.6 ft), and has lost 1 m (3.3 ft) from the surface through erosion. This eroded material, carried down as sediment by the great rivers and deposited over the lowlands, has formed the plains and deltas of northern India, Bangladesh and Pakistan.

The raising of the great mountain wall has made the Himalayas a very effective climatic barrier, excluding the Indian monsoon winds and so preventing warm, moist air from moving to the north. The Plateau of Tibet is consequently one of the coldest and driest places on Earth. Because of the low rain- and snowfall, there are fewer glaciers on the northern slopes of the Himalayas. On the snow-covered southern slopes the effects of glaciation, the continuing rise of the land and the high rainfall result in a very unstable landscape, with rockfalls and landslides an ever-present hazard.

Disappearing islands

Pollution of the atmosphere could cause one of the world's most low-lying countries to vanish completely.

The 1,800 small coral islands and sandbanks making up the Maldives are built on the crowns of old volcanoes that have been submerged beneath the Indian Ocean. Starting 595 km (370 mi) southwest of India's southern tip, the islands extend 764 km (475 mi) to the south. None of them rises to more than 1.8 m (6 ft), and they are protected from violent storms only by the coral reefs that form a ring around the islands.

The islands are threatened by the rise in sea level that is expected to result from the global warming taking place because of the greenhouse effect. Gases such as carbon dioxide trap the heat radiating from the surface of the Earth and keep it warm. However, an extra 5 billion tonnes of carbon dioxide enters the atmosphere every year from the burning of fossil fuels such as coal and oil. Scientists predict that the temperature of the Earth will rise by between 1.5°C and 4.5°C (2.7–8.1°F) in the next 50 years or so. This will cause the polar ice caps to recede, releasing more water into the oceans. Furthermore, the water in the oceans will expand as it warms. The resulting rise in sea level will be sufficient to threaten or even completely inundate the 200 and more islands that are inhabited by some 250,000 people.

Doomed to extinction, the islands of the Maldives are likely to disappear beneath the rising Indian Ocean as global warming continues.

A CHECKERBOARD LAND

DIVERSITY ON A GRAND SCALE · FROM INLAND DESERT TO THE HUMID EAST · A SHIFTING ENVIRONMENT

China, the third largest country in the world, covers almost a quarter of Asia and dominates the continent's eastern coast. The coastline extends some 23,000 km (14,000 mi); its land boundaries total over 19,000 km (12,000 mi). Well over half the land, to the west and northwest, comprises harsh deserts and grassy plains (steppes). The world's coldest and highest plateau and its loftiest mountains can be found in Tibet. China has some of the most remote and empty places on Earth, including places most distant from the oceans; it also has the largest population of any country. The great majority of the one billion Chinese live in the east, where the climate is wetter and more hospitable on the north China plains, the terraced hillsides and the lush paddy fields of the Red Basin and coastal lowlands to the east and south.

COUNTRIES IN THE REGION

China, Taiwan

LAND

Area 9,562,904 sq km (3,690,246 sq mi)
Highest point Mount Everest, 8,848 m (29,028 ft) highest on earth
Lowest point Turfan depression, −154 m (−505 ft)
Major features Plateau of Tibet, Himalayas, world's highest mountain chain, Red and Tarim basins, Takla Makan and Gobi deserts, river plains in east

WATER

Longest river Chang, 5,900 km (3,722 mi)
Largest basin Chang, 1,827,000 sq km (705,000 sq mi)
Highest average flow Chang, 32,190 cu m/sec (1,137,000 cu ft/sec)
Largest lake Qinghai, 4,460 sq km (1,721 sq mi)

CLIMATE

	Temperature °C (°F)		Altitude
	January	July	m (ft)
Lhasa	−2 (28)	15 (59)	3,658 (11,988)
Hami	−12 (10)	28 (82)	738 (2,421)
Guangzhou	13 (56)	28 (82)	63 (201)
Beijing	−5 (23)	26 (79)	51 (167)
Shanghai	3 (38)	28 (82)	45 (148)
Harbin	−20 (−4)	23 (73)	172 (564)

	Precipitation mm (in)		
	January	July	Year
Lhasa	0.2 (0.01)	142 (5.6)	454 (17.9)
Hami	2 (0.1)	6 (0.2)	33 (1.3)
Guangzhou	39 (1.5)	220 (8.6)	1,681 (66.2)
Beijing	3 (0.1)	197 (7.7)	683 (26.9)
Shanghai	44 (1.7)	142 (5.6)	1,129 (44.4)
Harbin	4 (0.2)	127 (4.9)	554 (21.8)

NATURAL HAZARDS

Large rivers in flood, earthquakes in interior and on Taiwan, typhoons in coastal areas, landslides

DIVERSITY ON A GRAND SCALE

The pattern of China's landscapes has been described as three giant steps that descend eastward to the sea, with the edge of each step marked by a mountain range. The westernmost step is the immense Plateau of Tibet, the "roof of the world", the highest and largest plateau on Earth. Extending over about 2.6 million sq km (1 million sq mi), it has an average height of 4,000 m (13,000 ft), with the snowcapped Kunlun Shan ranges to the north rising to 7,723 m (25,348 ft) at Wu-lu-k'o-mu-shih. To the south and west the world's greatest mountain range, the Himalayas, has many peaks over 8,000 m (26,000 ft), including the highest on Earth, Mount Everest. Northwest of the Himalayas lies the Karakoram Range, which includes glaciers among the longest outside polar regions. The Himalaya–Karakoram system has 96 of the world's 100 highest peaks.

The second step, to the north and east, is a vast checkerboard of plateaus and basins mostly 1,000–2,000 (3,300–6,600 ft) above sea level. It includes part of the Gobi desert, a plateau made of fine wind-blown loess deposits and the Tarim and Red basins. The region is bordered to the west by mountain ranges rising to well over 7,000 m (23,000 ft) in the Karakoram, the Pamirs and the Tien Shan, and to the north by the Altai.

The great plains of China form the third step, stretching east and southeast from the plateaus to the sea. They were created by the deposition of soils brought down by great rivers flowing from the higher land to the west such as the Huang (Yellow), Chang (Yangtze) and Xi rivers. The plains are generally below 500 m (1,600 ft), broken by occasional hills.

A complex history

The landscapes of China are the result of a complex geological past in which there have been several periods of earth movement lifting up the land. During the early Tertiary period, some 60 million years ago, much of present-day China consisted of low-lying areas with sediment-filled depressions. About 25 million years ago the mountains, plateaus and basins of today were formed. The boundaries between the three steps follow the lines of deep-seated faults in the crust. This tectonic activity continues: there are frequent earthquakes, particularly in the mountains of the west and on the island of Taiwan.

The Plateau of Tibet has continued to be lifted up through most of the last 2 million years (the Quaternary). During this period the prevailing northwesterly winds lifted dust from what is now the Gobi desert and surrounding areas and deposited it in northern China to form huge loess deposits that cover 600,000 sq km (230,000 sq mi). In many areas several layers of loess of different ages can be distinguished. Erosion of the uplifted areas also led to vast amounts of sediment being carried by rivers to the great plains of China, where the alluvial deposits accumulated to depths as much as 1,000 m (3,300 ft).

Humans in the landscape In the cultivated parts of eastern and southern China the landscapes have been altered over many centuries by the work of humans. The fertile slopes of this valley in Yunnan, southwest China, have been terraced to make them flat enough for cultivating rice.

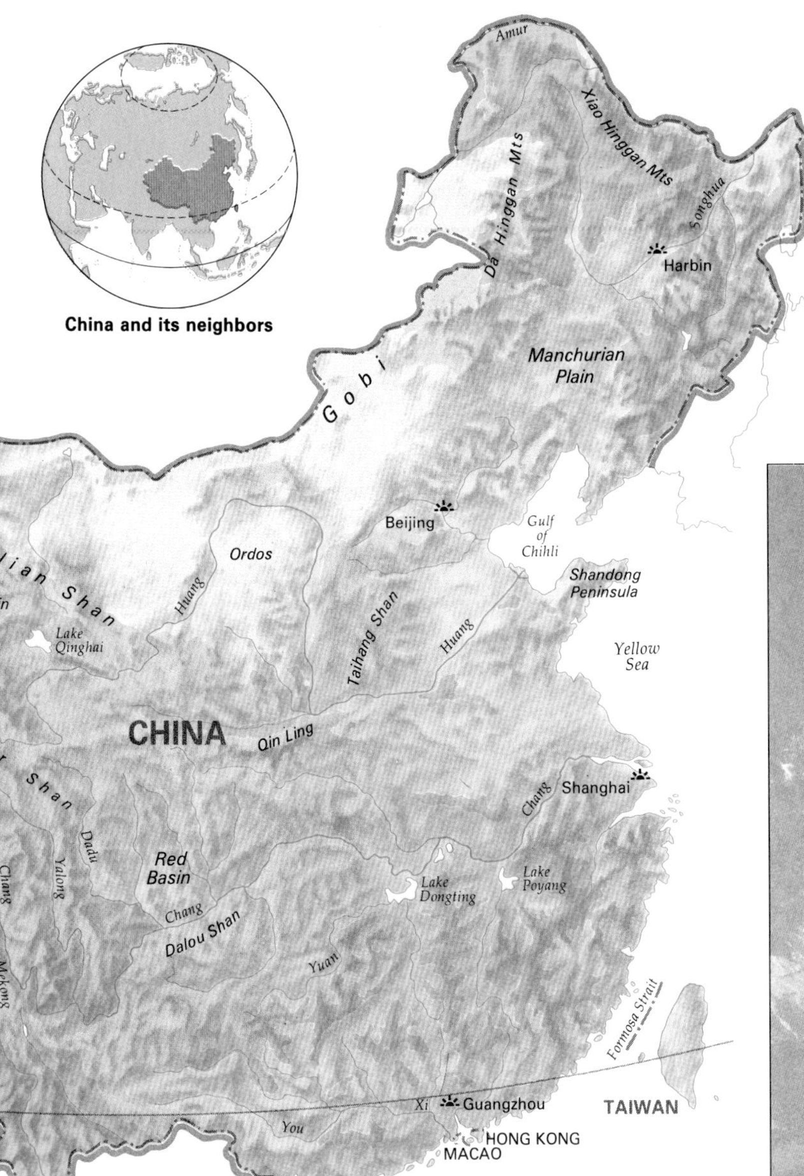

Map of physical zones, China includes the most remote areas of the world as well as some of its most populous. It is a region of many contrasts, notably between the high mountains, plateaus and inland basins of the remote, dry western interiors, and the fertile plains, hills and valleys of the more humid north, east and south toward the Pacific Ocean.

The basin of Dzungaria in northwest China is over 2,200 km (1,370 mi) from the Yellow Sea, but it is only 200–500 m (650–1,650 ft) above sea level. The basin is enclosed by the high, snowcapped Tien Shan and Altai mountain ranges. On the arid grasslands rainfall is slight. The few permanent rivers all drain inland, and are fed by the snowmelt of spring and summer.

FROM INLAND DESERT TO THE HUMID EAST

The variety of climates in China today reflects differences in altitude, distance from the Equator and distance from the sea. There are three major climatic zones, and they correspond to the three giant steps in relief. The Plateau of Tibet has a harsh, cold, dry climate, which has earned it a reputation as "Earth's third pole". The high altitude of the mountain wall to the south prevents warmer air from moving north. The average January temperature ranges from −10 to −15°C (14 to 5°F). Above 4,000 m (13,000 ft), the average altitude of the plateau, there is no frost-free season throughout the year.

In the remote arid steppes and deserts of northwest China, far from the ocean and therefore from rain-bearing winds, annual precipitation rarely exceeds 100 mm (4 in). There are marked seasonal extremes of temperatures.

The monsoon effect

The lowlands and hills of eastern monsoon China occupy almost half the country. Here temperature varies greatly with latitude: the northeast lies in a temperate zone; farther south there are both warm-temperate and subtropical zones; southern Yunnan in the southwest and also the island of Hainan are tropical. Throughout eastern China the monsoon brings high rainfall in summer. The monsoon cycle dominates the seasons. In the fall the Asian landmass cools much more rapidly than the surrounding ocean to the south and east, forming a center of high atmospheric pressure (anticyclone). Cold, dry northerly winds (the winter monsoon) blow outward over much of China. When the land warms in spring a great current of warm humid air penetrates inland from the south and east, bringing rain. This summer monsoon accounts for most of the rainfall throughout China.

Landscapes and climates

The landscapes of China fall broadly into the same three zones as climate and relief. Snow, glaciers, and a subsurface layer of permanently frozen earth (permafrost) are typical of the mountains and high plateaus of Tibet. Most rivers drain inland, and there are many lakes in the closed basins.

Northwest China is an arid land, of undulating plateaus and of great inland basins enclosed by high mountains. To the north of the Takla Makan desert in the great Tarim river basin lies China's lowest and hottest place, the Turfan depression. It lies 154 m (505 ft) below sea level, and in summer temperatures may exceed 47°C (117°F). Turfan has an oasis, one of a number in the northwest, supporting some cultivation. Wind erosion and dust storms are common and there are large areas of sandy and stony (gobi) desert. To the northwest desert gives way to grassy steppes and then to the remote Altai mountain range. Here too the rivers drain inland and many of the numerous lakes are salty.

Compared with Tibet and the northwest the landscapes of eastern China have been formed to a greater extent by rain and rivers and by deep weathering of the rocks. The mountains and basins, hills and plains of eastern China stretch from northeast to southwest, in a series of

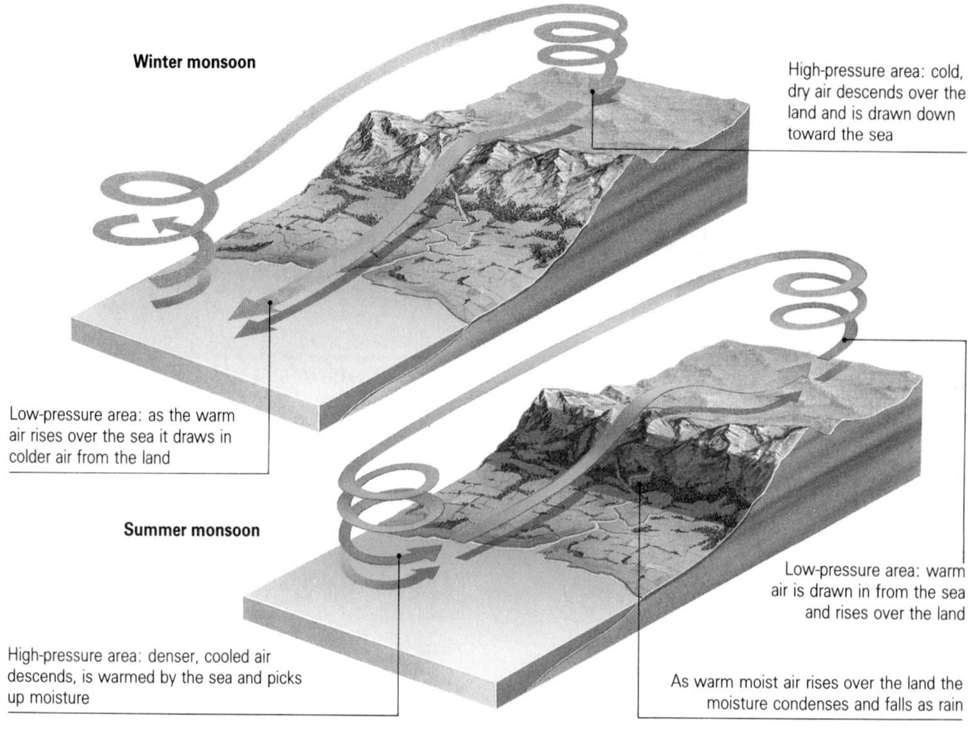

Monsoons are seasonal winds that blow predominantly from the northeast for six months and from the southwest for another six months. In summer warm, moist air is drawn in from the ocean by low atmospheric pressure over the land. This air rises, cools and condenses in the form of rain. In winter the pattern is reversed: the land cools more quickly than the sea, and dry, cold air flows out from the continent.

China's karst landscapes Stretching across south-central China is a stunning landscape of razor-sharp limestone peaks. They are weathered remnants of a vast block of limestone, which has been dissolved by rainwater made acid through carbon dioxide taken up from the atmosphere or soil. Between the peaks the flat valley floors are formed by the deposition of sediments.

A HANDMADE LANDSCAPE

Agriculture has been practiced in China for more than 7,000 years, and during this time the land has been shaped by the people who work on it. The fertile eastern areas are where nine-tenths of China's one billion people live, and most of them still make their living from the land. Here the human activity ranks alongside climate, topography and rock type as a major influence on the landscape. Both the terraces on the loess plateau in the north and the terraced rice paddies of southern China have been constructed entirely by hand – a major feat of earthmoving that has modified the land surface over large areas. In some parts terraces serve simply to increase the area of flat land available for growing crops; in others they also help to conserve the soil by reducing erosion.

At present almost half the total cropland of China is irrigated, and in several areas the proportion exceeds four-fifths. About one quarter of the world's total irrigated land is in China. To collect and distribute water for irrigation, intricate systems of reservoirs, wells, canals, ditches and sluices have been built over thousands of years. These human works have become an integral part of the landscape.

depressed and raised belts created by gigantic movements of the Earth's crust.

Moving from west to east, the first of these belts is formed by the Inner Mongolian plateau south of the Gobi, the Ordos plateau, the Red Basin and the eastern Yunnan plateau in the southwest, followed by the Da Hinggan, Taihang and south-central Dalou mountain systems. The second belt consists of the cultivated plains of northeast and northern China and the middle of the Chang valley, followed by the mountains of the northeast, the Shandong peninsula and ranges parallel to the coast in the southeast. Across the sea from the lowlands of the south coast, mountainous Taiwan, with extensive plains in the west, represents the third belt of raised land. Old shorelines show that the island's west coast has risen over 800 m (2,600 ft) in the last million years.

Many of the high rugged mountains are composed of ancient resistant crystalline rock, whereas the lower mountains, hills and basins are sedimentary rock. The type of underlying rock is an influence on the landscape, which is then shaped by the action of the wind and water. In the north, for example, is the unique and highly eroded landscape of the loess plateau, where fine wind-blown deposits lie up to 300m (1,000ft) deep. In the southwest the spectacular karst towers and flat-floored valleys are produced by rainwater percolating through and dissolving limestone. The distinctive outcrops that remain can be found over an area of more than 1 million sq km (400,000 sq mi).

Granite rocks are to be found widely throughout China, particularly in the center and south, where the characteristic dome shapes are covered with a reddish crust that is easily eroded. The rounded granite hills of the southeastern lowlands are the product of deep chemical weathering, aided by the hot, humid climate. Large areas are also covered by sediments, known as red beds; these include sandstones, conglomerates and shales in central and southern China.

Half of China was once forested. In the humid east the natural cover ranges from coniferous taiga forest in the north to tropical rainforest in the south. Agricultural crops have replaced many of the trees, and only one-eighth of the land is now covered by forest. To the northwest there are grassy steppes; in desert regions the vegetation is limited to scrub.

A SHIFTING ENVIRONMENT

The natural processes that shape China also pose major problems for people. Soil erosion threatens cultivated land and desert margins advance into agricultural areas. Landslides can destroy buildings and disrupt the transport network, floods may spill out over the densely populated plains, and typhoons frequently rage across coastal areas.

Erosion of the land

Nearly one-third of China has been subject to severe soil erosion in recent decades – part of a global phenomenon. Wind erosion, which has caused about half the soil loss, is particularly severe in the arid northwest. Water erosion, affecting large areas of farmland in eastern China, has posed a more serious problem, particularly for humans. In total, the country was losing about 5 billion tonnes of soil each year.

China covers some 6.5 percent of the world's land surface; its rivers carry about 15 percent of all the sediment taken to the oceans. Much of the soil eroded by rivers from the farmland is transported downstream, where it silts up reservoirs and irrigation systems. Effective soil protect-

A scarred landscape The loess plateau of north China, shown here near Datong, Shanxi Province, lies between arid north-central China and the fertile plains to the east. It is the largest and thickest expanse of loess in the world. The plateau is spectacularly dissected by streams and rivers, and is almost treeless. It owes much of its character to the soft nature of the loess soil.

Erosion and riverborne sediment The natural erosion of the soft loess deposits is exacerbated by the lack of vegetation, by the intense summer rain and by agricultural activity. Sediment yield is expressed as the average amount of sediment washed to a river from an area of 1 sq km (0.4 sq mi); a yield of 1,000 tonnes represents a lowering of the land surface by 1 mm.

SOIL EROSION – A WORLDWIDE PROBLEM

Erosion is a natural process. All soils undergo some erosion by wind and/or water, but usually the rate of soil loss does not exceed that of soil formation. However, if the natural balance is disrupted, for example by removal of the vegetation cover, by excessive grazing or by cultivation, erosion may increase. Then soils become progressively thinner and can even disappear. Erosion in one place means deposition of eroded material in another, bringing its own set of benefits and problems.

The depletion rate of the Earth's soil is increasing. Recent estimates have suggested that each year the world is losing from its croplands about 23 billion tonnes more soil than is being formed. This means that soils are being reduced worldwide by 0.7 percent a year, or 7 percent each decade. Other estimates suggest that since settled agriculture began about 10,000 years ago some 430 million ha (1 billion acres) of productive land have been destroyed by the effects of soil erosion.

tion measures have become essential to slow down these powerful erosive processes, with their potentially disastrous consequences for millions of Chinese.

Unstable slopes

In many areas of southern China the underlying granite and volcanic rocks have undergone deep chemical weathering. The warm, wet climate accelerates the process and weathering to depths of up to 60m (200ft) is common. Rock material is loosened by the weathering process, and on slopes it is highly susceptible to landslides, particularly at times of heavy rainfall. Along the south China coast more than 250 mm (10 in) of rain may fall in one day; most landslides happen after such a storm, when the saturated ground becomes unstable. Excavations to build foundations or roads can reduce the stability of slopes still further.

Landslides can cause considerable damage and disruption. In crowded Hong Kong, where there is a major landslide about once every two years on average, they can be triggered by exceptionally heavy rains. On 18 June 1972, for example, more than 650 mm (26 in) of rain fell in three days. One of the resulting landslides caused a number of buildings to collapse, and 67 people were killed. Landslides may well become a more frequent hazard in southern China as the pace of urban development and pressure to build on steep slopes increases.

Typhoons in coastal areas

In China a hurricane is known as a typhoon, named from the Chinese word for a strong wind. Typhoons are moving cells of intensely low pressure that occasionally strike the coastal areas of east Asia. Because the atmospheric pressure changes considerably over short distances, typhoons are accompanied by extremely high winds that may reach 160 km/h (100 mph) and can generate exceptionally high sea waves. Typhoons are also associated with torrential rain, causing severe flooding.

On average China experiences about eight typhoons a year, most of which usually take place between July and November. Typhoons can bring havoc and disaster to the coastal areas where they strike, whereas they die out rapidly inland and have little effect.

The island of Taiwan is hit by some four typhoons each year. During the 1950s about 200 people were killed and more than 13,000 houses destroyed each year. Since then the inhabitants of Taiwan have become better prepared for the periodic violent onslaught of wind and rain, and typhoons inflict less damage, but they remain a great hazard to fishing boats and coastal shipping.

China's Sorrow

It is said that the Huang river has both claimed more lives and caused more human suffering than any other natural feature on Earth. According to historical sources the river, often known as China's Sorrow, has broken its banks more than 1,500 times since about 600 BC, and has changed its course on 26 occasions.

One of the most serious floods took place in September 1887, when flood-water breached the dike near the city of Zhengzhou some 600 km (370 mi) from the sea, and an immense lake, the size of Canada's Lake Ontario, was formed. When the water receded vast spreads of sand covered the countryside to depths of 2–3 m (6–10 ft). Well over a million people perished from drowning and, in the aftermath, from starvation and disease. Only with enormous human effort was the river returned to its original course.

The actual course of the Huang is bound up with human activity. In 1938 a Guomindang general sought to stem the advance of Japanese troops on Zhengzhou by ordering the main south dike to be breached: the invaders were halted as the river rushed southeast, but nearly 900,000 Chinese drowned. It took seven years to restore the northeast course.

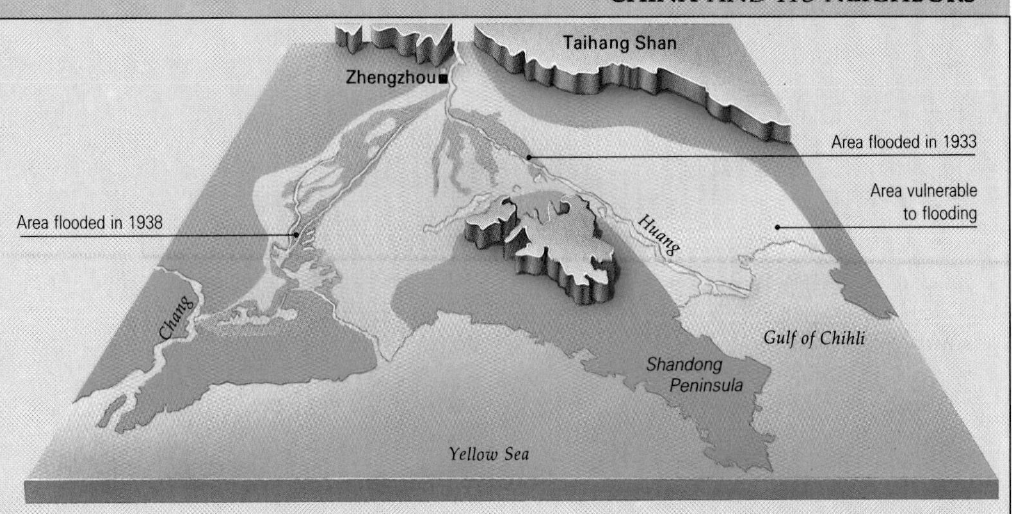

The wandering Huang set its present course northeast to the Yellow Sea in 1861. In major floods, such as the one that took place in 1933, the river burst its raised banks. In the 1938 disaster the Huang rushed southeast in the direction of what had been its former principal course.

Fertile valleys and plains are made of sediments released when the Huang in flood slows down on leaving its channel. The vast north China plain exceeds the state of Oregon in size and supports over 100 million Chinese.

GREAT RIVER SEDIMENT

The Huang carries a greater load of sediment than any other river in the world, but for the size of the area they drain, other rivers transport more material.

River	Millions of tonnes carried per year	Tonnes per year sq km (sq mi)
Huang, China	1,600	1,080 (420)
Ganges, India/Bangladesh	1,455	1,520 (590)
Brahmaputra, India/Bangladesh	726	1,370 (530)
Chang, China	501	250 (95)
Indus, Pakistan	436	260 (100)
Mississippi–Missouri, USA	365	150 (58)
Amazon, Brazil	363	143 (55)
Irrawaddy, Burma	299	700 (270)
Mekong, China/Southeast Asia	170	435 (167)
Nile, Egypt	111	37 (14)

China's other great river, the Chang, is also prone to flooding. During the 1933 floods the water rose in places by 30 m (100 ft). Nearly 40,000 sq km (15,000 sq mi) of land were submerged and more than 3.7 million people died. In 1954 some 10 percent of China's farmland was flooded.

The fertile floodplains

Large rivers dominate landscapes in eastern China. The lower reaches of the Huang, Chang and Xi rivers cross large areas of lowland that are frequently flooded. These fertile flood plains support some of the highest population densities in the world. It is unavoidable that human lives will be lost when so many people live on land where flooding is a major and frequent natural process.

Although the headwaters of both the Huang and the Chang lie in the mountains to the west and the rivers receive large volumes of water from melting snows in the spring, most floods result from heavy rain during the summer monsoon. Typhoons can also bring torrential rain during the summer months.

The huge load of yellow sediment that the Huang carries, and that gives the river its name, is the major cause of the floods. At the point where it leaves the loess plateau the Huang carries some 1.6 billion tonnes of sediment each year, the highest sediment load of any river in the world. The silt carried by the river can make up as much as 46 percent of its total flow weight at this point.

During its journey across the north China plain to the sea, a distance of over 800 km (500 mi), the Huang falls only about 100 m (330 ft) – just 12.5 cm for every kilometer or 3 in for every mile. The slow-moving river dumps vast amounts of sediment, about one-third of its load, during this part of the journey. As a result, the river channel and adjacent floodplain have been built up several meters above the level of the surrounding land. In a major flood the river spills out onto the lower land, inundates vast areas and changes its own course. The sediments deposited are hugely beneficial to plant life because the soil is so fertile. The process is also what attracts people to cultivate the land – but the very flood that brings fertility and richness to the land can bring disaster to the people whose lives it benefits.

For over 4,000 years the Chinese have devoted great efforts to controlling the floods by building dikes along the mudbanks, called levees. The mudbanks have formed naturally when floodwaters overflow the channel, slow down and deposit some of their coarser sediment load. A "Great Wall along the river" lines both banks of the Huang for some 1,400 km (870 mi). Since the 1960s major dams and dikes have been built, and reservoirs and canals have been constructed both upstream and alongside the main channel to hold and control the waters. Like the smaller-scale terraced rice paddies, channels and water gates of the hills and inland basins, they are a well-established part of the landscape in eastern China.

Construction and destruction

The Great Wall of China is hailed as one of the great creative achievements of mankind, but as in many parts of the world, the destructive achievements are even more impressive.

At the time the Great Wall began to be built in 221 BC, forests covered more than 30 percent of China. The demand for farmland to grow crops and graze animals has been so great that much of the natural vegetation cover has been removed. Forests now cover only about 5 percent of the country. The natural vegetation prevented what has become one of China's most serious problems – soil erosion. The fertile but easily eroded wind-blown loess soils of the interior are most at risk; it is estimated that 5 billion tonnes of loess is washed into rivers each year. Gullying is very common, as the slightest channel or depression, even a tire track, can set in train disastrous erosion.

Trees and other vegetation once protected the soil from the direct impact of heavy rain and also held the soil together, preventing channels from becoming established. Runoff is now much greater, causing erosion and flooding. Terracing the slopes and massive replanting schemes will eventually help to solve the problem, but the soil that has been lost will take hundreds of years to replace.

A ribbon in the landscape As it traces the crests of steep valleys and gullies to the northeast of Beijing, the Great Wall of China appears almost insignificant in this vast landscape.

A PATTERN OF ISLANDS

CLIMATE AND THE LAND · THE ISLAND CHAIN · LANDFORMS AND LANDSCAPE

The peninsula and islands of Southeast Asia extend over 6,000 km (3,700 mi) between the eastern end of the Himalayas and the continent of Australia to the south. They include over 20,000 tropical islands – the greatest archipelago on Earth. The loops and curves of the mountainous island arcs are bounded by the Java trench of the Indian Ocean and the Philippine trench of the Pacific. Here two great areas of the planet's crust – the Indo–Australian plate and the landmass of Asia – are in slow collision. Mantled by the dry teak forests of Burma, the luxuriant green rainforests of Indonesia and the eucalyptus and gum trees of Timor lie some of the world's fastest changing landscapes, in a region that is frequently subject to many of the Earth's more violent hazards – typhoons, tropical rains, earthquakes and volcanoes.

COUNTRIES IN THE REGION

Brunei, Burma, Cambodia, Indonesia, Laos, Malaysia, Philippines, Singapore, Thailand, Vietnam

LAND

Area 4,495,392 sq km (1,735,221 sq mi)
Highest point Hkakabo Razi, 5,881 m (19,296 ft)
Lowest point sea level
Major features mountain chains and flood plains, deltas of great rivers in north of region, mountainous and volcanic islands of Malaysia, Indonesia, Philippines, world's largest archipelago

WATER

Longest river Mekong, 4,180 km (2,600 mi)
Largest basin Mekong, 811,000 sq km (313,000 sq mi)
Highest average flow Irrawaddy, 12,660 cu m/sec (447,000 cu ft/sec)
Largest lake Tonle Sap, 10,000 sq km (3,860 sq mi)

CLIMATE

	Temperature °C (°F) January	July	Altitude m (ft)
Rangoon	25 (77)	27 (81)	23 (75)
Ho Chi Minh City	26 (79)	27 (81)	10 (33)
Manila	25 (77)	28 (82)	15 (49)
Cameron Highlands	18 (64)	18 (64)	1,449 (4,753)
Singapore	26 (79)	27 (81)	10 (33)
Djakarta	26 (79)	27 (81)	8 (26)

	Precipitation mm (in) January	July	Year
Rangoon	3 (0.1)	580 (22.8)	2,618 (103.1)
Ho Chi Minh City	6 (0.2)	242 (9.5)	1,808 (71.2)
Manila	18 (0.7)	253 (10.0)	1,791 (70.5)
Cameron Highlands	168 (6.6)	122 (4.8)	2,640 (104.0)
Singapore	285 (11.2)	163 (6.4)	2,282 (89.8)
Djakarta	335 (13.2)	61 (2.4)	1,755 (69.1)

Rainfall of 1,170 mm (46 in) in 24 hours has been recorded at Baguio in Luzon, Philippines

NATURAL HAZARDS

Typhoons and floods, earthquakes and volcanic eruptions, landslides and mudslides

CLIMATE AND THE LAND

Southeast Asia is a tropical region. The islands straddle the Equator, but the Philippines and the continental peninsula extend farther north, so the region's climates are far from uniform. The areas nearest the Equator have a climate that is hot, wet and humid all year round. They include Sumatra, the Malay Peninsula, Borneo, Sulawesi, Halmahera and Mindanao. The annual rainfall averages over 2,000 mm (80 in), with at least 100 mm (4 in) every month, up to half of which comes in short, heavy downpours. During the day the temperature in the shade is 25–30°C (77–86°F), but just before dawn it falls to about 16–18°C (61–64°F) – hence the saying "in the tropics, winter comes every night".

Away from the Equator climates show the influence of the seasonal changes in winds (monsoons) and their accompanying summer rains. Both mainland Southeast Asia and the northern islands of the Philippines experience their wettest months from June through August. In the northern Philippines and in Vietnam, Laos and Cambodia rainfall patterns are strongly influenced by typhoons. Java and other islands to the south of the Equator have their heaviest rainfall from December through February.

Rainstorms and erosion

Typhoons are tropical storms, known elsewhere as hurricanes or cyclones. They travel from the western Pacific on erratic westerly courses toward the Asian mainland. The ferocious storms bring extreme rainfalls. Areas near the coast feel the effects most strongly. On Luzon, the main island of the Philippines, a fall of as much as 1,170 mm (46 in) in one day has been recorded. Even in the typhoon-free equatorial areas rain can be very heavy, with falls of 200 mm (8 in) in one day recorded in the Malay Peninsula.

The seasonal variation of rainfall causes the great rivers of mainland Southeast Asia, the Irrawaddy, Salween, Chao Phraya and Mekong, to have their peak flow in summer. They spill over their banks in the lowland and delta courses and inundate the ricefields.

The sudden heavy downpours can cause severe erosion on the ground. However, it is the same abundance of water, coupled with the continuously high temperatures, that encourages prolific growth of vegetation, which helps to retain the soil. Wherever rocks are weak, the streams and rivers erode them. The collapse of river banks, often triggered by fallen trees, and sand and silt churned up from the river bed are the main causes of the muddiness in the great rivers of this part of the world. This mud builds up the deltas and extends the great coastal mangrove swamps farther out to sea on many Southeast Asian coasts.

The patterns of climate help determine the types of vegetation. In the equatorial areas high humidity all year encourages the growth of the world's richest rainforest. A monthly rainfall of 100 mm (4 in) or more is usually enough in the tropics to keep plants growing all year round, allow water to reach rivers and wash out chemicals from the soil. In the monsoon areas to the north and south, where there is a sequence of dry months with under 60 mm (2 in) of rain, there are deciduous woodlands, including teak, and some savanna grassland vegetation.

Phra Nang, a limestone island off the southwest coast of Thailand. Southern Thailand is part of the Malay Peninsula. The peninsula is fringed with thousands of tiny islands. Phra Nang was shaped by the chemical weathering of its limestone rock. The limestone is dissolved by rainwater that has become slightly acid by absorbing carbon dioxide from the atmosphere or from the soil. The water attacks the rock through the joints and bedding planes, producing distinctive "karst" landforms, named after the plateau in Yugoslavia famed for its weathered limestone scenery. In humid climates, such as that of the tropical Malay Peninsula, the rate of weathering is increased.

A meeting of continents

Landscapes in Southeast Asia are being shaped under the influence of a warm, wet climate. The underlying structure was formed when parts of the ancient continent of Gondwanaland moved north and pushed into the continent of Asia.

When the crustal plate of India collided with Asia farther west some 40 million years ago it created a complex area of folding that extends into the peninsula of Southeast Asia. In the north the headwaters of the Irrawaddy, Salween, Chao

Map of physical zones Southeast Asia has a complex pattern of peninsulas and numerous islands lying in warm tropical seas.

Limestone towers dominate the landscape around Khao Yai in west-central Thailand. This area contains over a hundred such peaks, spectacular landforms created by chemical weathering of the limestone.

Phraya and Mekong rivers run south close together and parallel. So too do the upper courses of the Brahmaputra and Chang of India and China respectively. Mountain chains run between the rivers rather like the fingers of a hand. They extend east into China and southeast to form the Annam Highlands of Vietnam. To the south the mountains and plateau of east Burma continue into the Malay Peninsula. To the southwest the Arakan Yoma extends down the western side of Burma. On the border with China lies Hkakabo Razi, at 5,881 m (19,298 ft) the region's highest peak. Away from the extreme north, only the summits of the Arakan Yoma exceed 3,000 m (10,000 ft).

Along the lower parts of the valleys are the great plains where most of the people of Southeast Asia live. The plains of the lower Red river and of the Mekong, the Chao Phraya central depression in Thailand and the Irrawaddy basin of Burma mostly consist of fertile river muds and sands that have been deposited in the last 2 million years.

THE ISLAND CHAIN

On the islands of Southeast Asia mountain slopes rise steeply from the sea. Parallel deep-sea troughs separate neighboring islands to north and south. The trench parallel to the coast of Java drops to 7,455 m (24,460 ft) below sea level, the deepest abyss of the Indian Ocean. This marks where the edge of the Indo-Australian plate of the Earth's crust is sliding under the Eurasian plate. On the eastern edge of the archipelago the Philippine trench of the Pacific Ocean is even deeper, plunging to as much as 10,497 m (34,444 ft).

Building the landscape

The islands of Java, Sumatra, Bali and Lombok and most of Borneo are peaks on an undersea extension of the Eurasian plate known as the Sunda shelf. Dry land once joined them and the continental peninsula of Southeast Asia when sea levels were lower during the last great ice age, 35,000 to 10,000 years ago, and a glacier topped Kinabalu, Borneo's highest peak at 4,094 m (13,432 ft).

In the south and east the pattern of smaller islands is much more complex. The long peninsulas of Sulawesi are different fragments of the Earth's crust that slid against one another, were twisted around and then became attached as a result of various plate movements. Mindanao, which is made of three such fragments, was also formed in this way.

Island Southeast Asia is a young landscape, much of which is still being created and built up by tectonic and volcanic activity, such as the great eruption that destroyed the island volcano of Krakatau between Sumatra and Java in 1883. Indonesia is the most active volcanic region in the world, and Java alone has over 50 active volcanoes out of the country's total of 150. Earthquake activity is also unmatched, most quakes taking place along the outer edges of the island arcs, for example along the Mentawai Islands on the southwestern flank of Sumatra.

The steaming volcano of Mount Bromo rises above the clouds in eastern Java. Mount Bromo is famous for its frequent activity, but it is only one of many mountain volcanoes in the region. Java alone has some 50 volcanoes. The slopes are frequently covered with tropical forest vegetation, and the lava and ash produce fertile soils that attract agricultural development despite the instability of the land.

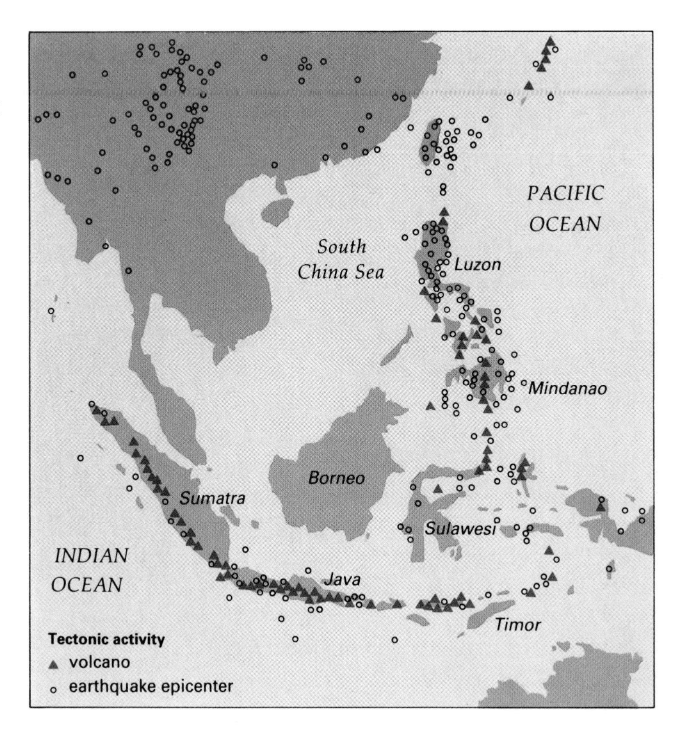

Earthquake and volcanic zones When earthquake epicenters and volcanoes are plotted on a map they show the major zones of tectonic activity and earth movement. These correspond to the boundaries of the Earth's plates. Southeast Asia lies where the Indo-Australian and Pacific/Philippine plates meet the extension of the Eurasian plate known as the Sunda shelf. Earthquake epicenters take place near the crust surface in ocean trenches, and at greater depths both under the island arcs and toward the mainland.

THE WORLD'S LARGEST ISLANDS

The Malay Archipelago comprises the islands of Southeast Asia and is the largest island group in the world. It includes three of the world's 10 biggest islands. Including Greenland, the North American Arctic contains four of them. Antarctica and Australia are regarded as continents, and are therefore excluded from the list.

	Area	
	sq km	sq mi
Greenland, Denmark	2,175,000	839,780
New Guinea, Indonesia/ Papua New Guinea	808,510	312,085
Borneo, Indonesia/ Malaysia	757,050	292,220
Madagascar	587,041	226,658
Baffin Island, Canada	476,070	183,760
Sumatra, Indonesia	473,700	182,900
Honshu, Japan	230,455	88,955
Great Britain, United Kingdom	229,870	88,730
Ellesmere Island, Canada	212,690	82,100
Victoria Island, Canada	212,147	81,910

Volcanic activity is now concentrated along the inner edge of the Asian plate: in central north Sumatra, the western mountains of central and southern Sumatra, and from west to east across Java into Bali, Lombok, Sumbawa, Flores and smaller islands. An arc of small volcanic islands stretches from Timor northeast to Ceram. There are parallel volcanic chains on the west coast of Halmahera and from the northern tip of Sulawesi to Mindanao. In west-central Mindanao there is a broad belt of volcanoes, while in the south of the island the peak of the Mount Apo volcano at 2,954 m (9,690 ft) forms the highest point of the Philippines.

A little farther north, Canlaon in northern Negros is now the only active volcano in the Negros–Parnay–Samar–Leyte group of islands, which contains many extinct or possibly dormant volcanoes. Luzon has several active volcanic areas; in the south the present crater of Taal forms a small island in the center of the lake that shares its name. The lake occupies a vast caldera, produced by an explosive eruption in prehistoric times that blew the top off the volcano. An eruption in 1911 killed 1,200 people; when the volcano erupted again in 1965 most of the 2,000 people squatting on the island were killed. In the far southeast of Luzon, Mount Mayon has a beautiful cone 2,421 m (7,943 ft) high, renowned as one of the most perfectly symmetrical volcanoes in the world.

THE RISE AND FALL OF CORAL REEFS

Sections of ancient coral now exposed many tens of meters above sea level have helped geologists to reconstruct the history of earth movements in Southeast Asia. The reefs are remnants of the old shorelines, providing traces of the system of coral reefs that developed in the past around many of the tropical islands of Indonesia strung out to the east of Java and Bali.

The great, relatively recent movements of the Earth's crust have had a dramatic effect on the land surface in some parts of these islands in Southeast Asia. In the center of the island of Timor, for example, reefs formed less than a million years ago have been lifted by as much as 1,300 m (4,260 ft) at a rate of 13 cm (5 in) per hundred years. To both the west and the east the rate of uplift lessens; at Tanimbar Island, more than 400 km (250 mi) to the east, there are no traces of the reefs being raised.

A great barrier reef on the edge of the stable Sunda shelf provides further evidence of old shorelines. Most of the shelf is now submerged, but when sea levels were lower 30,000 years ago it was exposed. The reef runs for more than 500 km (310 mi) off the southeast coast of Borneo. It is believed to have grown from a fringing reef that developed along the exposed eastern edge of the shelf. It did not grow fast enough to keep pace with the rising sea level, so it is now much farther offshore and most of it is submerged.

LANDFORMS AND LANDSCAPES

The climate and the upheavals of the Earth's crust have created a patchwork of landscape types in Southeast Asia. It is a region with a variety of scenery, though similar landforms are found in widely separated parts of the region. For example, the spectacular limestone tower karst scenery of the Red river delta in northern Vietnam, northwest Malaysia and the adjacent Langkawi Islands can also be found in scattered parts of Borneo, in Cebu in the Philippines and among the Indonesian islands, especially Java.

The dominant influences in many areas are the frequent earthquakes and volcanic eruptions that happen along the arc from Burma through Indonesia to the Philippines. Individual volcanoes erupt, earth tremors trigger landslides, earthquakes cause seismic sea waves (tsunami) that can destroy beaches, and extremely heavy monsoon rains and typhoons cause disastrous flooding. Riverbeds are often strewn with boulders, and there are great banks of gravel and sand where they emerge from the mountains onto the edge of the coastal plains.

A region of rugged terrain

The volcanic areas have the most fertile soils of all Southeast Asia. In Java, Bali and much of the Philippines the lower slopes are intricately terraced for rice cultivation. Higher up the slopes, much of the original forest has been replaced with plantations. Elsewhere patches of forest remain, as on Sumatra and many of the smaller islands.

The sandstones and shales that were folded and tilted by the collision of the tectonic plates form another kind of rugged terrain, particularly extensive in Borneo. Typically a succession of deep river valleys is overlooked by precipitous slopes, sometimes separated from another set of hills and valleys by a gently undulating plateau or plain. The land is covered with tropical rainforest.

Older rocks form the backbone of the most stable highlands, such as those of the Malay Peninsula, whose succession of granite peaks continues through the islands of Singapore, Bangka and Belitung into western Borneo. Except where used for farming, as in the cool, damp Cameron Highlands in the center of the southern part of the peninsula, the ranges are generally still forested. The summits, often shrouded in mist, rise to 2,000 m (6,500 ft) or more. From the steep-sided slopes boulders are released by the weathering of blocks of granite. These roll down and accumulate in the narrow river channels, where they are gradually broken down into sand. Where the rivers begin to enter the lowlands their channels widen, and they meander from side to side, creating a broad, flat-floored valley where rice may be cultivated.

Long-eroded landscapes

In stable areas the land surface has been worn down over long periods, and the uplands are low plateaus or gently undulating hills and valleys. Chemical weathering has penetrated deeply into the rocks, and some of the soils formed on them have a hard layer where iron has accumulated (laterite or ferricrete).

West of the mountains of Vietnam these uplands comprise sandstones, in places overlain by basalt forced between the rocks during mountain-building about 20 million years ago, when the South China Sea was opening up.

The major low tableland in continental Southeast Asia is the southern plateau of southern Thailand. It resembles a tilted saucer, with a low, mountainous rim that separates it from the rest of the peninsula to the south and west. The plateau is one of the drier areas of Southeast Asia, with

The Irrawaddy floodplain The Irrawaddy flows south through central Burma and enters the sea through numerous river mouths. The waters are swollen both by mountain snowmelt and by the monsoons. The huge delta has been built up with river silts, producing fertile soil in which rice is widely grown.

less than 1,000 mm (25 in) of rain each year. Its sandy soils are thin and poor. Deciduous forests were once widespread, but with the help of irrigation rice is now cultivated. The plateau of southern Vietnam also has poor soils. The tablelands of Java are often limestone areas with dry surfaces, but there are water supplies below ground. The largest cave in the world, the Sarawak Chamber, is in a limestone area in northwest Borneo. It is 700 m (2,300 ft) long and an average 300 m (980 ft) wide and 70 m (230 ft) high.

Valleys, plains and deltas

In the upper reaches of the Irrawaddy, Chao Phraya and Mekong rivers, and on the eastern margin of the mountains of Sumatra, there are old flood plains and river terraces well above the level of modern river channels. Their soils are easily irrigated by diverting tributary streams. The dry plain of the Irrawaddy in Burma lies in a rain shadow, cut off by mountains from the southwest monsoon; the upper Chao Phraya plain of Thailand is much less dry.

There are two major types of landscape in the geologically more recent flood plains. On the deltas of the Irrawaddy, Chao Phraya, Mekong and Red rivers the land is intensively managed for rice cultivation. Along the coasts of eastern Sumatra, the Malay Peninsula and Borneo there are forests and mangrove swamps. The great deltas are built up of river silt that is often rich in nutrients, so the soil is very fertile. In the coastal swamps, on the other hand, decaying vegetation builds up peat deposits that even when drained are acidic and unsuitable for many crops without further treatment.

PLANTS AND THE LAND

Some habitats are so limiting that they can accommodate only a single plant species. In the lush unspoiled tropical rainforest that is the natural vegetation of Southeast Asia, however, there are over a hundred different plant species in a single hectare (2.5 acres). The diverse plant community supports a host of birds, animals, insects, fungi, bacteria and other microorganisms.

In northern Burma, Laos and Thailand the luxuriance of the subtropical and temperate vegetation is limited by cold at high altitudes and by drought in the dry season. These deciduous forests include teak, but in drier areas they may be replaced by scrub and bamboo.

At the opposite end of the region, in Timor, the length of the dry season and the island's closeness to the plant species of Australia result in a dry tropical flora of eucalyptus and gum trees. Nearby, in western Sumbawa, the dry lowlands up to 800 m (2,600 ft) were originally covered with deciduous forest. Above 800 m the higher rainfall supports evergreen rainforest. On the dry summits of the mountains at 1,500 m (4,900 ft) this becomes stunted montane forest. Here the effects of seasonal drought and altitude can result in contrasting vegetation on different sides of the same mountain.

Individual rock types have developed characteristic vegetation. A special heath forest (kerangas) grows on the sandy beach ridges of the eastern Malay Peninsula. Behind the ridges, river outlets may be blocked and create waterlogged areas that are occupied by freshwater swamp forest. At the mouths of major rivers and along the coasts of eastern Sumatra and southern Borneo there are great mangrove swamps, the tropical equivalent of the salt marshes of temperate zones. Eventually a mangrove swamp may become dry and high enough to develop into coastal swamp forest.

Southeast Asia's sliding slopes

Landslides play an important part in shaping the landscapes of Southeast Asia. Landslides are triggered wherever there is enough energy to set loose material moving down steep slopes. Such energy is abundant in Southeast Asia in the form of earthquakes and prolonged torrential rain, and is often found in combination with steep mountain terrain.

Much of Southeast Asia lies in a zone where different plates of the Earth's crust have collided in geologically recent times and continue to do so, raising the land and creating steep slopes. Deep chemical weathering under the hot and wet conditions has provided a mantle of rotted rock several meters deep. Torrential rainfall is common both in the equatorial belt and in areas subject to typhoons, such as the Philippines. The slopes are already unstable, and once the soil and rotted rock have become saturated with water there may be so much pressure caused by the water in the tiny spaces between individual rock particles that they start to slide over each other. When this happens, part or all of the hillside begins to move downslope as a landslide.

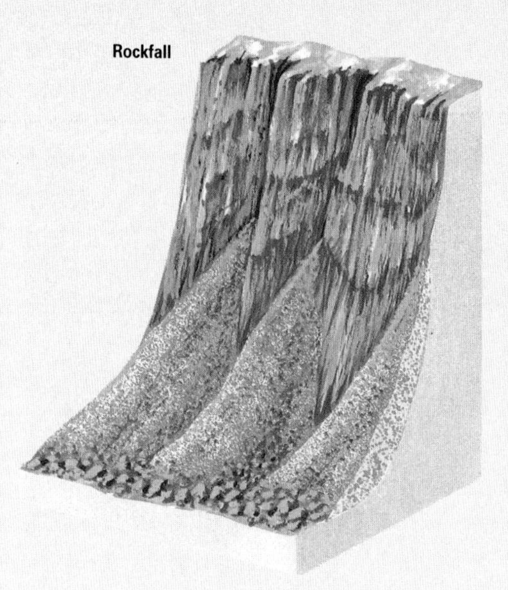

Forested slopes in mountainous Sabah, northeast Borneo. Where vegetation cover is removed by logging or during road or building construction, the soil is exposed to erosion and the risk of landslides. On the steep slopes of Southeast Asia torrential rain results in many landslides. Volcanic eruptions and earthquakes also trigger landslides.

Rockfalls and rockslides are the fastest types of mass movement – the downslope shiftings of weathered material. A rockfall (*left*) is a free fall of fragments on a mountainside or other precipitous slope. In high mountains, fragments accumulate as scree at the foot of slopes. In a rockslide (*below*), masses of rock slip down a slope. The term avalanche is used to cover both types of mass movement. Slides cause more damage to people than rock falls do because they take place on lower slopes, which are closer to human settlements.

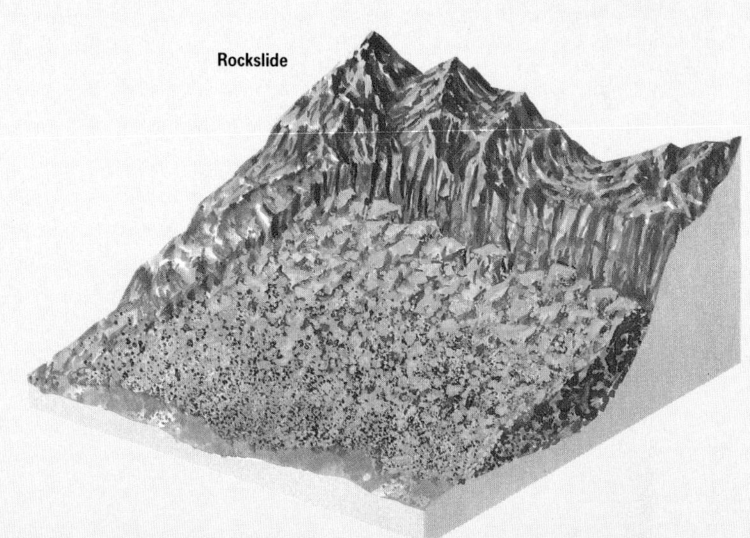

Causes and effects

A landslide can be set in motion and accelerated by vibration of the ground, as occurs in an earthquake. In Indonesia, especially around the large islands of Sumatra and Java, relatively few earthquakes happen on land compared with the frequent quakes that originate in the ocean trench offshore to the southwest. Nevertheless, most of the economic damage and social disruption on land is caused by earthquakes. One earthquake in 1962, in the steep-sided mountains near the city of Padang in central west Sumatra, caused numerous landslides and much damage to buildings and other structures. It also lowered the floor of a lake so places that were previously above the shoreline became submerged beneath 10 m (33 ft) of water.

Most large Indonesian cities have been built on relatively stable ground away from the main earthquake zone. Batavia, on the site of Djakarta in northwest Java, was devastated in 1699 by an earthquake that caused landslides in the catchment area of the Liwung river; great masses of mud and uprooted trees hurtled down into the river, causing widespread flooding and silting up of channels downstream in the city.

From the offshore islands and oceanic Java trench southwest of Sumatra and Java, the active zone of earthquakes passes through Flores in an arc east through Sulawesi to the Philippines. Earthquakes that trigger landslides are particularly strong close to the Philippine trench of the Pacific and in eastern Luzon. Weaker earthquakes occur in western Mindanao and southwestern Luzon.

Many landslides are also triggered by volcanic eruptions. Mount Mayon on Luzon is one of the most active volcanoes of Southeast Asia, with a peak period of 28 major eruptions between 1814 and 1928. Some of the landslides that develop form mud-and-lava mixtures (lahars) and pour into rivers, fill the channels and cause widespread flooding. The 1963 eruption of Goenoeng Agoeng, the sacred mountain of northeast Bali, started flows of mud, stones and boulders that buried the houses of a nearby village in up to 3 m (10 ft) of debris.

In the Philippines the effects of earthquakes and volcanic eruptions are often aggravated by heavy rain. Other landslides are associated simply with heavy rain on unstable slopes. Major typhoons, such as those that struck central Vietnam in 1953 and Manila in 1967, cause extensive landslides, as do heavy rains in equatorial areas, especially where hill slopes have been modified by cuttings for roads or excavation for building construction. One or two major storms may cause so many landslides and such high river flows that more eroded rock material is carried away by them than by all the rest of the rains during the year.

MAJOR LANDSLIDE DISASTERS

Torrential rainfall often triggers landslides, in Southeast Asia as elsewhere, but many are triggered by events some distance away. In 1920 earthquakes started the catastrophic landslides in the loess region of northern China that cost some 200,000 lives. Volcanic eruptions caused the landslides and mudflows of 1970 and 1985 in the Andes mountains of South America.

Year	Location	Deaths
1963	Italy	2,500
1966–67	Brazil	2,700
1970	Peru	21,000
1971	Peru	600
1972	Hong Kong	300
1972	United States	400
1981	Indonesia	500
1983	China	277
1985	Colombia	26,000
1985	Philippines	300

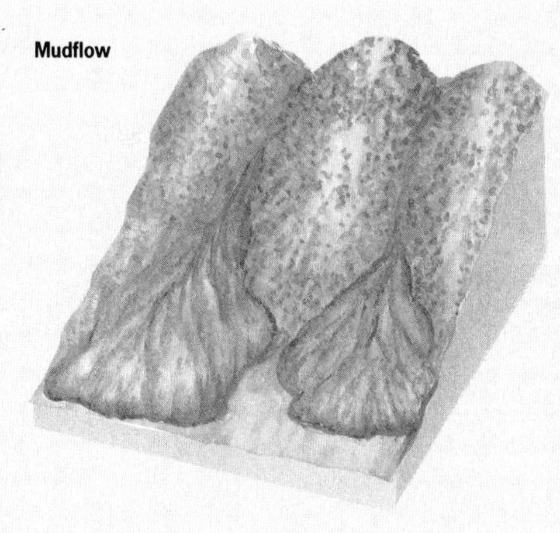

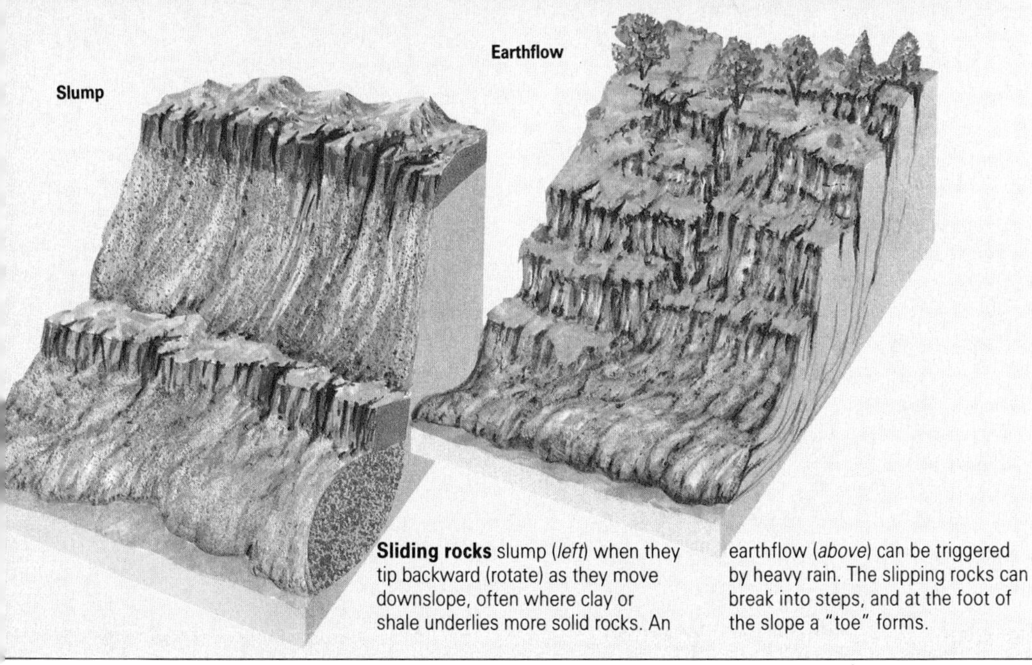

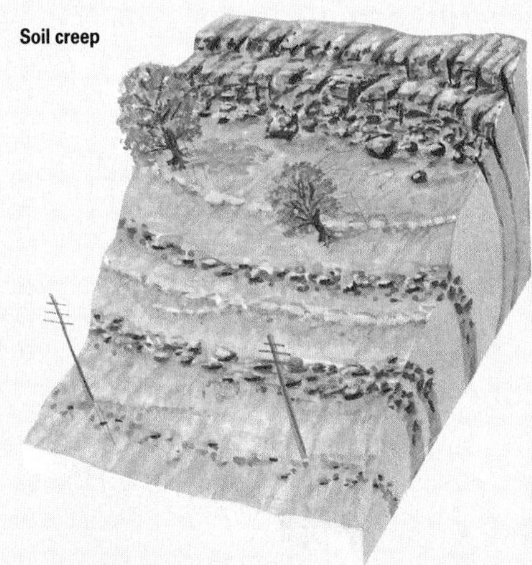

Sliding rocks slump (*left*) when they tip backward (rotate) as they move downslope, often where clay or shale underlies more solid rocks. An earthflow (*above*) can be triggered by heavy rain. The slipping rocks can break into steps, and at the foot of the slope a "toe" forms.

In a mudflow (*top*) the saturated earth flows down even a gentle slope. The consequences can be catastrophic. Soil creep (*below*) is much slower; imperceptible to the eye, it is revealed by displaced rock layers, trees, soil and walls.

A HAZARDOUS ENVIRONMENT

A LAND OF MOUNTAINS · A MONSOON CLIMATE · ISLANDS AND COASTS

A wilderness of scenic mountains that is interspersed by small, intensively used lowlands is a landscape common to the Japanese archipelago and the peninsula of Korea. Nearly 4,000 islands make up the archipelago, while the Korean peninsula projects some 1,000 km (620 mi) south from China toward the Japanese island of Kyushu. Japan and Korea lie in a frontier zone between Eurasia, the world's largest landmass, and the Pacific, the world's largest ocean. They have a monsoon climate of hot, humid summers and cold winters. Typhoons and heavy flooding regularly strike the region. Korea is an old and stable land; most of Japan's island chains mark the boundaries between plates of the Earth's crust. Japan still lives with the threats posed by frequent earthquakes, volcanic eruptions and seismic sea waves.

COUNTRIES IN THE REGION

Japan, North Korea, South Korea

LAND

Area 601,241 sq km (232,141 sq mi)
Highest point Mount Fuji, 3,776 m (12,388 ft)
Lowest point sea level
Major features mountains of moderate height cover most of Japan and Korea; over 3,900 islands in Japan

WATER

Longest river Yalu, 810 km (503 mi)
Largest basin Yalu, 63,000 sq km (24,000 sq mi)
Highest average flow Yalu, 526 cu m/sec (19,000 cu ft/sec)
Largest lake Biwa, 695 sq km (268 sq mi)

CLIMATE

	Temperature °C (°F) January	July	Altitude m (ft)
Wonsan	−4 (25)	23 (73)	37 (121)
Seoul	−5 (23)	24 (75)	86 (282)
Sapporo	−6 (21)	20 (68)	18 (59)
Niigata	2 (36)	24 (75)	4 (13)
Tokyo	4 (39)	25 (77)	6 (20)

	Precipitation mm (in) January	July	Year
Wonsan	29 (1.1)	273 (10.7)	1,310 (51.6)
Seoul	17 (0.7)	358 (14.1)	1,258 (49.5)
Sapporo	111 (4.4)	100 (3.9)	1,136 (44.7)
Niigata	194 (7.6)	193 (7.6)	1,841 (72.5)
Tokyo	48 (1.9)	146 (5.8)	1,563 (61.5)

NATURAL HAZARDS

Earthquakes and associated sea waves (tsunami) in Japan, floods, typhoons, landslides

A LAND OF MOUNTAINS

Japan and Korea are dominated by mountains. More than three-quarters of the land is above 250 m (820 ft) and has slopes steeper than 15 degrees. In Japan, most of the highest peaks are volcanic in origin. They are found in the mountains of central Honshu, where the most famous and highest mountain, Mount Fuji, last erupted in 1707–8. Volcanic deposits cover a quarter of Japan, whose 60 active volcanoes account for over one-tenth of the world's total. The volcanic areas also produce many hot springs. Glaciation has not had much effect on the scenery except in central Honshu and Hokkaido.

The rest of Japan's mountains are great slabs of lifted terrain, rarely above 1,500 m (4,900 ft). They have been folded and faulted under the stress of great earth movements, and deeply cut by rivers to form V-shaped valleys separated by narrow ridges.

In Korea also the mountains rarely rise above 1,500 m (4,900 ft). They are the product of a much older period of mountain-building and are part of the raised rim of the Eurasian continent. The peninsula is divided from the Asian mainland by the Nangnim Mountains and the Chang-pai range on the border with China. They contain most of Korea's peaks. The highest peak, Mount Paektu (2,744 m/ 9,003 ft), is an extinct volcano. The lower Taebaek range runs south down the length of the peninsula, close to the east coast.

The mountains of Korea and Japan are clad with forests that soften the underlying ruggedness of the terrain. In Japan forest cover changes from subtropical evergreen hardwoods, including cedar, laurel and myrtle in the south, through deciduous hardwoods such as ash, beech and oak, to coniferous softwoods, chiefly fir and spruce, in the north.

Map of physical zones Japan and Korea are comparable in their size, mountainous character and position extending north–south beside the Sea of Japan. Korea belongs to the Asian landmass, while Japan lies in a zone of dramatic tectonic disturbance in the Earth's crust.

Lowlands and intermediate zones

The lowlands account for about one-eighth of Japan and one-fifth of Korea. The typical lowland is a plain formed of sediments deposited by streams and rivers situated in a coastal bay, inlet or a basin between mountains.

Because the mountains frequently extend down to the sea, most lowlands are small and there is no continuous lowland belt, particularly in Japan. The largest lowland in Japan is the Kanto Plain in eastern Honshu, covering an area of 7,000 sq km (2,700 sq mi); it accommodates the huge metropolitan area of Tokyo. In Korea the most extensive plains are in the southwest.

A third type of landscape is the zone of gravel and sand deposits laid down, in the form of a fan or cone, by each turbulent mountain stream as it leaves its upland valley. The gravel is deposited first, at the edge of the upland area, the sand where the fan merges into the lowlands. Often the streams are so close together that their fans form a continuous

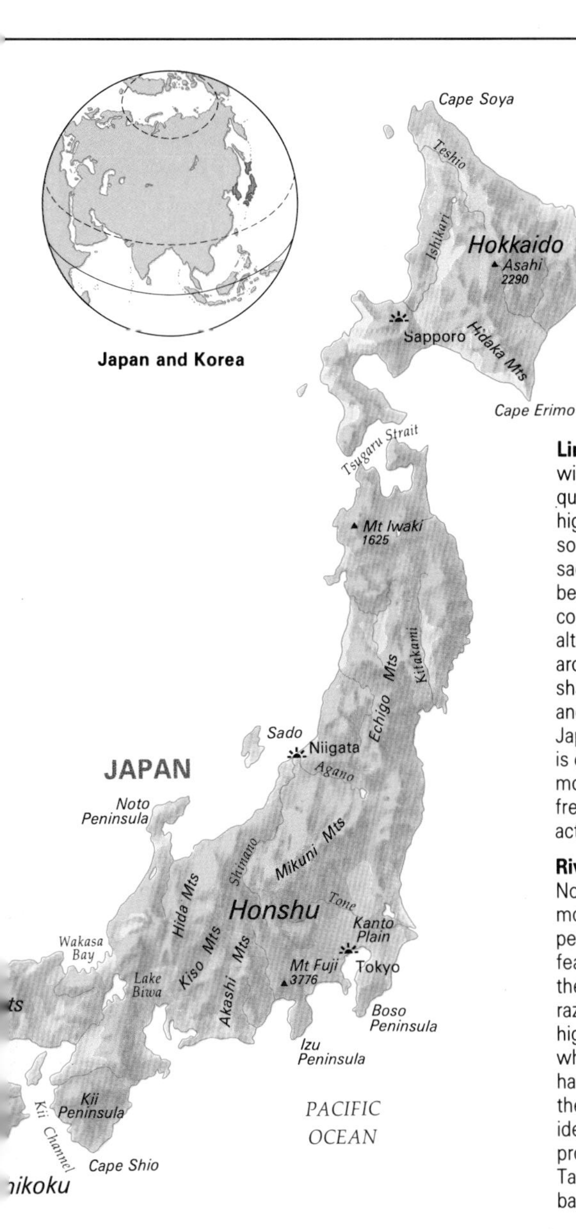

Linking heaven and earth High winds carry snow away from the quiescent volcanic cone of Japan's highest mountain, Mount Fuji in south-central Honshu. Japan's sacred mountain is one of the most beautiful examples in the world of a composite volcanic cone built up of alternating layers of ash and lava around its central vent. Its classic shape has been a source of spiritual and artistic inspiration to the Japanese over the centuries. Japan is constantly affected by earth movements, accompanied by frequent earthquakes and volcanic activity.

River deep, mountain high North and South Korea are largely mountainous countries although no peaks exceed 3,000 m (10,000 ft). A feature of mountain landscapes in the northeast of South Korea are the razor-back ridges shown here. The highest mountains are in North Korea where short, swift-flowing rivers have eroded deep valleys through the ancient igneous rocks, creating ideal locations for damming rivers to provide hydro-electric power. The Taebaek range in the east forms the backbone of the peninsula.

slope at the foot of the hills. On the Kanto Plain tongues of such flood (diluvial) deposits extend over large areas.

Nowhere in Japan or Korea is far from the sea, so rivers are short. By far the longest is the Yalu, much of whose 800 km (500 mi) course provides part of the border between North Korea and the Manchuria region of China; Korea's second longest river, the Han, is nearly half that length. In Japan the longest river, the Shinano, is just 367 km (228 mi) long; the Tone, which drains much of the Kanto Plain, flows for 322 km (200 mi). Both rivers are on the island of Honshu.

The rivers have steep gradients, particularly in mountainous Japan, and in full spate they have great power to erode. The valleys between the mountains are deeply cut; in the middle zone the river courses, choked with gravel and sand, are braided into numerous channels; on the plains, except during periods of flood, the rivers are single channels. Flooding along the lower courses causes much damage.

A MONSOON CLIMATE

Most of Japan and all of Korea are located in the middle latitudes between 30° and 45°N on the eastern edge of the Eurasian landmass. During the course of a year, a zone of conflict between cold polar and warm tropical air masses passes twice over the region, establishing a pattern of six seasons.

From polar to tropical climates

The mechanism behind the seasonal pattern is the monsoon, a twice-yearly reversal of pressure and wind systems. In winter the Eurasian interior is dominated by high pressure; as a result cold northwesterly blasts of polar continental air push out toward the Pacific over Korea and Japan. As the days get longer the continental landmass warms up. The high-pressure system gives way to low pressure, a reversal that results in southeasterly winds of tropical maritime air.

Spring is a relatively long season of unsettled weather from late February to mid-June. There is a short rainy season (the *baiu*) from mid-June to mid-July, when the unsettled front between polar continental and tropical maritime air masses passes northwest over Japan; the conflict yields great amounts of rain. High summer, with its hot, humid weather, lasts until early September, when the frontal zone retreats southeast back across Japan, giving rise to another short rainy season (the *shurin*). The transition back to the dominant winter monsoon is quite rapid, so the fall is relatively short (mid-October to mid-November).

In January, when most of Hokkaido in the north has an average temperature of −6°C (21.2°F), southern Kyushu enjoys an average temperature of 6°C (42.8°F). The January average temperature range in peninsular Korea, from −19°C to 6°C (−2.2° to 42.8°F), occurs over a north–south range of only nine degrees. The abruptness of this temperature change is

STORMS FROM THE PACIFIC

Typhoons are a major natural hazard of the northwest Pacific. These revolving storms of the tropics bring very high winds and torrential rainfall. They strike principally between July and November and disrupt the normal seasonal weather pattern. The number in any one year varies between 3 and 30; they are more frequent in Japan than Korea.

Typhoons originate out over the ocean as small cells of low pressure that develop into upward-spiraling wind patterns, fueled by the energy released by warm, moist, tropical sea air. Typhoons reaching Japan curve first northwest and then northeast in front of prevailing winds, before dissipating their energy over colder sea or land.

The winds sometimes reach speeds up to 200 km/h (125 mph) and can cause enormous damage, destroying houses, bringing down power lines, flattening the ripening rice, and downing tree crops. Heavy rain, often with over 300 mm (12 in) falling in 24 hours, causes flooding and landslides. Flooding in coastal areas is also caused by the huge waves that are whipped up by the wind, particularly when they are funneled into deep inlets and bays.

CHINA
JAPAN
July
August
September
October
November
June
May
April
February
December
40°
30°
20°
10°
120°
130°
140°
Typhoon starts as a cell of low-pressure air over the tropical Pacific Ocean

Typhoon paths Typhoons originate over the warm waters of the southwest Pacific in unstable atmospheric conditions. They are moved to the west by the trade winds but curve to the right, following northwesterly and then northeasterly tracks before they dying out over the land or the cooler waters of the north. The western and southern coasts of Kyushu suffer most typhoon damage.

Precious coastal lands About four-fifths of Japan is mountainous. Flat land is highly populated and intensively cultivated. These paddies are overlooked by Aso, Kyushu's highest peak. Aso is one of many active volcanoes in Japan and has the largest crater in the world, 24 km (15 mi) long and 16 km (10 mi) wide.

unequaled anywhere else in the world.

Winters are distinctly more severe in Korea, and temperature contrasts are much less marked in summer. In August the north–south difference at sea level is 6°C (11°F) in Japan but only 2°C (4°F) in Korea. Hokkaido experiences a cool-summer continental climate, northern Honshu and most of Korea a warm-summer continental climate, and the rest of Japan and the southern extremity of Korea a humid subtropical climate.

Rainfall and sunshine

Annual precipitation in Korea ranges from 600 mm (24 in) in the northeast to more than 1,500 mm (60 in) in the south. In Japan there are three main areas where precipitation is above average (more than 2,000 mm/80 in): the Pacific side south of Tokyo, with its high mountains facing the summer onshore winds; the Sea of Japan side north of about the same latitude, and the mountains of central Honshu.

Over the eastern half of Hokkaido, eastern Honshu north of Tokyo, and the coastlands of the Inland Sea between Honshu, Shikoku and Kyushu there is relatively little rain (less than 1,000 mm, 40 in). Ocean currents influence rainfall. The cold Oyashio Current along the northern Pacific coast diminishes rainfall but elsewhere around the shores of Japan and in southernmost Korea the warm Kuroshio Current from the south encourages precipitation. Rainfall is heavier in summer, when it is generally greater on the eastern sides of both Japan and Korea due to the southeast monsoon.

In winter the pattern is reversed in Japan, but remains broadly the same in Korea. Northwesterly winds from Asia pick up moisture as they move across the Sea of Japan and its branch of the warm Kuroshio Current; they then rise up over the mountains, where most of the moisture condenses and falls as snow. On the Pacific side of the mountains the winter weather is altogether different. The winds, having lost their moisture and strength, give clear, sunny days and frosty nights. Tokyo receives an average of 189 hours of sunshine in January, compared with only 56 hours at Niigata on the Sea of Japan coast, barely 250 km (155 mi) away, where rainfall is up to four times as heavy.

In most parts of Japan and Korea water is a major factor in shaping the landscape. On exposed peaks the freezing and thawing of water in winter helps to break up rocks. Water lubricates the movement of weathered rock downhill, often in the form of landslides. Rivers and streams erode their banks, carrying debris from uplands to lowlands. During winter the strong onshore northwesterly winds help build up extensive sand dunes along the western coasts of Japan.

ISLANDS AND COASTS

Japan is both an island nation and a land of islands. The four main islands of the 3,900-island chain – Honshu, Hokkaido, Kyushu and Shikoku – account for 98 percent of the total land area.

Island arcs

The configuration of the islands traces a number of quite clearly defined arcs. The Honshu arc provides the crescent-shaped backbone for the largest island, and to it are linked at least five other arcs. To the north, the former Japanese islands of Sakhalin and the Kurils mark two arcs that converge on Hokkaido; Shikoku has its own arc; the Ryukyu Islands arc runs from Kyushu to the southwest through Okinawa to Taiwan, while the Bonin arc defines the alignment for the Izu and Ogasawara islands, which extend due south of Tokyo into the Pacific.

These arcs coincide with the margins of plates of the Earth's crust; the Bonin arc, for example, follows the the boundary between the Pacific and Philippine plates. Many of the islands have been produced by volcanoes. Some remain active, such as the volcano on Miyake, an island of the Izu Peninsula, which suddenly erupted on 4 October 1983, causing the 4,500 inhabitants to be evacuated.

Rising and sinking coasts

Japan has a very long coastline relative to its area: for every 22 sq km (8.5 sq mi) of land there is 1 km (0.6 mi) of coastline. (For comparison the ratio in the British Isles, for example, is 31 to 1.) The shape of the coastline reflects the rock structure, faulting and changes in sea level. Over thousands of years the action of the sea tends to reduce irregularities by infilling bays and cutting back headlands, creating great sweeps of smooth coastline.

The Pacific coast of Honshu north of Tokyo is mainly a rising coastline. Raised beaches are widespread. These range in height from a little above the present high-tide line to over 300 m (1,000 ft) above present sea level. On the coasts of eastern Hokkaido and those edging the mountains of northeastern Honshu the land has subsided, and the rising sea level has accentuated the indentations of the coast in these areas.

Most of the coastline of the southwest, including much of Kyushu and the Inland

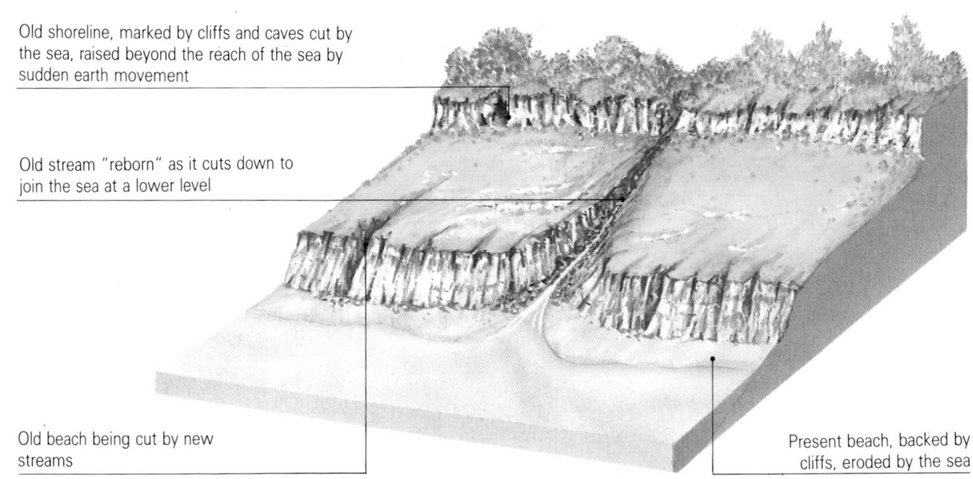

Fossil coasts The relationship between the level of the land and that of the sea rarely remains static for long. In Japan, raising of the land has meant that old coastlines and their features are removed from the sea that created them. The shore becomes a stranded wave-cut platform, and cliffs and caves formed by the sea are abandoned. They gradually become less steep through erosion and because there is no further undercutting by the sea. Meanwhile the sea starts the whole process again at a lower level.

RECLAIMING THE WETLANDS

There are few lowlands in Japan and Korea today that have not, at least in part, been reclaimed at least in part from the water. Reclamation of coastal and riverine wetlands has a long history in the region, possibly dating back to the 9th century.

In many cases reclamation has simply involved banking up rivers and draining adjacent marshes; in some other instances lakes and ponds have been drained. In recent years some 16,000 ha (40,000 acres) of land have been recovered from Japan's second largest lake. Coastal lagoons, tidal marshes, bayheads and creeks have also been reclaimed. Until the 20th century land was drained mainly to supplement the meager amount suitable for agriculture. Today the driving force is urbanization. Nowhere in the region is that pressure greater than around Tokyo Bay, where it considerably modifies the shoreline.

An estimated 110,000 ha (270,000 acres) of new land have been created, largely by infilling, in order to provide additional space, not just for Tokyo but also for the other sizeable cities that together make up a huge metropolitan area accommodating some 26 million people. The land has been put to a variety of uses, including new port installations, heavy industrial development, housing, commercial growth and airport expansion.

The Inland Sea is a narrow arm of the Pacific bounded by Shikoku and Kyushu islands, and at its widest is 65 km (40 mi) across. It occupies land that dropped and became flooded by the sea. Hundreds of small islands are all that can now be seen of the hills.

Sea, is subsiding and has a highly indented coastline. The whole of the Inland Sea has subsided; rocky inlets and rectangular bays bounded by faults have developed, and former hilltops now jut above sea level as islands. Sediments from the land are being deposited in many of the bays and inlets, leading to the formation of marshes. Elsewhere the structural grain of the land runs at right angles to the coast, so rising sea levels have produced long inlets and promontories, including harbors and peninsulas on the western coasts of both Kyushu and Shikoku.

Along most of the Sea of Japan coast of Honshu the land has risen. Along the Pacific coast from Tokyo to the Inland Sea it has subsided. Faults in the rocks cut across the trend of the coast, which is broken into a series of impressive bays (from east to west, Tokyo, Sagami, Suruga and Ise) and promontories (Boso, Miura and Izu). In contrast, long stretches of the Japan Sea coast are elevated, and are backed by wide belts of beach ridges and dunes marking the transition to relatively wide coastal plains. Here the forces of sea, wind and river continue to work together to smooth the coastline by infilling bays or depositing bars of sand, shingle or mud behind which lagoons have formed.

Elsewhere in Honshu the coast is generally smooth because it runs parallel to a fault system. Only occasionally, as at Wakasa Bay on the west coast, do the faults cut across the coastline. The Noto peninsula, with its indented coast, is another exception. The large island of Sado, a little farther to the northeast, may once have been the outer end of a similar promontory that has since been cut by the eroding action of the sea.

East–west contrasts in Korea

The Korean east coast is monotonous, smooth and precipitous, while the south and west coasts are indented and offer plenty of good natural harbors. On the western or Yellow Sea side of the peninsula there are also many islands (over 2,000), but unlike the islands of Japan none is of volcanic origin. The only three volcanic islands in Korea lie off the other coasts: Cheju to the south and the smaller and more isolated Ullung and Tok out in the Sea of Japan.

Earthquakes: part of everyday life

On 1 September 1923 Japan experienced its most devastating earthquake. In a matter of minutes 100,000 people were dead and over 300,000 buildings had been destroyed. So great was the force of the earthquake, centered on Sagami Bay 90 km (56 mi) to the south of Tokyo in southeastern Honshu, that the floor of the bay split. The south coast was lifted by nearly 2 m (6.6 ft) and the north coast was correspondingly depressed. In places the land moved horizontally up to 4 m (13 ft) either side of a fault. Even in distant Tokyo the land was raised 10 cm (4 in) and moved horizontally 20 cm (8 in).

The eventual toll of 142,807 included many deaths caused not by falling buildings but by the fires started by overturned stoves. Giant waves (tsunami), generated by the vibrations of the earthquake rocking the sea floor, washed ships inland and then swept people and houses out to sea.

Five thousand tremors a year

The Great Kanto Earthquake, as it is known, was an extreme event. However, in Japan earthquakes are a feature of everyday life. In any one year there may be as many as 5,000 earthquake tremors in Japan, though most go unnoticed. The city of Gifu in central Honshu holds the record for earthquakes detectable without instruments (those that measure 2 or above on the Richter scale): 516 in one year. Tokyo averages about 150.

Earthquakes are a way of releasing the stress energy that is built up mainly as the plates of the Earth's crust converge or slide past each other. Earth movements raise the land surface in Japan about 4.5 m (15 ft) every thousand years, compared with, for example, about 7.5 m (25 ft) in California on the west coast of the United States. (A similar estimate made for the Himalayas, where the Indian subcontinent meets the landmass of Eurasia, is 2m (6.6 ft). The history of quake activity is written into the landscapes of Japan, where there are faults visible on slopes, including coasts, and river courses have been changed.

There are two types of earthquake in Japan: those that take place where the Pacific and Philippine plates dip beneath the Eurasian plate (the subduction zone), and those at the rigid continental margin of Eurasia and along transform faults, where plates slide past each other without crust being built or destroyed. Quakes of the second type are experienced almost everywhere in Japan and are the more frequent. The shockwaves start at a point (the focus) close to the surface. The earthquakes affect both geological structures and landscape features such as hill slopes and sea cliffs. Because of their violence they can cause great havoc.

Tsunami and landslides

Earthquakes in the subduction zone are large scale and have deeper foci. Their strength is often so great that they cause considerable disturbances in the Earth's crust. Because they tend to take place offshore, they generate tsunami. These menacing seismic sea waves travel at great speed (160–1,100 k/h, 100–700 mph). They are about 1 m (3.3 ft) high over deep water, but greatly increase in height in shallow waters, and can momentarily raise the sea level in bays or estuaries by as much as 30 m (100 ft).

The destructive potential of the tsunami is increased by the fact that so many Japanese live and work in low-lying areas near the coast. The plight of some vulnerable areas is increased by land subsidence. This is caused both by water being pumped up from artesian wells beneath the surface and by the sheer weight of the built-up area compacting the ground.

Earthquake shock waves can shake loose entire mountainsides. Landslides can move by as much as 4 km (2.5 mi) downslope; where slopes are steep and the material is lubricated by a high water content, they can move very fast, with devastating effects.

It is easier to predict where a quake may strike than when. Measurements – of movement in rocks, strength and frequency of minor earthquakes, ground level and tilting, water level in wells and in the sea, magnetic fields, and quantities of radon gas – may help. However, some earthquake zones, such as those off the coast of Japan, lie in the ocean abyss, and are much more difficult to monitor. Many earthquakes take place without warning; even when there are warning signs it is difficult to fit them into a pattern that helps prediction. Scientists in other countries have joined Japan's quest for a way to predict earthquakes, but a universally applicable method of doing so has still to be developed.

KILLER EARTHQUAKES

In Japan's Great Kanto Earthquake of 1923 many people were victims of sea waves (tsunami) or fires associated with the event; landslides and famine swell the death toll of other quakes. On the magnitude scale devised by Charles Richter in 1935, a difference of one denotes a tenfold difference in energy released: an earthquake of magnitude 8 is 10 times more powerful than one of magnitude 7.

Year and location	Magnitude	Deaths
1976 Tangshan, China	8.2	240,000
1920 Gansu, China	8.6	200,000
1927 Tien Shan, China	8.3	200,000
1923 Tokyo, Japan	8.3	142,000
1908 Messina, Italy	7.5	85,000
1932 Gansu, China	7.6	70,000
1970 Northern Peru	7.7	70,000
1955 Quetta, Pakistan	7.5	50,000
1939 Erzincan, Turkey	7.9	40,000
1988 Armenia, USSR	6.9	25,000
1960 Agadir, Morocco	5.8	10,000

Siting, design and materials used in buildings can make the difference between life and death in areas that are affected by earthquakes. Shock waves may cause the disintegration of load-bearing pillars and walls. Where settlements are built on relatively recent deposits of sedimentary materials, such as clay, a major threat is soil liquefaction triggered by an earthquake after heavy rain. Earthquake intensity is the degree to which a quake is felt by people and causes damage to buildings. Buildings on soft soil suffer more damage than those built on rock, so intensity may vary at identical distances from the epicenter. Earthquake intensity is measured on the Mercalli scale, while the Richter scale is used to measure the power of the stress energy released.

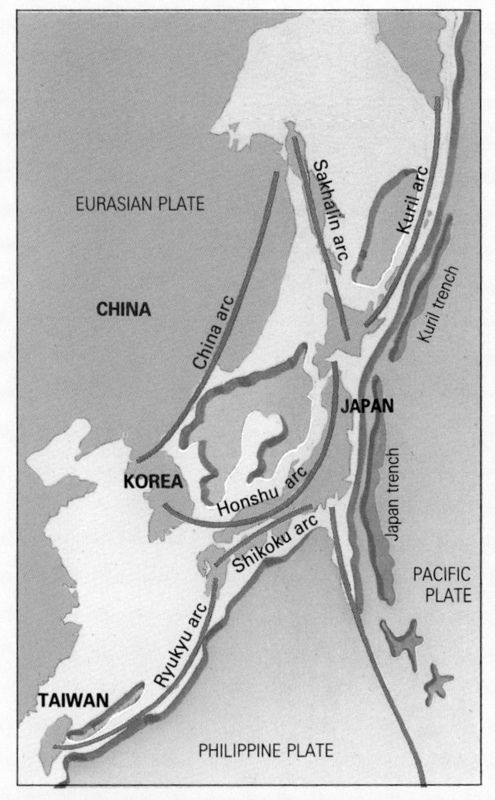

Volcanic island arcs and ocean trenches show where the Philippine and Pacific plates of the Earth's crust dip below the Eurasian plate. They are part of the "Ring of Fire", a belt of intense tectonic activity that encircles the Pacific.

Quake-triggered landslide A hillside, house and road sheared off in a landslide caused by an earthquake at Otaki village in the mountains of southeast Hokkaido.

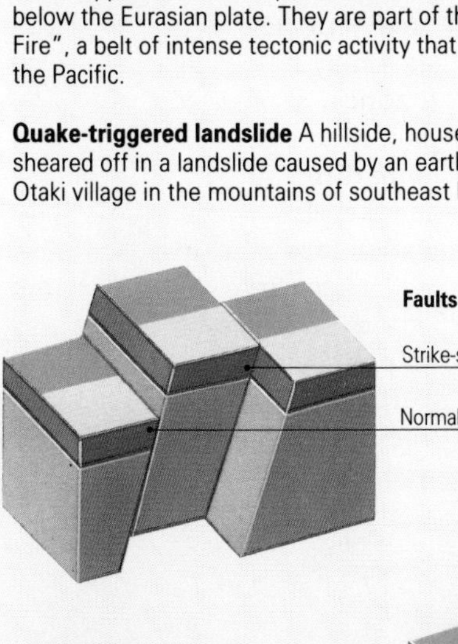

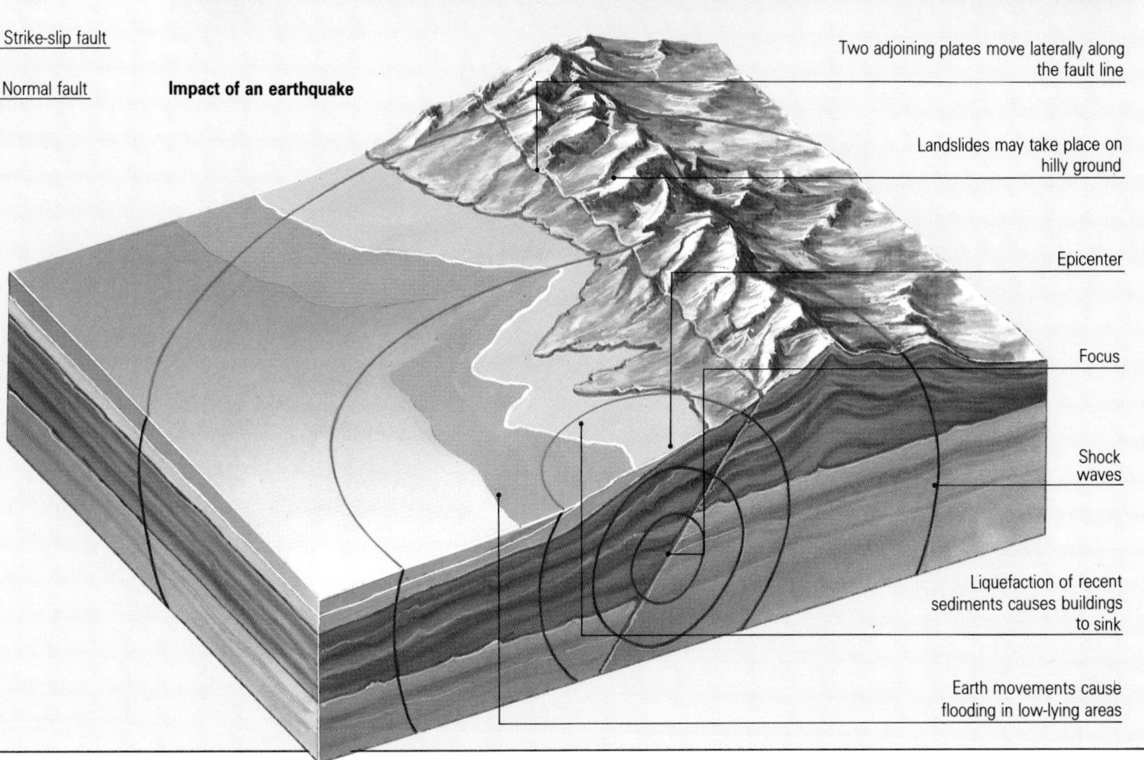

Faults and earthquakes spread out from the focus of an earthquake. Here stress energy is suddenly released as blocks of rock, deformed when portions of the Earth's crust converge or slide past each other, move to a new position. The shock waves trigger landslides on steep slopes and cause water- saturated sediments to liquefy. One block of rock can move several meters in relation to its neighbor, wreaking havoc that may include flooding. In a "normal" fault *(above)* stretching rocks break vertically along a steep fault plane. A strike-slip or tear fault involves horizontal shearing along a vertical plane.

LANDS OF THE SOUTHERN SEAS

ENVIRONMENTS OF THE ANTIPODES · AUSTRALASIAN LANDSCAPES · ANTARCTICA

Originally all part of an ancient supercontinent, Gondwanaland, Australia, New Zealand, Antarctica and the southwest Pacific islands have moved slowly away from each other over millions of years. Volcanic activity is a feature of the Pacific "Ring of Fire", and generates the steaming lakes, geysers and bubbling mud pools of North Island, New Zealand. Spectacular coral reefs and atolls have developed in the warm seas of the Pacific Ocean. They include Australia's Great Barrier Reef. Climates in the region range from equatorial in New Guinea through wet tropical in northern Australia to the arid and semiarid desert interior that forms a major part of an ancient plateau. Antarctica in the south is the coldest place on Earth. It is also the driest continent, though it holds almost nine-tenths of the planet's ice.

COUNTRIES IN THE REGION

Antarctica, Australia, Fiji, Kiribati, Nauru, New Zealand, Papua New Guinea, Solomon Islands, Tonga, Tuvalu, Vanuatu, Western Samoa

LAND

Area 22,748,000 sq km (8,783,000 sq mi)
Highest points Vinson Massif, Antarctica, 5,140 m (16,863 ft), Mount Cook, New Zealand, 3,764 m (12,349 ft)
Lowest point Lake Eyre, Australia, –16 m (–52 ft)
Major features Australia, 7,686,848 sq km (2,967,909 sq mi) is the world's lowest continent, Antarctica, 14,245,000 sq km (5,500,000 sq mi) the highest and coldest with the greatest ice sheet; also New Guinea (part), New Zealand, Pacific islands

WATER

Longest river Murray–Darling, 3,780 km (2,330 mi)
Largest basin Murray–Darling, 1,072,000 sq km (414,100 sq mi)
Highest average flow Murray, 400 cu m/sec (14,000 cu ft/sec)
Largest lake Eyre, 8,900 sq km (3,400 sq mi) when flooded

CLIMATE

	Temperature °C (°F) January	July	Altitude m (ft)
Darwin	28 (82)	25 (77)	27 (89)
Perth	23 (73)	13 (55)	60 (197)
Alice Springs	28 (82)	12 (54)	548 (1,797)
Auckland	20 (68)	11 (52)	38 (125)
Sydney	22 (72)	12 (54)	42 (138)

	Precipitation mm (in) January	July	Year
Darwin	409 (16.1)	1 (0.04)	1,661 (65.4)
Perth	8 (0.3)	183 (7.2)	873 (34.4)
Alice Springs	39 (1.5)	10 (0.4)	252 (9.9)
Auckland	46 (1.8)	139 (5.5)	1,323 (52.1)
Sydney	102 (4.0)	101 (4.0)	1,214 (47.8)

World's lowest recorded temperature, –89.2°C (–128.6°F), Vostok, Antarctica

NATURAL HAZARDS

Drought, storms and flooding, bush fires, extreme cold in Antarctica

ENVIRONMENTS OF THE ANTIPODES

Antipodean environments are many and varied. They include the high equatorial mountains of New Guinea, the cold desert of Antarctica, thousands of islands – small tropical volcanic islands and coral atolls in the Pacific Ocean, and the sub-polar islands of the Southern Ocean – and the ancient island continent of Australia.

The Australian continent

Deserts and plains form the largest natural environment in Australia. The vast desert interior, fringed by semiarid plains of scrub and grassland, stretches from the Gulf of Carpentaria to the Great Australian Bight, and from the eastern mountains to the west coast. Rainfall from the monsoons in the north fails to reach inland, and in the east the mountains shield the interior from the Pacific winds, so it is very dry.

Temperatures in the interior average more than 30°C (86°F) in January, with extremes of 39°C (102°F). Daily maximum temperatures of over 38°C (100°F) have been recorded for 160 consecutive days in the Marble Bar area west of the Great Sandy Desert in the northwest.

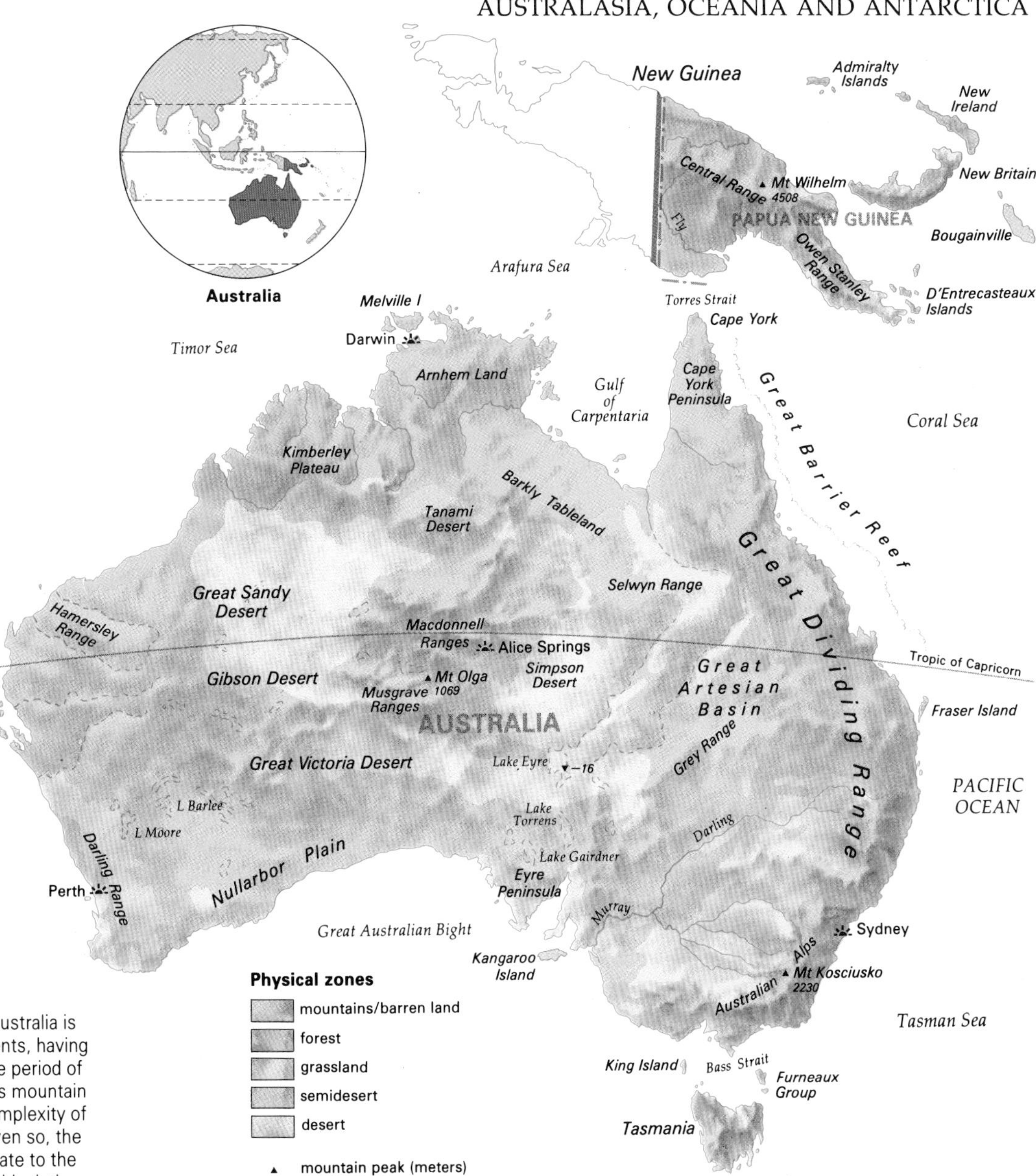

Map of physical zones Australia is the flattest of the continents, having escaped the recent Alpine period of mountain-building. It lacks mountain ranges of the size and complexity of the Alps or Himalayas. Even so, the arid interior owes its climate to the eastern highlands, which block the path of moisture-bearing winds from the Pacific.

A drowned coastline Mount Ruby dominates the coast of Tasmania's Southwest National Park. Following the end of the last glaciation, the rising sea flooded the valleys to form rias. Tasmania is a mountainous island; in the west are several parallel ridges and valleys, while to the east lie a series of plateaus often studded with lakes. By contrast, in the northwest and northeast and in the low South Esk river valley there are extensive plains.

The Devil's Marbles in Northern Territory, Australia, are a product of the extreme desert temperatures. When the outer layer of the rock is intensively heated by the Sun it expands and pulls away from the cooler layer below. During rare desert storms the rocks undergo rapid chilling. Different minerals expand and contract at different rates and the rocks start to disintegrate. The process is accelerated by chemical weathering as air and water gain access. Eventually the outer layers peel off, like the skin of an onion (exfoliation).

The eastern highlands rise gently from central Australia to a series of high plateaus and mountains that run the length of the east coast from Cape York to Tasmania. They include the mountains of the Great Dividing Range and of the Australian Alps, which rise to heights of 1,000–2,000 m (3,300–6,600 ft) in the south. Winds blowing onshore from the Coral Sea to the northeast produce locally high rainfalls and subtropical climate in the northern half of the eastern highlands. The highlands are flanked by a narrow, fertile coastal strip where agriculture is concentrated and many people live.

The savannas of the northern tropics extend east from about halfway along the northwest-facing coast. This area receives seasonal rain from the monsoon winds. In summer, when the winds blow onshore, there is heavy rainfall; in winter they blow offshore, and are so dry that they create drought conditions. Darwin's average annual rainfall of 1,611 mm (65.4 in) falls almost entirely between November and April. Temperatures in the area average 28°C (82°F), with little seasonal variation.

The southwest tip of Australia around Perth and Albany, and the southern coastal areas to the east of the Great Australian Bight, have a Mediterranean climate, with warm, dry summers and cool, wet winters. The average annual temperature is 17°C (63°F); it ranges from about 23°C (73°F) in summer to 11°C (52°F) in winter. The annual rainfall average in Perth is 873 mm (34.4 in); in Adelaide it is 533 mm (21 in), with most rain falling in the winter.

Islands of the Pacific

The Pacific climates are mostly tropical and wet with seasonal monsoons, but between latitudes 5°N to 5°S an equatorial climate prevails. Mountain ranges can modify local climates. In New Guinea the central mountain ridge reaches 4,508 m (14,789 ft) at Mount Wilhelm, and this

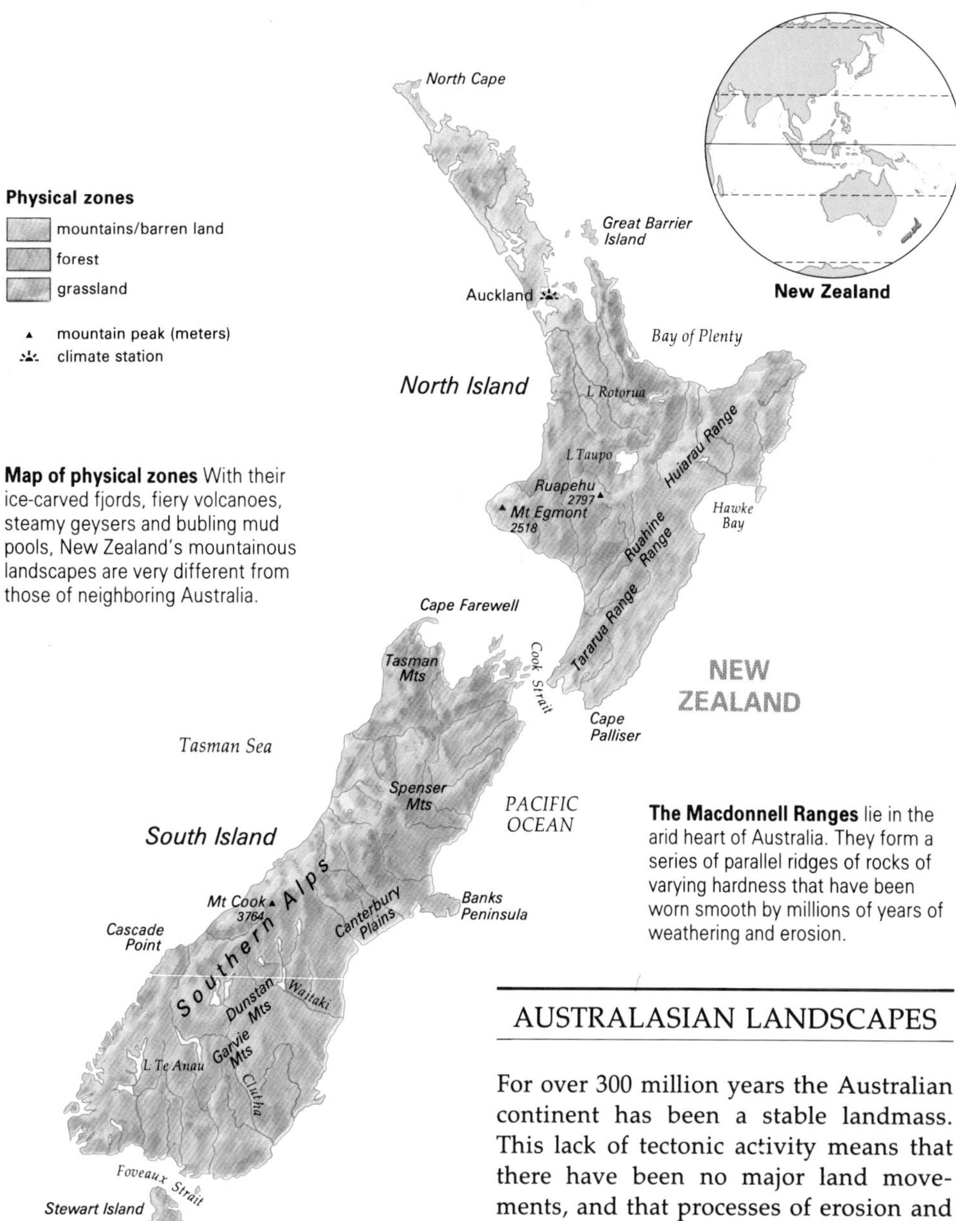

Map of physical zones With their ice-carved fjords, fiery volcanoes, steamy geysers and bubling mud pools, New Zealand's mountainous landscapes are very different from those of neighboring Australia.

The Macdonnell Ranges lie in the arid heart of Australia. They form a series of parallel ridges of rocks of varying hardness that have been worn smooth by millions of years of weathering and erosion.

affects the rainfall in the upland areas north and south of the central range. Rainfall can average 4,000 mm (160 in), with extremes of over 10,000 mm (400 in) in places. Even though New Guinea is close to the Equator, there may be frosts above 2,200 m (7,200 ft), and snow may settle above 4,000 m (13,000 ft).

In South Island, New Zealand rainfall is affected by the Southern Alps. The mountains rise to 3,764 m (12,349 ft) at Mount Cook, and are crossed by the dominant westerly winds. On the windward, western side the annual rainfall is over 7,000 mm (280 in), whereas in the rain shadow on the eastern side it can be as little as 300 mm (12 in). The maritime influence also affects temperatures, which average 15–21°C (60–70°F) in the summer and 4–10°C (40–50°F) in winter. The climate in New Zealand varies with latitude, ranging from subtropical in the north to wet temperate in South Island.

AUSTRALASIAN LANDSCAPES

For over 300 million years the Australian continent has been a stable landmass. This lack of tectonic activity means that there have been no major land movements, and that processes of erosion and deposition, rather than of mountain-building, have been responsible for shaping the land.

The most striking feature of Australia is its flatness. The western plateau that comprises three-fifths of Australia rises only 230–460 m (750–1,500 ft). This huge shield plateau is 600 million years old and is one of the Earth's original geological plates. It contains granites, metamorphic and hardened sedimentary rocks, many of which are now covered in desert sand. Deserts such as the Great Sandy Desert in the west and the Great Victoria Desert in the south extend into the interior of the plateau. These are areas of sand dunes, stony deserts and salt lakes, with only occasional waterholes and rivers to intersperse the dryness. When rain reaches these areas it usually comes in torrential downpours, bringing desert plants and animals to life for a short season, and filling the salt lakes and gorges.

This arid plateau is broken only by a few mountain Ranges, such as the Macdonnell Ranges. The flatness of the land highlights other individual mountains such as Ayers Rock (867 m/2,845 ft) and Mount Olga (1,069 m/3,507 ft).

The central lowlands

Between the western plateau and the eastern highlands is a series of lowland plains and basins based on sedimentary rock. These central lowlands rarely rise above 150 m (500 ft), and some areas, such as the inland drainage basin of Lake Eyre, are below sea level. The lowlands are arid and uniform expanses, broken occasionally by riverbeds and waterholes that are usually dry. Water is available but it is mostly stored deep within the ground in artesian basins. These have been likened to huge underground saucers that collect the water flowing beneath the Earth's surface. The largest is the Great Artesian Basin in the north. Wells to tap its valuable resource have to be sunk to depths of some 300–600 m (1,000–2,000 ft).

THE GREAT BARRIER REEF

The Great Barrier Reef is a vast complex of coral islands and reefs, stretching for about 2,000 km (1,240 mi) along the Queensland coast of northeast Australia. It covers an area of about 270,000 sq km (104,000 sq mi) and is the largest living structure on Earth. It can even be seen from the Moon. Of the 2,500 reefs, more than 2,000 are concentrated in a band 50–60 km (30–37 mi) wide along the margin of the continental shelf. Some individual reefs are capped with islands, ranging from sandspits uncovered only at low water to wooded cays and mangrove swamps. Many of the islands rise only slightly above sea level while others, such as Whitsunday and Magnetic islands, can rise to about 1,000 m (3,300 ft), and are relics of an earlier continental landmass.

The reef is formed by colonies of coral polyps, small animals rather like sea anemones. The corals secrete a skeleton of calcium carbonate, and it is their skeletons that make up the bulk of the reef. Many other animals such as fish, starfish and sea urchins graze on the corals, producing a fine coral sand. This and limestone-secreting algae help to cement the reef together.

Coral reefs form very slowly. They need a temperature close to 25°C (77°F), normal marine salinity, a solid foundation, and a supply of food brought in by waves and currents. Corals contain algae in their tissues that need light to live, so they can grow only in shallow, clear water such as the higher parts of the continental shelf. The continental shelf off the Queensland coast did not support reefs until 15,000 years ago, when the sea began to rise after the ice age. Since then, the growth of the reefs has been interrupted several times by temporary sea-level retreats and extensive erosion. Today the maximum known thickness of reef resting on base rock is 150 m (500 ft).

Australia's most important rivers, the Murray and Darling and their tributaries, flow from the eastern highlands, across the southeastern lowlands to the Great Australian Bight.

The Great Dividing Range

The name given to the eastern highlands, the Great Dividing Range, is in fact misleading, because in many places the area is very flat. Nevertheless the Great Divide, the watershed between east- and west-flowing rivers, does exist.

The eastern highlands are formed from Paleozoic rocks that are about 300 million years old. They include limestones, sandstones, quartzite, schists and metamorphic dolomite that have been faulted and folded into mountain blocks and plateaus. Rivers and streams have eroded and dissected the highlands to create deep valleys, gorges and rugged hills, and this process is still in evidence today. The eastern edges of the hills form high escarpments, or cliffs, over which the rivers flow in some spectacular waterfalls before crossing the coastal strip to the sea.

Island landscapes

New Zealand is dominated by mountains. In South Island four-fifths of the land is taken up by the Southern Alps, a rugged range of snowcapped and glaciated peaks. The Canterbury Plains make up the remaining fifth of the island. This eastern coastal plain was formed by eroded material from the Alps, and is covered in tussocky grass that makes it ideal sheep-rearing country. Rivers flow down from the mountains, cutting deep valleys into the plains. North Island scenery by contrast, consists largely of forested hills, a volcanic plateau and downlands, with a much less extensive and lower mountain region.

The Pacific islands are usually grouped into three realms – Melanesia, Micronesia and Polynesia – based on their ethnic peoples. They can be also be grouped by their geological origins into coral islands, volcanic islands and continental islands. Coral islands are made of limestone rock formed from the skeletons and bodies of coral polyps. The island reefs can form in a more or less circular shape called an atoll, as at Kwajalein in the Marshall Islands, the largest atoll on Earth, which is over 274 km (170 mi) long.

Many Pacific islands are created by volcanic activity. Continental islands, such as New Guinea and New Caledonia, may have volcanic, sedimentary and metamorphic rocks, so there is a wide range of climate, soils and vegetation including grassland, high mountains, dense forests and mangrove swamps.

The South Alligator river valley near El Sharana in Arnhem Land, northern Australia. Here the high level of rainfall brought by the summer monsoon promotes lush, rapidly growing vegetation.

ANTARCTICA

Two-thirds of all the Earth's fresh water is held in Antarctica. Twice the size of Australia, this frozen continent contains ice and snow to depths of 3,500 m (11,500 ft) and more. It is also the world's highest continent – more than half of it is over 2,000 m (6,560 ft) above sea level. It is the coldest place on Earth, with temperatures so low that boiling water thrown into the air will spontaneously "explode" into ice crystals before reaching the ground.

The frozen continent

The presence of coal deposits in Antarctica reveals that it was not always a frozen wilderness. About 500 million years ago it was near the Equator, and had a warm, the wet climate. It formed part of a huge supercontinent known as Gondwanaland. Over millions of years Gondwanaland slowly broke up, and the continents as we know them were formed. They drifted apart, with Antarctica moving toward the South Pole. Today, 95 percent of Antarctica is covered by an ice sheet that conceals the landforms, leaving bare rock exposed on only a few mountain summits and sheltered coastal areas. What land is exposed shows the effects of erosion in this, the area of most active glaciation in the world.

The pressure of snow and ice on the ice sheet moves it slowly toward the sea, where it forms floating shelves such as the Ross Ice Shelf, an area the size of France. Ice breaks off the ice shelf and forms huge, flat-topped icebergs. Icebergs are also created when glaciers break up on meeting the sea.

Antarctica has two distinct regions. They are divided by the Transantarctic Mountains, which stretch from Victoria Land on the Pacific Coast to Queen Maud Land on the Atlantic Coast. Greater, or East, Antarctica is a high plateau area that covers two-thirds of the continent and rises to a height of 3,355 m (11,008 ft) at Mount Menzies. It is an old shield region containing Antarctica's oldest rocks, which date back to Precambrian times (over 600 million years ago). Lesser, or West, Antarctica includes Marie Byrd Land and the 1,280 km (800 mi) long Antarctic Peninsula. It includes the highest peak in the two southernmost continents and the Pacific islands, the Vinson Massif, at 5,140 m (16,863 ft). It also has a series of lower and younger mountain ranges, which are thought to be a continuation of the Andes range in South America.

The worst weather in the world?

The Antarctic environment is characterized by low temperatures, low precipitation and strong winds. High atmospheric pressure, giving clear winter skies, dominates the interior of the continent. Surface temperatures are controlled by the influences of latitude and altitude. Vostok, on the Greater Antarctic plateau, has an altitude of 3.5 km (11,155 ft) and an average annual temperature of −60°C (−76°F). At the coast the temperatures are warmer, ranging from 0°C (32°F) in summer to −20°C (−28°F) in winter; they can even rise above freezing point on the sheltered west coast of the Antarctic Peninsula. Despite having almost nine-tenths of the world's ice, Antarctica has little rainfall. At the South Pole there is only about 70 mm (2.8 in) of snow each year. This increases to a water equivalent of about 170 mm (6.7 in) toward the coast.

The lack of land between latitudes 40 °S and 65°S around Antarctica allows a vigorous westerly wind circulation to develop in the Roaring Forties and Howling Fifties. Easterly winds blow around the Antarctic coastline, but these are influenced by local winds.

A surface temperature inversion is a feature of the Antarctic lower atmosphere (troposphere) throughout the year, and particularly in winter. Air usually becomes colder with increasing height, but in Antarctica the air on the surface can be colder than that at greater heights. At the South Pole the air temperature at

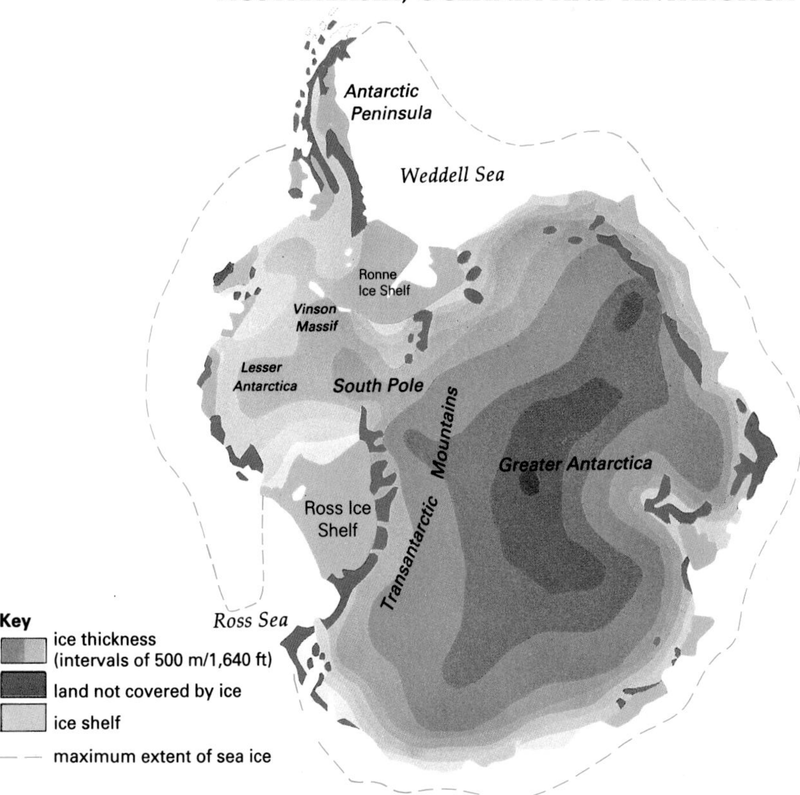

A frozen wilderness Despite being covered in ice to an average depth of 2,000 m (6,500 ft), Antarctica receives very little precipitation – it is a polar desert.

The coldest and highest continent in the world, Antarctica remained at the south pole when the other continents that once made up the great supercontinent of Gondwanaland drifted away to the warmer north.

914 m (3,000 ft) above the surface can be 30°C (54°F) higher than it is at the surface.

The strength of this inversion and the slope of the terrain creates local inversion winds of the type known as katabatic winds, which blow down slopes. In Antarctica these winds can blow ferociously and unceasingly for days or even weeks, and may extend well out beyond the edge of the continent.

By the end of March the Sun has set on Antarctica, and the continent is in freezing darkness for six months. Although the summer is very short the land receives more sunlight than equatorial regions do throughout the year. However, the ice sheet reflects most of the Sun's energy back into the atmosphere and, in so doing, plays a key part in the temperature balance of the planet.

The geographical separation of the Antarctic landmass after the breakup of Gondwanaland, and the freezing, dry climate have prevented many animals and plants from invading and colonizing it. There are over four hundred species of lichens but only two flowering plant species. The animals that live there, such as krill, fish, seals, penguins and whales, depend largely on the surrounding nutrient-rich oceans for their food. Antarctica is the only continent where there are no native peoples and where no civilizations have developed.

THE HOLE IN THE OZONE LAYER

The ozone layer in the stratosphere, the upper atmosphere, forms a protective shield around the Earth. It screens out 90 percent of the Sun's ultraviolet (UV) radiation. In 1982 scientists working for the British Antarctic Survey at Halley Bay discovered a reduction in concentration ("hole") in the ozone layer at altitudes of 12–27 km (7.5–17 mi).

Further research confirmed this phenomenon, and fears were voiced that this was the beginning of a global depletion of the layer. Life on Earth is delicately balanced, and an increase in UV radiation would affect global life cycles. It would decrease the productivity of crops, reduce photosynthesis by plankton in the oceans, change climates and increase the incidence of skin cancer in humans.

Chlorine from human-made chemicals is seen as the culprit. Chlorofluorocarbons (CFCs) are used in industrial processes, as aerosol propellants and refrigerator coolants, and they have been released into the atmosphere in increasing concentrations. CFC compounds persist in the lower atmosphere (troposphere), but they migrate slowly to the stratosphere, where UV radiation breaks them down and starts a chemical chain reaction releasing chlorine and destroying the ozone molecules.

The ozone hole is seasonal, developing only in the Antarctic winter and spring. Its size also changes. In October 1987 it was at its largest, an area the size of the United States, and had spread as far as southern Australia. There are signs that the hole has been less extensive since then. Further surveys now suggest that a similar process is taking place over the Arctic. Urgent measures are needed to reduce the release of CFCs into the atmosphere. At least one international agreement has been signed, and "ozone-friendly" aerosols have been developed.

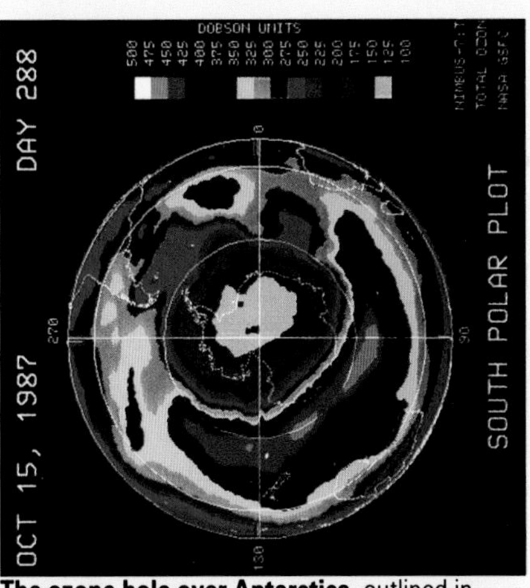

The ozone hole over Antarctica, outlined in white, seen from a satellite in October 1987. The Dobson units are a measure of atmospheric ozone. The deep blue, purple, black and pink area is the "hole".

The hot springs of New Zealand

Spectacular hot-water fountains, bubbling mud pools, and clouds of steam rising out of crevices in the land and even from drains in the street, are features of the volcanic plateau of New Zealand's North Island. The plateau rises steeply from the crater lake of Lake Taupo to the peaks of Tongariro (1,968 m/6,458 ft), Ngauruhoe (2,291 m/7,515 ft) and Ruapehu (2,797 m/ 9,175 ft). Volcanic ash, pumice and lava cover the plateau and a chain of active volcanoes forms the Taupo volcanic belt. This extends from the Tongariro National Park to the Bay of Plenty on the north coast, and the largest and hottest thermal springs are found in this area, particularly in the Rotorua and Taupo valleys.

Thermal springs are formed when underground water comes into contact with hot volcanic rocks. The water, under pressure from the mass of rock, soil and water above, can rise to temperatures of 260°C (500°F). This superheated water, which is lighter than the cooler water underground, rises to the surface through cracks and joints in the rock.

Thermal springs contain high concentrations of dissolved minerals, which build up in deposits around the surface vents. They form peculiar and colorful natural sculptures. These may be made of calcium carbonate, when they are called travertine, or of quartz, called geyserite.

Geysers and mud pots

Like thermal springs, geysers are formed by the contact of underground water with very hot volcanic rock. As the water boils steam is produced, and the pressure forces a steam explosion, sending a violent jet of water and steam up through channels in the rock.

Geysers can be extremely dramatic; the Pohutu geyser at Whakarewarewa reaches heights of over 30 m (100 ft). New Zealand holds the record for the world's highest geyser. It erupted out of a fissure near Mount Tarawera in 1901, reaching a height of 450 m (1,500 ft).

After a geyser has erupted the water settles and the cycle begins again. Some geysers erupt spasmodically, while others are very regular; they may erupt two or three times an hour or only every few weeks. When barometric pressures are low, the downward pressure is reduced and the geysers become more active.

Other features of the area are mud volcanoes and hot, bubbling mud pools. The mud is formed from solid particles and volcanic gases dissolved in the water. It oozes out from vents in the Earth's surface, creating mud cones. Small vents releasing steam (known as fumaroles) can also be found. The area is characterized by the smell of hydrogen sulfide gas.

Energy from the Earth

Humans have found a number of ways to harness the intense energy beneath the Earth's surface for domestic and industrial use. In the past New Zealand's Maori people once used this natural geothermal energy to cook their food. They dug holes in the ground and constructed earth ovens. Today local people sink bores into the ground to fill swimming pools and heat their homes. On a larger scale, bores are sunk to tap the steam, which is used to drive power station turbines. The world's first geothermal power station was built at Wairakei near Lake Taupo, and utilizes this valuable source of renewable energy to supply electricity to New Zealand's national grid.

Boiling fountains of mineral-rich water spout from the Earth at Pohutu geyser at Whakarewarewa, New Zealand. The hot water dissolves minerals from the rocks below, and deposits them as glistening terraces when the water evaporates.

A bubbling pool of boiling mud at Waiotapu, near Rotorua in New Zealand's North Island, indicates that hot volcanic rocks lie not far below the surface.

A classic coral island

The twin mountain peaks of Bora-Bora, Taimanu and Pahia look down on a fringe of coral protecting the blue waters of a lagoon. The sandy beaches are lined with palm trees. Bora-Bora is the tour operator's dream of a "paradise island".

The core of Bora-Bora is an extinct volcano, one of many scattered across the South Pacific. The volcanic center has been eroded by thousands of years of wind and rain, so only the hard rocks are left to give Bora-Bora its rugged skyline.

Such reef-fringed islands are formed when volcanic eruptions on the ocean floor force lava to pile up, forming an island that provides a base for coral to grow. Coral polyps require warm, shallow tropical seas. They use the calcium in seawater to form limestone skeletons. As new generations of polyps build on the skeletons of previous generations a reef is built up, fringing the old volcano.

If the volcanic island later subsides or the sea level rises, the coral growth may keep pace with the change, and a sheltered lagoon forms between the reef and the island. The volcano may eventually become totally submerged, and only a circular reef then remains above the level of the sea. These ring-shaped reef islands are called atolls.

Island in the sun The extinct volcano of Bora-Bora, in the Leeward group of the Society Islands, forms a palm-fringed island surrounded by coral reefs.

GLOSSARY

Acid rain
Rain or any other form of PRECIPITATION that has become more acid by absorbing waste gases (sulfur dioxide and nitrogen oxides) discharged into the ATMOSPHERE.
Adaptation
The process in which an organism gradually changes so that it becomes better suited to its ENVIRONMENT.
Air mass
A large mass of air that has consistent temperatures and humidity levels.
Albedo
The whiteness, or percentage of light energy, reflected from a surface. Snow and cloud have a high albedo, vegetation and sea a low albedo.
Alluvium
Fine SEDIMENT transported or deposited by flowing water.
Alpine
A TUNDRA-type environment found on a mountain above the tree line but beneath the limit of permanent snow.
Alpine mountains
High mountain ranges formed by folding by tectonic activity during the Alpine mountain-building period, which was at its peak in the Miocene period, when the Alpine–Himalayan belt was formed.
Anticline
A FOLD in which the rock strata curve upward; the opposite of SYNCLINE.
Anticyclone
A center of high ATMOSPHERIC PRESSURE; the opposite of CYCLONE.
Aquifer
A bed of rock that is capable of holding water.
Arête
A knife-edged ridge produced between two CIRQUES that have been eroded by ice.
Arid
Very dry; areas have little rainfall. When rain does fall it is often evaporated and also sinks underground, so vegetation is sparse.
Asthenosphere
The layer of the Earth's MANTLE below the LITHOSPHERE. The asthenosphere is not rigid, allowing the movement of PLATES above it.
Atmosphere
The gaseous layer surrounding the Earth. It consists of nitrogen (78 percent), oxygen (21 percent), argon (1 percent), tiny amounts of carbon dioxide, neon, ozone, hydrogen and krypton, and varying amounts of water vapor.
Atmospheric pressure
The pressure exerted by the atmosphere as gravity acts upon the air.
Atoll
A circular chain of CORAL REEFS enclosing a LAGOON but no island. Atolls form as coral reefs fringing a volcanic island; as sea levels rise or the island sinks a lagoon is formed.
Avalanche
A slide of snow and ice or rock debris.

Bar
An offshore strip of sand or shingle parallel to the coast. It can also refer to a submerged barrier of sand or gravel.
Barchan
A crescent-shaped sand dune with the horns of the crescent facing downwind. It has a steep leeward face and a gentle windward face.
Basalt
A fine-grained IGNEOUS rock. It has a dark color and contains little silica. Ninety percent of lavas are basaltic.
Base level
The lowest level to which a stream or river can erode its channel. This may be the sea or an inland drainage basin.
Beach
An accumulation of sand or shingle on a coast running parallel to a zone of breaking waves.
Biomass
The combined weight of all living organisms within a given area. Almost all the Earth's biomass is found on the land.
Biosphere
The part of the Earth and its ATMOSPHERE in which all organisms live.
Bornhardt
A high rock dome formed of an IGNEOUS rock such as granite.
Breccia
A rock composed of sharp-edged fragments of other rocks held together by natural cement.

Carbon cycle
A chain or cycle in which the ELEMENT carbon circulates through ecosystems.
Carbon dioxide (CO_2)
A gas that makes up about 0.33 percent of the ATMOSPHERE. Burning FOSSIL FUELS increases the amount of CO_2 in the atmosphere. See also GREENHOUSE EFFECT.
Carbonation
The solution of limestone rocks by rainwater made more acid by CARBON DIOXIDE from the ATMOSPHERE or the SOIL. Carbonation can produce KARST scenery.
Castle kopje
The Afrikaans name for the steep-sided, boulder-strewn hills found standing out above the plains in southern Africa.
Catena
The sequence of SOIL PROFILES across a valley, which often change as a result of different drainage conditions.
Chemical weathering
The chemical breakdown of rocks exposed to air or to water. OXIDATION and CARBONATION are chemical weathering processes.
Cirque
A steep-sided, bowl-shaped rock basin formed at the head of a valley by frost shattering and ice EROSION.
Compound
A chemical substance composed of two or more ELEMENTS that always occur in the same combination, e.g. water (H_2O).
Condensation
A process in which gases are changed into liquids. For example, water vapor in the ATMOSPHERE condenses to become rainwater.
Conglomerate
A rock formed from rounded fragments of other rocks cemented together. They are often composed of water-eroded rock material from mountains or coastlines.
Coniferous forest
A forest of cone-bearing evergreen trees with needle-like leaves. They are found in TEMPERATE zones and forests of the colder regions.
Continental climate
The type of climate found in the interior of continents, with a wide temperature range both daily and seasonally, and relatively low rainfall.
Continental drift
The theory that today's continents, formed by the breakup of prehistoric SUPERCONTINENTS, have slowly drifted to their present positions. The theory was first proposed by Alfred Wegener in 1912.
Continental shelf
An extension of a continent, forming a shallow, sloping shelf in the sea.
Coral reef
A barrier of limestone rock formed by the accumulation of the skeletons of millions of coral polyps. Reefs grow as polyps build on the shells of dead generations.
Core
The central mass of the Earth lying beneath the MANTLE. The core is made up of iron and nickel. It has a radius of 3,470 km (2,156 mi); the radius of the solid inner core is about 1,400 km (869 mi). The outer core is liquid.
Coriolis effect
The effect of the Earth's rotation, which alters the direction of a moving body toward the right in the northern hemisphere and to the left in the southern hemisphere.
Craton
A continental block that has been stable for a long period of geological time.
Crust
The outer layer of the Earth resting on the MANTLE. Continental crust underlies the continents and is about 40 km (24 mi) thick; oceanic crust underlies the oceans and is about 10 km (6 mi) thick.
Crystalline rock
Rocks, such as IGNEOUS or METAMORPHIC rocks, that are formed when molten material crystallizes or recrystallizes.
Cyclone
A center of low ATMOSPHERIC PRESSURE, the opposite of ANTICYCLONE. Tropical cyclones are known as HURRICANES and TYPHOONS.

Deciduous forest
A forest of trees or shrubs that shed their leaves each year, usually in the fall.
Delta
An accumulation of SEDIMENT at a river mouth. When the sediment blocks the river channel the flow of the river splits into many channels, known as distributaries, creating new routes for the water and its load.
Denudation
The total effect of WEATHERING and EROSION processes in wearing away the land surface.
Deposition
The laying down of material broken down by EROSION or WEATHERING and transported by the wind, water or gravity.
Desert
A very ARID area with less than 25 cm (10 in) rainfall a year. In hot deserts the rate of evaporation is greater than the rate of PRECIPITATION, and there is little vegetation.
Donga
The South African name for a wadi or dry watercourse. They normally occupy a narrow, steep-sided gully.
Drainage basin
The land drained by a river and its tributaries, the catchment area of a river. One drainage basin is separated from another by the WATERSHED.
Drought
A long period where rainfall is substantially lower than average.
Drumlin
An oval hill of deposited glacial TILL. Often found in groups, when they are described as a "basket of eggs" landscape.
Dune
A hill or ridge of sand that has been shaped by the wind. Dunes can be mobile, usually moving downwind. See BARCHAN and SEIF dunes.

Ecosystem
A community of plants and animals and the ENVIRONMENT or habitat in which they live and react with one another.
Element
A substance that cannot be resolved into simpler substances by chemical means.
Environment
The surroundings in which all animals and plants live.

Environmental impact assessment (EIA)
A study that assesses the impact a certain course of action will have on the environment. EIAs are used in planning and decision-making on new projects; in many countries EIAs are required by law.
Epicenter
The point on the Earth's surface directly above the focus of an earthquake.
Erg
The Arabic name for a sandy DESERT.
Erosion
The process by which exposed land surfaces are broken down into smaller particles or worn away by water, wind or ice.
Escarpment
A steep slope or cliff.
Esker
A narrow and often winding ridge of sand and gravel, many kilometers long. Eskers are formed when the meltwater of a GLACIER or ICE SHEET deposits its load either in ice tunnels or at the ice margins.
Eustasy
Eustatic change refers to a worldwide change in sea level.
Exfoliation
The weathering of a surface in which thin layers of rock flake away. It is caused by the expansion and contraction of rocks that have heated and cooled more on the surface than in the interior.
Extrusive rock
IGNEOUS rock formed by the solidification of MAGMA that has flowed out onto the surface from the lower CRUST or the MANTLE.

Fault
A fracture or crack in the Earth along which there has been movement of the rock masses relative to each other.
Fjord
A steep-sided inlet formed when a glaciated U-shaped valley is drowned by the sea.
Floodplain
A stretch of low, flat ground on one or both sides of a river channel. They are formed by SEDIMENTS deposited by the river.
Flow
River or stream flow is the amount of water flowing in the channel.
Fold
A bend in the rock STRATA caused by movements of the Earth's CRUST.
Fossil
The remains or traces of a once-living organism preserved in the Earth's crust.
Fossil dune
A dune that was formed in an area where the climate was once arid but has since become wetter. Vegetation stabilizes the dunes, which normally move in the wind.
Fossil fuels
Fuels such as oil, coal, peat and natural gas, formed from ancient organic remains.
Front
The zone where two different AIR MASSES meet.

Geological time-scale
The division of the geological history of the Earth. The principal division is into eras; eras are subdivided into periods, and periods are subdivided into epochs.
Geosyncline
A SYNCLINE extending over hundreds of kilometers, caused when SEDIMENTS accumulate and depress the CRUST below, allowing more sediments to accumulate on top.
Geothermal energy
Heat energy derived from the Earth's crust, shown by thermal springs, steam or dry hot rock.
Geyser
A fountain of steam and hot water that periodically erupts from the ground.
Glacial
A period of time in which GLACIERS and ICE SHEETS formed, spread and receded. Also called an ICE AGE.
Glaciation
The process of GLACIER and ICE SHEET growth and their effect on the landscape.
Glacier
A mass of ice formed by the compaction and freezing of snow and which shows evidence of past or present movement.
Glacioeustasy
Changes in the world sea level caused by the formation and melting of large bodies of ice such as GLACIERS and ICE SHEETS.
Glaciofluvial
The features of the landscape caused by meltwater at the end of an ICE SHEET or GLACIER.
Global warming
The increase in the average temperature of the Earth. This is currently being caused by the GREENHOUSE EFFECT.
Gondwanaland
The southern SUPERCONTINENT composed of present-day Africa, Australia, Antarctica, India and South America. It began to break up 200 million years ago.
Graben
A long, narrow block that has dropped between two parallel faults to form a trench-like depression. Also known as a RIFT VALLEY.
Greenhouse effect
The process in which radiation from the Sun passes through the atmosphere, is reflected off the surface of the Earth, and is then trapped by gases in the atmosphere. The buildup of carbon dioxide and other "greenhouse gases" is increasing the effect. There are fears that the temperature of the planet may rise as a result; this global warming is expected to have dire consequences.
Groundwater
Water held in spaces in rocks beneath the surface.
Gully
Deep, steep-sided V-shaped valley eroded by a fast-flowing stream well above its BASE LEVEL.

Hadley cell
The atmospheric circulation cell found in low latitudes. Air rises over the Equator creating low pressure, moves toward the poles, sinks over the TROPICS creating high pressure, and moves back to the Equator.
Hamada
The Arabic name for a bare, rocky desert.
Hanging valley
A tributary valley that ends abruptly at its junction with the main valley. The tributary valley was eroded less than the main valley by the GLACIERS that occupied them, and when the ice melted the tributary valley was left "hanging".
Hurricane
The name for a tropical CYCLONE found in the Caribbean and western North Atlantic.
Hydrolysis
The chemical interaction of water molecules and minerals to form new COMPOUNDS.

Ice age
A long period of geological time in which cold climatic conditions prevail and snow and ice sheets are present throughout the year. There have been many ice ages in the Earth's history. Also called a GLACIAL.
Ice cap
A dome-shaped GLACIER.
Ice sheet
A large area of glacial ice that moves slowly outward in all directions from a central point. Ice sheets cover the polar regions and can be thousands of meters thick.
Iceberg
A mass of ice floating in the ocean. Icebergs are formed when GLACIERS and ICE SHEETS break up on meeting the ocean.
Igneous
Rocks formed by the solidification of MAGMA.
Incised river
A river where the curves in the channel remain unchanged while the channel cuts vertically deeper.
Inselberg
A steep, isolated hill found in ARID or SEMIARID landscapes.
Interglacial
A period of milder global climate between GLACIAL periods when ICE SHEETS and GLACIERS retreat.
Intrusive rock
IGNEOUS rock formed when MAGMA solidifies beneath the surface of the Earth.
Island arc
Belts of volcanic islands produced by SUBDUCTION ZONES. Island arcs run parallel to ocean TRENCHES. There is usually great earthquake activity.
Isostasy
The state of balance thought to exist between lighter and heavier parts of the Earth's crust as they float on the denser material of the MANTLE.
Isotope
Atoms of the same element that contain the same number of protons but different numbers of neutrons. For example, uranium always has 92 protons but between 141 and 147 neutrons.

Jet stream
Flows of fast-moving west to east air at high levels in the atmosphere. They are found in the middle and subtropical latitudes.
Joint
A fracture or crack in rock along which there has been little or no movement.

Kame
A mound formed by deposition of GLACIOFLUVIAL sediments. A kame terrace has a flat top with steep sides and runs along a valley side; a kame moraine is a complex of mounds deposited along the margin of an ice sheet.
Karst
A bare limestone landscape formed by chemical weathering of the rock and named after a typical region in Yugoslavia. It features limestone pavements, caverns and gorges.
Kettle
A depression in a glacial deposit formed when a block of ice that was once part of the deposit melts. Kettle holes can fill with water, forming kettle lakes.

Lagoon
A shallow stretch of water lying behind a barrier such as a CORAL REEF, sand BAR or SPIT.
Landslide
The rapid MASS MOVEMENT of rock and earth on mountain or cliff slopes, usually caused when the land becomes saturated with water or is shaken by an earthquake.
Laterite
A rock-like layer of nodules found in tropical soils. Laterite consists mainly of aluminum oxides and hydroxides, iron, bauxite and limonite.
Latitude
The angular distance from the Equator to a given point on the globe. Lines of latitude are

imaginary circles around the Earth, measured in degrees north or south of the Equator.

Laurasia
The ancient northern SUPERCONTINENT, which broke up and formed present-day North America, Europe and Asia.

Lava
The molten rock produced from a volcano, or its later solidified form.

Leaching
The process by which water moves minerals from one layer of soil to another, or into streams.

Levee
A ridge of ALLUVIUM deposited at the side of a river by flood waters. As the river repeatedly floods the levees form high banks.

Lithosphere
The Earth's CRUST and the rigid upper part of the MANTLE above the ASTHENOSPHERE, forming the PLATES.

Loess
A fine-grained SEDIMENT, usually buff or yellowish in color, deposited after transport by the wind in dust storms. Loess is composed of particles of minerals such as quartz, feldspar and calcite. When it accumulates it makes fertile soil.

Longitude
The angular distance measured east or west from a certain point of the globe to a standard point. Lines of longitude are imaginary circles around the Earth, measured in degrees east or west of the Greenwich meridian.

Magma
Molten rock or material that originates in the lower CRUST or MANTLE. Solidified magma is known as IGNEOUS rock.

Mantle
The layer of the Earth that lies between the CRUST and the CORE. It is rich in silica rocks.

Mass movement
The movement of soil or rock just below the surface, as in a LANDSLIDE.

Meander
Winding bends in a river that form when a river flows around a curve. At the curve, the current is faster at the outer bank and erodes that bank. At the inner bank, the current slackens and deposits sediment.

Metamorphism
The alteration of the texture and/or composition of a rock by great heat or pressure without the rock passing through a molten phase.

Mineral
A naturally formed solid inorganic substance with a characteristic chemical composition and often a particular crystalline shape.

Mohorovicic discontinuity (Moho)
A major structural divide in the Earth that forms the boundary between the CRUST and the MANTLE.

Monsoon
Generally used to describe winds that change direction according to the seasons. They are most prevalent in southern Asia, where they blow from the southwest in summer, bringing heavy rainfall, and the northeast in winter.

Moraine
Rocks, gravel, clay and debris carried along by a glacier and deposited when the GLACIER melts.

Mountain-building
The process of tectonic activity when mountains are formed (see FAULT and FOLD). These periods are also known as orogenies.

Mudflow
The MASS MOVEMENT of material down a slope. Mudflows take place when soil becomes saturated with water, loosens and slides downhill lubricated by the water.

Nitrogen (N)
An element; an odorless, colorless gas that forms 78 percent of the ATMOSPHERE. Nitrogen is an essential nutrient for plants, and is circulated through the environment in the nitrogen cycle.

Oceanic ridge
A submarine ridge formed by earthquake or volcanic activity beneath the seabed at the boundaries between PLATES. It is usually associated with SEA FLOOR SPREADING.

Oxidation
The combination of oxygen with a metallic substance in MINERALS.

Ozone hole
The ozone layer is a band of enriched oxygen (O_3) found in the upper ATMOSPHERE. It filters out harmful ultraviolet radiation from the Sun. Certain chemicals, especially those containing chlorine, destroy the ozone; a hole appears seasonally above Antarctica, and over the Arctic the layer of ozone is becoming thinner.

Paleomagnetism
The study of the Earth's magnetic field in the geological past. The alignment of magnetic particles in rocks is examined to deduce the polarity of the Earth at the time the rocks were formed.

Paleontology
The study of the remains of extinct plants and animals found buried in SEDIMENTARY ROCKS.

Pangea
The SUPERCONTINENT that was composed of all the present-day continents and therefore included both GONDWANALAND and LAURASIA. It existed between 250 and 200 million years ago.

Pediment
The gently sloping plain found at the foot of steep slopes in ARID and SEMIARID areas.

Pediplanation
The processes of DENUDATION from streams and as sheetwash that form PEDIMENTS.

Periglacial
The areas on the fringes of ICE SHEETS and GLACIERS, and the processes of landscape development prevailing there.

Permafrost
Ground that is permanently frozen. The layer of soil above the permafrost melts in summer but is unable to drain through it. Permafrost is typically found in Arctic landscapes.

Permeable rocks
Rocks through which water can pass because they are either porous (see POROSITY) or have cracks through them.

Physical weathering
Also called mechanical weathering. Rocks are broken up by processes caused by the weather, such as freezing water and expansion and contraction, and by other organisms, such as tree roots.

Pingo
A dome of soil or gravel formed in TUNDRA landscapes when water below the surface freezes and expands, pushing up the land surface.

Planation
The processes of EROSION that wear away a land surface until it is almost flat. Planation surfaces may later be raised up and eroded again.

Plate tectonics
The movement of the PLATES of the LITHOSPHERE over the surface of the Earth.

Plateau
A large area of level, elevated land. When it is bordered by steep slopes it is called a tableland. If it has deep eroded valleys it is called a dissected plateau. Some plateaus have level STRATA, others are old PLANATION surfaces that have been lifted.

Plates
Rigid slabs of the Earth's CRUST and upper MANTLE that make up the LITHOSPHERE. Boundaries between the plates can be constructive, where plates are moving apart, or destructive, where one plate is forced under another along a SUBDUCTION ZONE.

Plutonic rock
IGNEOUS rocks that cooled down and slowly solidified deep within the Earth. Because cooling is slow the MINERALS have time to crystallize. The more slowly the rocks cool, the larger the crystals.

Polar
The polar regions lie within the lines of latitude known as the Arctic and Antarctic circles – 66° 32'N and 66° 32'S respectively. At this line of latitude the Sun does not set in midsummer.

Porosity
The extent to which rocks are able to absorb water. When they are underlain by non-porous rock, water can saturate the porous rock. The top of the saturated layer is known as the water table. A rock holding water in this way is known as an AQUIFER.

Precession of the equinoxes
The slow change in the time of the equinoxes caused by a cyclical wobble in the Earth's axis. A one-wobble cycle takes about 25,800 years.

Precipitation
Moisture that reaches the Earth from the ATMOSPHERE, including mist, dew, rain, sleet, snow and hail.

Pyramidal peak
A steep-sided angular mountain whose peak is formed when three or more CIRQUES cut back into it. Each CIRQUE is separated from the others by a steep ridge known as an ARETE. A classic example is the Matterhorn in Switzerland.

Radiation
The movement of radiant energy in a form such as heat. The Sun's energy is radiated into space, and some of this heats the Earth. The Earth's surface radiates heat back into the ATMOSPHERE, where some of it is absorbed and some re-radiated back to Earth.

Radioactivity
The RADIATION emitted from atomic nuclei. This is greatest when the atom is split, as in a nuclear reactor. There are three common types – alpha, beta and gamma – which have very different properties.

Rainforest
Usually known as tropical rainforest, and found in the equatorial belt where there is heavy rain and no marked dry SEASON. Growth is very lush and rapid. Rainforests probably contain half of all the Earth's plant and animal species.

Raised beach
An old shoreline, with beach deposits or wave-cut platform, that has been abandoned by the sea as sea level has dropped or the land has risen. Sometimes several are found one behind the other, giving a stepped appearance.

Reg
A large area of DESERT from which the sand has been blown away, leaving behind heavier stones; also known as stony desert.

Remote sensing
The use of unmanned equipment (such as a weather satellite) to monitor the environment and relay the information to a center where the data can be collated and analyzed.

Resistance
In geology, the ability of a rock to resist the processes of WEATHERING and EROSION. Resistant rocks frequently stand above less resistant rocks. SCARP and vale topography is found where there are alternating bands of resistant and less resistant rocks.

Ria
A river valley drowned by the sea. Ria coasts are highly indented, as the landmass has been partly submerged by the sea.
Rift valley
A long valley formed by the subsidence of a block between two parallel FAULTS. (See also GRABEN.)
Roche moutonée
A mass of rock in a GLACIATED valley over which ice travelled, creating a smooth, eroded slope on the side facing the glacier and a steeper, craggy slope on the side from which blocks were plucked.
Rock cycle
The cycle in which rocks are constantly being formed, broken down and reformed by landforming processes. Even the MINERALS in the rock are completely reformed when they are dragged back into the mantle along a SUBDUCTION ZONE or subjected to intense heat and pressure at the surface.
Runoff
Rainfall that flows across the land surface into streams and rivers. Delayed runoff is water that soaks into the ground and later emerges on the surface as springs.

Savanna
Also known as tropical grassland, savanna covers the areas lying between tropical RAINFOREST and hot DESERT areas. There is a marked dry SEASON, and too little rain even in the rainy season to support large areas of forest.
Scarp
An inland steep slope or cliff that cuts across the bedding planes of a rock. May be formed by a FAULT or by EROSION working on rocks of varying RESISTANCE at the surface.
Scree
An unsorted mass of rocks and boulders that accumulates at the bottom of a slope, having been broken off the rocks higher up by WEATHERING processes.
Sea floor spreading
The zone beneath an ocean or sea where adjacent PLATES move apart and new rock wells up from the MANTLE to fill the gap, building an OCEANIC RIDGE.
Seasons
Divisions of the year according to regular changes in the weather.
Sediments
Layers of rock, gravel, sand or silt – sometimes including organic material – that have been laid down by water, ice, wind or mass movement by water or ice, and cover the bedrock.
Sedimentary basin
A broad, deep hollow or depression filled with sedimentary rocks.
Sedimentary rocks
Rocks made up of layers of SEDIMENTS. The layers are called STRATA, and the rocks are often referred to as stratified rocks.
Seif
A type of sand DUNE forming a long, narrow ridge running parallel to the direction of the prevailing wind.
Seismic activity
Movements of the land, usually in the form of earthquakes, that take place near the edges of the Earth's PLATES. Over millions of years the cumulative effect can create high mountains and deep RIFT VALLEYS.
Semiarid
Areas between ARID deserts and better-watered areas, where there is sufficient moisture to support a little more vegetation than in the DESERT.
Sheetwash
Water that runs down a slope as a thin sheet or film rather than in channels. It can cause SOIL EROSION if the land is bare.
Snowline
The altitude at which snow cover is permanent. Below this line, winter snow melts in summer.
Soil
The unconsolidated, WEATHERED layer of material changed by soil-forming processes that lies immediately below the Earth's surface. It consists of mineral material and living and dead organic matter.
Soil creep
The slow, barely perceptible movement of soil downhill under the influence of gravity. The surface layer moves more rapidly than layers lower down.
Soil erosion
The loss of topsoil from an area as a result of EROSION by wind or water. Sheet erosion is caused by SHEETWASH, GULLY erosion by water flowing along channels.
Soil profile
The series of horizontal layers (horizons) in the SOIL, from the surface down to the bedrock.
Solar energy
Energy derived from the Sun. The Sun's energy is transformed into other types of energy such as wind and waves. Plants convert the Sun's energy into plant material by photosynthesis.
Solar system
The Sun and the numerous astronomical bodies that are bound to it by gravity, including the Earth and the other planets.
Solifluction
A more rapid form of SOIL CREEP, usually occurring where the surface layer becomes saturated with water.
Spit
A narrow bank of shingle or sand extending out from the shore, often found running part way across the mouth of a river. They are formed of material transported along the coast by currents, wind and waves.
Strata
The layers in SEDIMENTARY ROCK.
Stratigraphy
The branch of geology that studies the formation, composition, chronology and distribution of stratified rocks (see STRATA).
Stratosphere
The layer of the ATMOSPHERE between 12 and 50 km (7.4 and 31 mi) above the Earth's surface. The stratosphere lies above the TROPOSPHERE.
Subduction zone
A destructive PLATE margin, where one plate descends beneath another.
Subtropical
The area of the Earth's surface between the TROPICS and TEMPERATE zones. There are marked SEASONAL changes of temperature, but it is never very cold.
Supercontinent
A large continent formed by the fusion of smaller continents, or one that later breaks up to form smaller continents. The Earth's supercontinents were called GONDWANALAND, LAURASIA and PANGEA.
Sustainable development
Using the Earth's resources to improve people's lives without diminishing the ability of the Earth to support life today and in the future.
Syncline
A FOLD in which the rock bed curves downward; the opposite of ANTICLINE.

Taiga
Russian name given to the CONIFEROUS FOREST belt of the Soviet Union, bordering TUNDRA in the north and mixed forests and grasslands in the south. The name is generally applied to the northern coniferous forest belt as a whole.
Temperate
Areas between the warm TROPICAL and cold POLAR regions that have a mild climate.
Thermokarst
Landforms found in TUNDRA areas that resemble those normally associated with limestone (KARST scenery); they result with thermal changes in ground ice.
Till
Material carried along by a GLACIER and deposited when the ice melts; also called boulder clay.
Tornado
A narrow rotating column of air, usually with a funnel-shaped cloud around the vortex. The wind speed is very high, and if the column reaches the ground it can cause a lot of damage.
Transform fault
A fault caused when rocks move past each other in a sliding horizontal motion rather than a vertical one.
Transpiration
The process by which plants absorb water through their roots and return it to the ATMOSPHERE as water vapor.
Treeline
The greatest altitude or latitude at which the growing SEASON is long enough for trees to grow.
Trench
A long, narrow ocean floor valley found along a SUBDUCTION ZONE. They are formed as one PLATE is drawn down beneath another. Trenches are the deepest part of the oceans.
Tropics
The area of the Earth lying between the Tropic of Cancer (23° 30′N) and the Tropic of Capricorn (23° 30′S). They mark the line of LATITUDE farthest from the Equator where the Sun is directly overhead at midday in midsummer.
Troposphere
The lower layer of the ATMOSPHERE, up to 12 km (7.4 mi) above the Earth's surface. It contains the world's weather processes.
Tsunami
A very large wave caused by shock waves from an earthquake under the sea. As the wave travels it becomes larger and can cause great damage when it hits the coast. Often incorrectly called a tidal wave.
Tundra
The land area lying in the very cold northern regions of Europe, Asia and Canada, where winters are long and cold and the ground beneath the surface is permanently frozen.
Typhoon
The local name for a TROPICAL CYCLONE or HURRICANE occurring in the South China Sea and western Pacific Ocean. They bring exceedingly strong winds, heavy rain and rough seas, and can inflict great damage.

Water cycle
The movement of water as water vapor from open surfaces to the ATMOSPHERE by evaporation and TRANSPIRATION, and its return as PRECIPITATION, which is then transferred to lakes and oceans by river flow.
Watershed
The boundary line dividing two river systems. It is also known as a water-parting or divide, particularly in the United States, where the word watershed refers to a river basin – i.e. the area drained by a river and its tributaries.
Weathering
The decay and disintegration of rocks when exposed to the ATMOSPHERE. When moisture causes chemical changes the process is known as CHEMICAL WEATHERING; the breaking of the rock by mechanical means is called PHYSICAL WEATHERING.

Further reading

Banyard, P.J. *Natural Wonders of the World* (Orbis, London, 1981)
Clark, M.J., Gregory, K.J. and Gurnell, A.M. (eds.) *Horizons in Physical Geography* (Macmillan, London, 1987)
Coates, Donald R. *Geology and Society* (Chapman & Hall, London, 1985)
Goudie, A.S. and others (eds.) *The Encyclopaedic Dictionary of Physical Geography* (Blackwell, Oxford, 1988)
Goudie, A.S *The Human Impact on the Natural Environment*, 2nd edn, (Blackwell, Oxford, 1986)
Gregory, K.J *The Nature of Physical Geography* (Arnold, London, 1985)
Guinness Book of Records (Guinness Publishing, London, 1989)
Lambert, D. *The Cambridge Guide to the Earth* (Cambridge University Press, Cambridge, 1988)
The New Encyclopedia Britannica, 15th edn (Encyclopedia Britannica, Chicago, 1989)
Paxton J. (ed.) *The Stateman's Year-book 1988–1989* (Macmillan, London, 1988)
Reader's Digest Book of Natural Wonders (Reader's Digest, Pleasantville, 1985)
Reader's Digest Guide to Places of the World (Reader's Digest, London, 1987)
Rudloff, W. *World Climates* (Wissenschaftliche Verlagsgesellshaft, Stuttgart, 1981)
Showers, V. *The World in Figures* (Wiley, New York, 1973)
Simmons, I.G. *Changing the Face of the Earth* (Blackwell, Oxford, 1989)
Smith, D.G. (ed.) *The Cambridge Encyclopedia of Earth Sciences* (Hutchinson, London, 1985)
Smith, P.J. (ed.) *Hutchinson Encyclopedia of the Earth* (Hutchinson, London, 1985)
Strahler, A.N. and Strahler, A.H. *Elements of Physical Geography* (Wiley, New York, 1984)
The Times Atlas of the World (Times Books, London, 1988)
Webster's New Geographical Dictionary (Merriam-Webster, Springfield, 1984)
Whittow, J.B. *Disasters: The Anatomy of Environmental Hazards* (Allen Lane, London, 1980)
Whittow, J.B. *The Penguin Dictionary of Physical Geography* (Penguin, Harmondsworth, 1984)

Acknowledgments

Picture credits

Key to abbreviations: AD Art Directors Photo Library, London; **AGE** AGE Fotostock, Barcelona; **APL** Aspect Picture Library, London; **BCL** Bruce Coleman Ltd, Uxbridge, Middlesex; **E** Explorer, Paris; **FSP** Frank Spooner Pictures, London; **GC** Gerald Cubitt, S. Africa; **GSF** Geoscience Features Picture Library, Wye, Kent; **HL** Hutchison Library, London; **JH** John Hillelson Agency, London; **M** Magnum, London; **NHPA** Natural History Photographic Agency, Ardingly, Sussex; **NSP** Natural Science Photos, Watford, Herts; **OSF** Oxford Scientific Films, Long Hanborough, Oxon; **PEP** Planet Earth Pictures, London; **RHPL** Robert Harding Picture Library, London; **SGA** Susan Griggs Agency, London; **SPL** Science Photo Library, London; **TCL** Telegraph Colour Library, London; **VE** Victor Englebert, Colombia; **WPL** Wilderness Photographic Library, Kendal, Cumbria.

b bottom; l left; r right; t top.

6–7 Zefa **8–9** EOSAT/British Aerospace **10–11,12** SPL/NASA **16–17** JH/Georg Gerster **18–19** AGE **21** David Pratt **24–25** GC **24b** PEP/John Lythgoe **27** HL/Jesco von Puttikamer **29** Vadim Gippenreiter **31** **Dr** Hans Bibelriether **35** AD **40–41** AGE **41b** BCL/M. Freeman **44–45** E/M Moisnard **46–47, 47r** Fred Bruemmer **48t** TCL/Boulton-Wilson **48b** E/Lenars **49** AGE **50–51** Swift Picture Library/Mike Read **51** B. Klingwall **52–53** NSP/J.W Warden **54–55** SPL/David Parker **55** RHPL **56–57** TCL/Hans Blohm **56b** SGA/Nathan Benn **58–59** RHPL/Carl Young **59** Zefa **60** FSP **60–61** Frank Lane Picture Agency/W. Carlson **61** SPL/J.G. Golden **62–63** AD/Ron Watts **64** TCL/Bill Brooks **65** Zefa **66–67** AGE **67** AD/Anne Griffiths **68, 69** Zefa **70** SPL/NASA **70–71** SGA/Nathan Benn **72–73** OSF/M. Fogden **74** OSF **75** OSF/M. Fogden **77** PEP/A. Mounter **78** Zefa **79** BCL/Ken Balcomb **80** FSP/Richards **82–83t** E/F. Gohier **82–83b** VE **84–85** E/Edouard **84b** TCL/Adrian Warren **86** South American Pictures/Tony Morrison **87** E/Krafft **88–89** VE **90–91t** Husmo Foto **90–91b** TCL/S.J. Allen **92** Zefa **93** E/Philippe Beuzen **94–95t** Naturfotograferna/Claes Grundsten **94–95b** BCL/Norbert Schwirtz **96** Zefa **96–97** Naturfotograferna/Claes Grunsten **98–99** BCL/N. Schwirtz **100–101t** APL/Rob Moore **101–101b, 102** T. Woodcock **103t** PEP/J. and G. Lythgoe **103b** TCL **105, 106** BCL/Geoff Doré **106–107** OSF/Peter O'Toole **108–109, 110** Zefa **111b** AD **112** E/Michel Castéran **112–113** E/Monique Claye **113** Tony Stone Worldwide **114–115** AD **116–117** ACE Photo Agency/Geoff Smyth **119** NHPA/Picture Box **121t** E/D. Clément **121b** E/Eric Saint **122–123t** Benelux Press/Art Deco **122–123b, 123r, 124–125** NHPA/Picture Box **125b, 126–127, 128** AGE **129** HL/B. Régent **130** HL/R. Francis **130–131** E/F. Gohier **132** E/Roy **132–133** E/Jean-Paul Nacivet 134–135, 135r Zefa **136** APL/Bob Davis **137** Zefa **138–139** AD **138b** Romano Cagnoni **140** FSP **142–143** Agence Top/J.P. Charbonnier **144** TCL/Greg Stott **145** TCL/JL **146, 147** Zefa **149t** AGE **149b** GSF **150–151** EISLF/D. Gros **151t, 151b** EISLF/E. Wengi **152–153** E/P. Weisbecker **154–155** Zefa **155** Tim Sharman **156–157** E/Louis-Yves Loriat **157** Tim Sharman **158** GSF **158–159** David Williamson **161** OSF/Tony Allen **162, 162–163** Vadim Gippenreiter **164t** EOSAT/British Aerospace **164b** Dr A. Knystautas **165** E/Roy **166–167** Vadim Gippenreiter **167t** Ardea/Masahiro Iijima **168–169** Vadim Gippenreiter **170–171** RHPL/Ian Griffiths **172** E/Nicholas Thibaut **173** BCL/Prato **175** Earth Satellite Corporation **176–177** GSF **177tl** Zefa **177br** E/Louis-Yves Loriat **178, 179, 180, 181, 182–183, 184** VE **184–185** JH/Georg Gerster **186–187t** RHPL **186–187b** OSF/R. Packwood **189** WPL/John Noble **190–191t** E/P. Cheuva **190–191b** E/C. Delu **192** Survival Anglia/D. and M. Plage **192–193** Zefa **194–195** GC **195** E/Krafft **196** E/J. Lenars **197** GC **198–199** AGE **199** Zefa **200** GC **201** EOSAT/British Aerospace **202** GC **204** SGA/Adam Woolfitt **204–205** Foto Features **206–207** WPL/John Noble **207** E/D.V. Boutin **208–209** GC **210–211** ACE Photo Agency/Paul Steel **213l** E/Eric Morvan **213r** APL/Peter Carmichael **215** M/Hiroji Kubota **216–217** JH/Georg Gerster **217** SGA/Mike Andrews **218–219** Wu Shouzhuang **220–221** JH/Georg Gerster **222** Chapel Studios **223** OSF/Alastair Shay **224–225** RHPL/Sassoon **226–227** RHPL/Grieves **228–229** International Centre for Conservation Education/Sylvia Yorath **231t** APL/L. Okuda **231b** JH/R. Michaud **232–233** Zefa **234–235** NHPA/Orion Press **236, 237** The Press Association/Jiji Tsushin **238–239t, 238–239b, 240–241, 241** Ardea/J.P. Ferrero **242–243** AGE **243b** SPL/NASA **244, 245** BCL **246–247** BCL/Nicholas DeVore

The map on pp 38–39 is based on one in Starkel, L. "Man as a cause of sedimentological changes", *Anthropogenic Sedimentological Changes* (L. Starkel, ed.), *Striae* vol. 26, Uppsala, 1987

Editorial, research and administrative assistance
Jill Bailey, John Baines, Helen Burridge, Barbara James, Shirley Jamieson, Susan Kennedy, Amanda Kirkby, Dr N. J. Middleton, Sarah Rhodes, Huw Rowlands

Artists
Julian Baker (The Maltings Partnership), Janos Marffy, Mick Saunders

Cartography
Maps drafted by Euromap, Pangbourne

Index
Barbara James

Production
Clive Sparling

Typesetting
Brian Blackmore, Catherine Boyd, Peter Macdonald Associates

Color origination
Scantrans pte Limited, Singapore

CONVERSION TABLE

1 *millimeter* (mm) = 0.0394 inches (in)
1 *meter* (m) = 3.281 feet (ft)
1 *kilometer* (km) = 0.621 miles (mi)
1 *hectare* (ha) = 2.471 acres
1 *square kilometer* (sq km) = 0.3861 square miles (sq mi)
1 *cubic meter* (cu m) = 35.315 cubic feet (cu ft)
1 *degree Celsius* (°C) = 1.8 degrees Fahrenheit (°F)
Celsius temperature = (Fahrenheit temperature − 32) × 5/9
1 *tonne* = 1.1023 short tons = 0.9842 UK tons

1 *inch* = 25.4 millimeters
1 *foot* = 0.3048 meters
1 *mile* = 1.609 kilometers
1 *acre* = 0.4047 hectares
1 *square mile* = 2.59 square kilometers
1 *cubic foot* = 0.0283 cubic meters
1 *degree Fahrenheit* = 0.55 degrees Celsuis
Fahrenheit temperature = (9/5 × Celsius temperature) + 32

1 *billion* = 1 thousand million = 1,000,000,000

INDEX

Page numbers in *italics* refer to captions, maps or tables

F

G

H

I

J

K

L